Scanning Electron Microscopy, X-Ray Microanalysis, and Analytical Electron Microscopy

A Laboratory Workbook

Scanning Electron Microscopy, X-Ray Microanalysis, and Analytical Electron Microscopy

A Laboratory Workbook

Charles E. Lyman
Lehigh University
Bethlehem, Pennsylvania

Dale E. Newbury
National Institute of Standards and Technology
Gaithersburg, Maryland

Joseph I. Goldstein
Lehigh University
Bethlehem, Pennsylvania

David B. Williams
Lehigh University
Bethlehem, Pennsylvania

Alton D. Romig, Jr.
Sandia National Laboratories
Albuquerque, New Mexico

John T. Armstrong
California Institute of Technology
Pasadena, California

Patrick Echlin
University of Cambridge
Cambridge, England

Charles E. Fiori
National Institute of Standards and Technology
Gaithersburg, Maryland

David C. Joy
University of Tennessee
Knoxville, Tennessee

Eric Lifshin
GE Corporate Research and Development
Schenectady, New York

Klaus-Ruediger Peters
The University of Connecticut Health Center
Farmington, Connecticut

Plenum Press • New York and London

Library of Congress Cataloging-in-Publication Data

Scanning electron microscopy, X-ray microanalysis, and analytical
   electron microscopy : a laboratory workbook / Charles E. Lyman ...
   [et al.].
      p.   cm.
   Includes bibliographical references.
   ISBN 0-306-43591-8
   1. Scanning electron microscopes--Laboratory manuals. 2. X-ray
microanalysis--Laboratory manuals. 3. Electron microscopy-
-Laboratory manuals.   I. Lyman, Charles E.
QH212.S3S33 1990
502'.8'25--dc20   90-33608
                                                              CIP

© 1990 Plenum Press, New York
A Division of Plenum Publishing Corporation
233 Spring Street, New York, N.Y. 10013

All rights reserved

No part of this book may be reproduced, stored in a retrieval system, or transmitted
in any form or by any means, electronic, mechanical, photocopying, microfilming,
recording, or otherwise, without written permission from the Publisher

Printed in the United States of America

Foreword

During the last four decades remarkable developments have taken place in instrumentation and techniques for characterizing the microstructure and microcomposition of materials. Some of the most important of these instruments involve the use of electron beams because of the wealth of information that can be obtained from the interaction of electron beams with matter. The principal instruments include the scanning electron microscope, electron probe x-ray microanalyzer, and the analytical transmission electron microscope. The training of students to use these instruments and to apply the new techniques that are possible with them is an important function, which has been carried out by formal classes in universities and colleges and by special summer courses such as the ones offered for the past 19 years at Lehigh University.

Laboratory work, which should be an integral part of such courses, is often hindered by the lack of a suitable laboratory workbook. While laboratory workbooks for transmission electron microscopy have been in existence for many years, the broad range of topics that must be dealt with in scanning electron microscopy and microanalysis has made it difficult for instructors to devise meaningful experiments. The present workbook provides a series of fundamental experiments to aid in "hands-on" learning of the use of the instrumentation and the techniques. It is written by a group of eminently qualified scientists and educators.

The importance of hands-on learning cannot be overemphasized. It is only by experience that one acquires familiarity and only by familiarity that one acquires the confidence to attack problems using the manifold techniques now available to the researcher. This workbook provides the only comprehensive series of experiments devoted to a wide range of imaging and analytical experiments. I was particularly pleased to see a number of experiments devoted to some of the important but less frequently used techniques, such as channeling contrast, magnetic contrast, low-voltage SEM, and stereo microscopy. I was also pleased to see experiments on wavelength-dispersive spectrometry. Although the majority of students may not have an opportunity to use this technique in their later work, they should be familiar with it now in order to evaluate, in proper perspective, the work that they may do with energy-dispersive spectrometry.

I expect this workbook to be highly successful and widely adopted. But I would like to add one caution to the reader: do not be misled into thinking that, because you have done an experiment on a technique, you fully understand it. In fact, many of the experiments in this book could have been expanded into several more experiments. I hope that this workbook is only the beginning and that it will be followed by additional experiments in the not too distant future.

> David B. Wittry
> Professor of Materials Science and Electrical Engineering
> University of Southern California
> Los Angeles, CA 90089-0241

Preface to the Student

These laboratory exercises were developed over several years at the short courses on scanning electron microscopy, x-ray microanalysis, and analytical electron microscopy held annually at Lehigh University. Laboratories 1-7 cover basic techniques that should be mastered before attempting the more advanced SEM and x-ray microanalysis techniques of Laboratories 8-23. Laboratories 24-27 cover techniques used in analytical electron microscopy in the transmission electron microscope. The specimens required for Laboratories 1-27 can be prepared using step-by-step guidelines given in Laboratories 28-30. Often the recommended specimens may be substituted with others that are more relevant to the work at individual companies or institutions.

This laboratory workbook is a companion volume to three textbooks on electron microscopy published by Plenum:

Scanning Electron Microscopy and X-Ray Microanalysis by J. I. Goldstein, D. E. Newbury, P. Echlin, D. C. Joy, C. E. Fiori, and E. Lifshin, Plenum Press, New York, 1981 (designated *SEMXM* throughout this workbook),

Advanced Scanning Electron Microscopy and X-Ray Microanalysis by D. E. Newbury, D. C. Joy, P. Echlin, C. E. Fiori, and J. I. Goldstein, Plenum Press, New York, 1986 (*ASEMXM*),

Principles of Analytical Electron Microscopy, edited by D. C. Joy, D. Romig, Jr., and J. I. Goldstein, Plenum Press, New York, 1986 (*PAEM*).

Each laboratory exercise illustrates one or more of the principles and techniques discusssed in these textbooks. For state-of-the-art techniques such as environmental SEM, additional explanatory material has been given within the narrative of the laboratory. To obtain the most benefit from these laboratories, the student is encouraged to read the recommended sections of the appropriate textbook and the narrative of the laboratory before sitting down to the instrument. There is an important learning experience to be derived from performing all parts of these labs on an instrument in your own laboratory rather than simply examining the worked examples in the solutions section. It should be possible to complete each laboratory in less than half a workday after suitable preparation of the specimens and alignment of the instrument.

The publication of this book would not have been possible without the instrumentation and technical assistance provided by the following manufacturers of electron microscopes and x-ray detection equipment: Amray, Cambridge Instruments, CamScan USA, EDAX International, ElectroScan, Gatan, NSA Hitachi Scientific Instruments, International Scientific Instruments, JEOL (USA), Kevex Corp., Link Analytical, Microspec, Philips Electronic Instruments, Princeton Gamma-Tech, Tracor Northern, VG Microscopes, and Carl Zeiss.

The authors wish to thank the many students and colleagues who have contributed to these laboratories including V. P. Dravid, D. R. Liu, S. M. Merchant, J. R. Michael, C. Robino, H. E. G. Rommal, C. Russo, J. Sutliff, K. S. Vecchio, and M. Zemyan. We also appreciate the efforts of our reviewers for various sections of the book: R. M. Anderson, E. L. Hall, and D. B. Wittry. Thanks also go to our technical staff, D. W. Ackland, D. C.

Calvert, J. Kerner, and K. Repa, who prepared most of the specimens used in these labs. Finally, special thanks go to our short course coordinator, Sharon Coe, for her devotion to the production of this workbook.

C. E. Lyman

Bethlehem, Pennsylvania

Contents

Part I

SCANNING ELECTRON MICROSCOPY AND X-RAY MICROANALYSIS

Laboratory 1	Basic SEM Imaging	3
Laboratory 2	Electron Beam Parameters	8
Laboratory 3	Image Contrast and Quality	16
Laboratory 4	Stereo Microscopy	22
Laboratory 5	Energy-Dispersive X-Ray Spectrometry	27
Laboratory 6	Energy-Dispersive X-Ray Microanalysis	33
Laboratory 7	Wavelength-Dispersive X-Ray Spectrometry and Microanalysis	42

Part II

ADVANCED SCANNING ELECTRON MICROSCOPY

Laboratory 8	Backscattered Electron Imaging	51
Laboratory 9	Scanning Transmission Imaging in the SEM	55
Laboratory 10	Low-Voltage SEM	57
Laboratory 11	High-Resolution SEM Imaging	61
Laboratory 12	SE Signal Components	67
Laboratory 13	Electron Channeling Contrast	73
Laboratory 14	Magnetic Contrast	78
Laboratory 15	Voltage Contrast and EBIC	81
Laboratory 16	Environmental SEM	86
Laboratory 17	Computer-Aided Imaging	90

Part III
ADVANCED X-RAY MICROANALYSIS

Laboratory 18	Quantitative Wavelength-Dispersive X-Ray Microanalysis	99
Laboratory 19	Quantitative Energy-Dispersive X-Ray Microanalysis	108
Laboratory 20	Light Element Microanalysis	117
Laboratory 21	Trace Element Microanalysis	122
Laboratory 22	Particle and Rough Surface Microanalysis	127
Laboratory 23	X-Ray Images	132

Part IV
ANALYTICAL ELECTRON MICROSCOPY

Laboratory 24	Scanning Transmission Imaging in the AEM	139
Laboratory 25	X-Ray Microanalysis in the AEM	143
Laboratory 26	Electron Energy Loss Spectrometry	148
Laboratory 27	Convergent Beam Electron Diffraction	153

Part V
GUIDE TO SPECIMEN PREPARATION

Laboratory 28	Bulk Specimens for SEM and X-Ray Microanalysis	159
Laboratory 29	Thin Specimens for TEM and AEM	172
Laboratory 30	Coating Methods	180

SOLUTIONS TO LABORATORY EXERCISES

Laboratory 2	Electron Beam Parameters	189
Laboratory 3	Image Contrast and Quality	197
Laboratory 4	Stereo Microscopy	204
Laboratory 5	Energy-Dispersive X-Ray Spectrometry	207
Laboratory 6	Energy-Dispersive X-Ray Microanalysis	213
Laboratory 7	Wavelength-Dispersive X-Ray Spectrometry and Microanalysis	219

Laboratory 8	Backscattered Electron Imaging	227
Laboratory 9	Scanning Transmission Imaging in the SEM	232
Laboratory 10	Low-Voltage SEM	234
Laboratory 11	High-Resolution SEM Imaging	242
Laboratory 12	SE Signal Components	251
Laboratory 13	Electron Channeling Contrast	263
Laboratory 14	Magnetic Contrast	275
Laboratory 15	Voltage Contrast and EBIC	279
Laboratory 16	Environmental SEM	287
Laboratory 17	Computer-Aided Imaging	296
Laboratory 18	Quantitative Wavelength-Dispersive X-Ray Microanalysis	309
Laboratory 19	Quantitative Energy-Dispersive X-Ray Microanalysis	316
Laboratory 20	Light Element Microanalysis	330
Laboratory 21	Trace Element Microanalysis	335
Laboratory 22	Particle and Rough Surface Microanalysis	343
Laboratory 23	X-Ray Images	352
Laboratory 24	Scanning Transmission Imaging in the AEM	365
Laboratory 25	X-Ray Microanalysis in the AEM	373
Laboratory 26	Electron Energy Loss Spectrometry	381
Laboratory 27	Convergent Beam Electron Diffraction	389

Index 401

Part I
SCANNING ELECTRON MICROSCOPY AND X-RAY MICROANALYSIS

Laboratory 1

Basic SEM Imaging

Purpose
This first laboratory is designed to acquaint the beginning SEM operator with the steps for taking a micrograph. The steps are described without reference to a particular instrument. Please consult the manufacturer's operation manual or an instructor before proceeding.

Equipment
Any SEM with all automatic features switched off, Polaroid film.

Note: With the broad range of SEM instrumentation encountered throughout the field, it is difficult to write a general procedure which is applicable to all instruments. Many of the simple operations described below have been automated or eliminated in the most modern instruments. However, since many older instruments are still encountered, these simple procedures are well worth consideration.

Specimen
"Old-type" zinc oxide copy paper or "bond" paper coated with gold-palladium (see Laboratory 31) or any conductive specimen with fine surface roughness that cannot be seen with a light optical microscope.

Time for this lab session
Twenty minutes provided that the following text of the lab has been read in advance.

1.1 Identifying the Parts of the SEM

The SEM consists of two major parts. An electronics console provides the switches and knobs for adjusting the intensity of the image on the viewing screen, focusing, and photography. The electron column is the business end of the SEM, where the electron beam is created, focused to a small spot, and scanned across the specimen to generate the signals that control the local intensity of the image on the viewing screen. To understand how all this works, we must consider the parts of the electron column in more detail (see Figure 1.1).
Electron Gun. The electron gun, at the top of the column, provides a source of electrons. Electrons are emitted from either a white-hot tungsten or lanthanun hexaboride e or via field emission and accelerated down the evacuated column. There are three separate electrical parts of the gun: the tungsten filament that emits electrons, the Wehnelt cylinder that controls the number of electrons leaving the gun, and the anode that accelerates the electrons to a voltage selectable from 1 kV to 30 kV (1 kV=1000 volts). A vacuum is necessary because electrons can travel only very short distances in air.

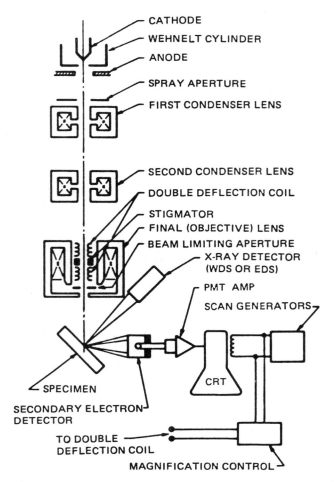

Figure 1.1. Schematic diagram of the parts of the electron column.

Electron Lenses. Two or three electron lenses are used to demagnify the electron beam to a small spot about 5-50 nm in diameter (5-50 nm = 50-500 angstroms = 0.005-0.050 µm = five millionths to fifty millionths of a millimeter) from a crossover diameter more than a thousand times larger located inside the electron gun. The lens closest to the gun is called the condenser lens while the lens closest to the specimen is called the final lens or the objective lens. The function of the objective lens is to move the smallest cross-section of the beam up and down until it meets the specimen surface. This corresponds to a focused image.

Scanning System. The image is formed by scanning the beam across the specimen in synchronism with a beam scanning inside the cathode ray tube (CRT) of the viewing screen. The scan coils to deflect the beam across the specimen are usually located inside the objective lens.

Liner Tube. Modern SEMs usually have a liner tube from the gun to the objective lens so that the lenses can be located outside the vacuum and so that the tube can be easily removed for cleaning.

Objective Aperture. This is a platinum disk with a small hole (~100 µm diameter) in it often located inside the objective lens (beam limiting aperture in Figure 1.1). Its function is to limit the angular width of the electron beam in order to reduce lens aberration effects and to improve depth-of-field in the image (more about this in Laboratory 2).

Specimen Chamber. This large evacuated space below the objective lens contains the specimen stage with all its motions, the electron signal detectors, the x-ray detector, and a pumping line to the main vacuum pump.

Everhart-Thornley Electron Detector. The Everhart-Thornley (E-T) detector is the most common type of detector for electron signals. The E-T detector is sensitive to both types of emitted signal-carrying electrons: backscattered electrons and secondary electrons. Backscattered electrons are sufficiently energetic to directly excite the detector. Low-energy secondary electrons are drawn toward the detector by the +300 V on the collector wire-mesh screen at the front. Once inside the screen the collected electrons are then accelerated to about +12 kV and collide with a scintillator producing light. The light is amplified by a photomultiplier tube (PMT) to produce an electrical signal that eventually modulates the intensity of the viewing CRT.

Vacuum System. For most SEMs the vacuum is produced by an oil diffusion pump backed by a rotary mechanical pump. The oil diffusion pump uses a stream of hot oil vapor to strike gas molecules in the vacuum and push them toward the mechanical pump which then expels the gas molecules from the system. Since the diffusion pump can only operate after a vacuum has been created, the mechanical pump also must be used to preevaluate or rough-pump the specimen chamber before the valving is switched to allow the diffusion pump to evacuate the specimen chamber.

Observe: Parts of the SEM. Find all of the parts of the SEM listed above. Open the specimen chamber door (as described below) to see the parts that are inside.

1.2 Inserting a Specimen

Venting the Specimen Chamber. Before the specimen chamber can be opened to atmosphere for inserting a new specimen, the electron beam must be shut off and the "specimen change" or "air" button must be activated. Once the specimen chamber has been vented with dry nitrogen or air, a new specimen may be inserted into the specimen stage. Note that many instruments make use of a sample exchange airlock to minimize the volume which must be brought up to atmosphere while changing the specimen. Such a system preserves the vacuum in the specimen chamber itself and allows faster pumpdown. In fact, a good vacuum is so important that anything placed in the evacuated areas of the microscope, including the specimen, should be handled only with lint-free plastic gloves.

Conductive Specimens. During imaging, electrons are continually bombarding the specimen and would eventually build up a negative charge on the areas of the specimen under the beam. This charge, if large enough, could deflect both incident and emitted electrons and ruin the image. To prevent this effect all specimens must be electrically conductive so that the current deposited onto the specimen has a path to an electrical ground through the specimen stage. Metals are already conductive but ceramics, polymers, and biological materials are not. The usual remedy is to coat the specimen surface with a thin layer of a conducting substance such as gold-palladium or carbon in a sputter coating or evaporative coating device. In fact, for the highest-quality SEM images, all specimens (even metals!) should be coated. Ensure that a good conductive path exists from the coating on the specimen to ground by means of a path to the specimen stub made with conducting paint.

Specimen Stage Motions. The stage holding the specimen is usually capable of tilt, elevation, and rotation motions in addition to x and y translations for changing the field of view. The specimen is usually tilted about 30°-45° toward the E-T detector to increase electron signal collection. Adjustment of the elevation or "z motion" is used to control the working distance, the distance from the objective lens to the specimen. Short working distances are used to achieve the best image resolution while long working distances improve depth-of-field and low-magnification operation (see Laboratory 2). The rotation motion is useful for aligning image features with a particular picture direction and for sequentially examining several specimens by rotating a turntable holding many specimen stubs.

Observe: Stage Motions. Insert the specimen on the stage and observe the effects of each stage motion. Tilt the specimen toward the E-T detector by 30° and center the x-y translation controls.

Pumping the Specimen Chamber. Close the stage door and evacuate the specimen chamber by pressing the "pumpdown" button. Wait 2-5 min for the "vac ready" light to come on, indicating that the vacuum is adequate for operation.

1.3 Obtaining an Image

Step 1: Assuring Adequate Vacuum. After the "vac ready" light comes on, confirm that a scanning raster is just visible on the viewing screen, and turn on the E-T detector/PMT, the high voltage (20 kV), and the filament heating circuit. (Many SEMs automatically switch these subsystems on after a specimen change.)

Step 2: Saturation of the Electron Gun. Three different types of electron sources are encountered: the tungsten hairpin (most common), LaB_6 (often used), and field emission (occasionally used). Each source requires a different procedure for proper operation. For a tungsten hairpin, slowly turn up the filament heating knob until the intensity on the viewing screen begins to increase. Observe the effect of this increase on the signal meter or the waveform monitor as well as in the general increase of the screen intensity. With a conventional tungsten filament you will observe a drop in intensity, the "false peak," as you heat the filament. When the intensity does not increase with continued increases in filament heating, you have reached saturation and should back off slightly for operation. *It is very important not to overheat the tungsten filament because this can cut the filament lifetime drastically.* Normal tungsten filament life is 40-100 hr, but lifetimes can be reduced to less than 10 hr if the filament heating is set too high.

Step 3: Contrast and Brightness. If an image is apparent on the screen, adjust the "contrast" or "PMT voltage" control to obtain a pleasing contrast. Adjust the "brightness" control bring the screen intensity to a suitable viewing level. These operations may have to be repeated after focusing.

Step 4: Focusing the Image. Many instruments have separate adjustments for "spot size" and "focus." "Spot size," the minimum possible size of the beam at the specimen, is controlled by the action of the condenser lens, which is the first lens below the electron gun. The smaller the spot size, the finer will be the detail which can be observed on a suitable specimen, which is generally a desirable goal. However, a serious consequence of reducing the spot size is that the beam current falls as the inverse square of the beam diameter. A lower beam current will result in a noisier image with poorer contrast sensitivity. Select a medium value for the spot size to obtain an adequate amount of current in the beam.

"Focus" is controlled by the action of the objective lens, which is the final lens in the column. Using the medium or fine control of the objective lens bring the image of the specimen into focus, reserving use of the coarse control for situations when the medium control cannot bring the image into focus. Use the stage motions to find an area of interest.

Notice that the image may go out of focus in the new area. The image may be refocused using either the objective lens or the z-drive.

Step 5: Optimizing the Image. SEM images are subject to defects which degrade their quality and which become apparent at high magnification. Astigmatism is the most common defect encountered. To look for this defect, find an area of the specimen with fine random detail, switch to a high magnification (10,000x or above), and focus the image as well as possible. As you rotate the objective lens knob back and forth on either side of the focus position, notice whether the image appears to line up or stretch out in two orthogonal directions (90° apart). If this is the case, the image is astigmatic and will never come into sharp focus without an astigmatism correction. Set the focus to a position midway between the stretched images where the image detail appears randomly oriented but still out of focus. Using only one stigmator control sharpen the image, then refocus. Now stigmate with the other stigmator control, refocus, and repeat until a sharp image is obtained (see *SEMXM*, Figure 2.15). Stigmating correctly takes some practice so do not be discouraged if your first picture is not perfect!

Step 6: Selecting the Magnification. Unlike light optical microscope images, after the SEM image is focused at a high magnification it is in focus at all lower magnifications since the objective lens setting has not changed. High image magnifications in the SEM mean that the scan raster on the specimen is very small, but that the image information is magnified to the size of the viewing screen or the photograph. Select an image magnification of about 1000x.

Step 7: Taking a Photo. Finally, adjust the "contrast" and "brightness" at the viewing screen or the waveform monitor for proper photography conditions. Note that the contrast and brightness for the photographic record tube are independently adjusted and should be correctly set in advance for the type of film in use. These "camera settings" should not be touched in normal operation. If the camera exposure settings are not correct, see Section 1.4.

Insert a piece of Polaroid film all the way into the camera on the "load" setting, pull the sleeve out until it stops, push the "photo" button on the SEM console, and wait for the slow single frame scan to expose the film (about 30-60 sec). When the photo scan is finished, push the film back into the camera until it stops. Move the camera lever to "process" and pull the film out with one firm, steady motion. After waiting the appropriate amount of time for the film to develop, separate the outer halves of the paper sleeve to view the photo. If the film is of the positive-negative (P/N) type, soak the negative in sodium sulfite solution (read the instructions on the box) before washing and drying.

Looking Ahead. You have now completed the basic tasks of SEM operation, but for a fixed set of microscope operating parameters. In the following lab sessions you will learn to gain control over these microscope parameters for special situations such as high resolution, high depth-of-field, maximum image contrast, and x-ray microanalysis.

Laboratory 2

Electron Beam Parameters

Purpose

This laboratory demonstrates: (1) electron gun saturation and alignment; (2) the measurement of beam current, beam size, and beam convergence; (3) the concept of electron gun brightness; and (4) the effects of these parameters on depth-of-field and resolution. More details and references can be found in *SEMXM*, Chapter 2.

Equipment
1. SEM with an Everhart-Thornley (E-T) electron detector. Switch off all automatic features.
2. Electrically isolated specimen stage and an electrical feed-through to outside the specimen chamber.
3. Picoammeter (electrometer with high impedance) to measure beam current.

Specimens
1. Faraday cup mounted on a specimen stub for measuring beam current.
2. Fractured silicon wafer on a carbon block for measuring beam size and beam convergence. For measuring small beam sizes (<100 nm) a piece of commercial silver mesh is useful.
3. Fracture surface of a metal or other specimen with surface relief of about 1 mm.

Time for this lab session

Three-hour lab session: Do Experiments 2.1, 2.3-2.9 (200-μm aperture only and use alternative α calculation for 8 and 9).

Four-hour lab session: Do Experiments 2.1-2.12.

2.1 Gun Saturation and Alignment

Proper saturation and alignment of the conventional tungsten hairpin electron gun are essential for optimal performance of the SEM. Without proper setup of the electron gun there may not be enough current in the scanning electron beam to produce acceptable pictures.

Insert the Faraday cup into the specimen stage and pump down the specimen chamber. When the vacuum is adequate for operation, turn the high voltage to 20 kV and heat the filament by slowly turning up the filament knob. For proper operation, the filament should be in a condition of saturation (no increase in beam current with increased filament heating). In this condition, small fluctuations in filament heating current have a negligible effect on the electron beam current.

Experiment 2.1: Filament Saturation. *Observe the changes in specimen beam current with the picoammeter as a function of applied filament heating current. Note the saturation level. Do not oversaturate.*

Can saturation be determined by measuring the secondary electron signal with the E-T detector instead? The emission current?

Note: The above procedure applies to a conventional tungsten hairpin filament. An LaB_6 source must be turned up more slowly according to the SEM manufacturer's instructions and only in the best possible vacuum. This source exhibits a less well-defined saturation. With a field-emission gun (FEG), electrons are extracted from a very sharp tungsten tip solely by the high electric field in front of the tip. There is no saturation condition for a field-emission gun. What limits the amount of current that can be drawn from an FEG?

Note on Computer Control: Many recent SEMs have computer-controlled startup and operation procedures. Many of the operator adjustments mentioned in these labs may actually be made automatically. It is suggested that the automatic option be turned off so that the purpose of these adjustments can be clearly understood.

Experiment 2.2: Gun Misalignment. *Intentionally misalign the electron beam by use of the external mechanical gun adjustment or the electromagnetic gun alignment controls. Observe again the variation of electron beam current versus applied filament heating current (or electron extraction voltage if the source is an FEG). How has this relationship changed from Experiment 2.1?*

Now the gun is misaligned. What is a good strategy for finding the proper alignment? A well-aligned gun is essential for proper SEM operation. See your SEM operator's manual for the gun alignment procedure (and lens alignment procedure, if necessary). The gun alignment process may need to be repeated about 15-20 min after the gun heats up.

2.2 Beam Current

The electron beam current entering the specimen (often called the probe current) typically ranges from picoamperes (pA) to microamperes (µA), depending upon the choice of aperture size and first condenser lens strength (spot size knob). The specimen absorbs part of this current and reemits the rest. Because of the emission of backscattered and secondary electrons, the specimen current that is measured on a given specimen is not an accurate representation of the beam current. An accurate measure can only be obtained by using a Faraday cup, a small blind hole of large depth drilled into a conducting material. A Faraday cup can be made by placing an electron microscope aperture, usually made of platinum or molybdenum, over a hole in a small block of nonmagnetic metal (see Figure 2.1).

The beam current is measured by placing the beam in the hole, either by directing a spot beam into the image of the hole or imaging at high magnification so that the image of the aperture hole appears to completely engulf the screen. The beam electrons injected into the hole, as well as backscattered and secondary electrons, are trapped with great efficiency by the Faraday cup providing an accurate measure of the beam current.

Experiment 2.3: Beam Current and Emission Current. *For a 200-µm real objective aperture (or equivalent virtual objective aperture, see note below), set a beam current of 0.1 nA, by adjusting the first condenser lens. Measure the beam current with the Faraday cup.*

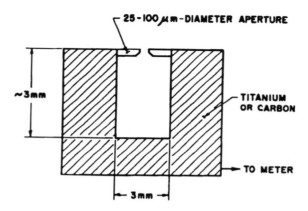

Figure 2.1. Cross-sectional diagram of a Faraday cup to be mounted on an SEM specimen stub.

Note the condenser setting and the emission current in the data box below. If time permits, repeat for a beam current of 10 nA (or as high as possible if the SEM has an FEG).
  Note: A virtual objective aperture is an aperture placed at a position before the objective lens which limits the beam angle in a manner similar to a real objective aperture. However, the size of the virtual aperture may be quite different.

Aperture	Condenser setting	Beam current	Emission current
200 µm		0.1 nA	µA
200 µm		10 nA	µA

How does the beam current at the specimen compare with the emission current at the gun? Where did the rest go? Note that the emission current did not change for the high-beam-current setting because no change was made to the gun parameters. Name three ways to change the emission current. Why would you want to change the emission current?
  Note: Often the emission current of the electron gun is labeled incorrectly on the SEM control panel as the "beam current." It is more correct to refer to the current reaching the sample as the "beam current" or the "probe current."
  Note: It is important to keep the same aperture for Experiments 2.3-2.6.

Experiment 2.4: Effect of Lens Settings on Beam Current. Vary the strength of the condenser lens (spot size) to obtain a 100-fold change in beam current, keeping the objective lens fixed. Measure three levels of beam current and plot the results versus condenser setting. Similarly, vary the strength of the objective lens (focus), keeping the condenser lens fixed, and observe the beam current. Plot the results.

Condenser lens	Beam current	Objective lens	Beam current

Why is there such a difference in the influence of these two lenses on the beam current?

2.3 Beam Size

The electron beam diameter (probe diameter) can be measured by sweeping the beam across a sharp, electron-opaque edge and observing the rise or fall in signal as a function of beam position (Figure 2.2). An arbitrary measurement of the beam diameter may be taken as the horizontal distance (corrected for image magnification) between the 10% and 90% signal levels since these signal levels are easy to determine on a graphical presentation of the intensity waveform during a linescan. This measurement is not difficult to make for probes of about 0.1-µm diameter and larger with the fractured silicon wafer specimen. The silver mesh specimen with its tiny needles is more useful for measuring smaller beam diameters, but this is still a very difficult experiment for probe sizes less than 0.1 µm. It is more important to understand the concept that the beam diameter (and other beam parameters) can be varied than to be able to measure the smallest beam sizes. In fact the beam size may be so small (e.g., 2-3 nm for an FEG) that this part of the laboratory may be impossible.

Experiment 2.5: Beam Diameter. *Insert the fractured silicon wafer specimen. Set the working distance to 10 mm. Rotate the line of the fracture to be perpendicular to the scan lines on the viewing screen. Tilt the specimen so that the edge of the silicon overhangs the carbon underneath. Using the above procedure, measure the beam diameter for the best focus and stigmation that can be obtained with a beam current of 0.1 nA. If time permits, repeat for a beam current of 10 nA.*

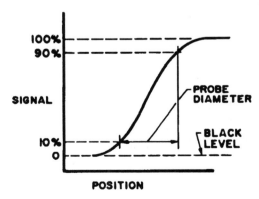

Figure 2.2. Schematic showing how to make an estimate of beam diameter from the signal waveform.

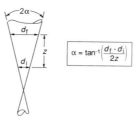

Figure 2.3. Schematic diagram of a longitudinal section of the electron beam mean focus and the equation for calculating the beam convergence.

Aperture	Condenser setting	Beam current	Beam diameter
200 μm		0.1 nA	
200 μm		10 nA	

Compare the ratio of beam diameters to the ratio of beam currents for the two operating conditions.

2.4 Beam Convergence

The beam convergence angle α (total angle = 2α) can be measured by the following method for any SEM, including those with a virtual objective aperture located above the final lens. (If your microscope has an objective aperture in the gap of the final lens, see the note at the end of this section for a simple alternative method of estimating α.) The convergence angle can be measured directly by measuring the same beam at two different working distances without changing the objective lens current (the focus). If d_i and d_f are the initial and final effective beam diameters and z is the vertical change in working distance (measured by the change in z on the stage), then the beam convergence may be calculated as shown in Figure 2.3. It should be noted that if the change in z is not large enough, the value of α will not be accurate.

***

Experiment 2.6: Beam Convergence. With a short working distance (exactly 5 mm) and with the same aperture and beam current as in the previous experiments (200-μm aperture and 0.1-nA beam current) measure d_i. Without touching the objective lens, increase z at least 1 mm, and measure d_f. Calculate the convergence angle, α. If time permits, repeat for the largest aperture available (e.g., 400 μm).

Aperture	d_i	d_f	Δd	z_i	z_f	Δz	Convergence
200 μm							
400 μm							

Will the convergence increase or decrease at a longer working distance?

Alternative method: For instruments with the beam limiting aperture in the gap of the final lens, an estimate of the beam convergence angle can be made using the formula $\alpha = R/W$ where R is the aperture radius and W is the working distance.

2.5 Brightness

The reason for using brightness, rather than the more obvious parameters of beam current or current density, is that brightness is a normalized parameter that permits valid comparisons between beam characteristics at different points in the SEM column or between different SEM columns. From the combined measurements of beam current i, beam diameter d, and beam convergence α, taken with the same aperture and condenser lens setting, the electron optical brightness β may be estimated from the brightness equation:

$$\beta = \frac{\text{Current}}{(\text{Area})(\text{Solid angle})} = \frac{4i}{\pi^2 d^2 \alpha^2} \quad \text{A/(cm}^2\text{ sr)} \tag{2.1}$$

Experiment 2.7: Brightness. *Calculate the gun brightness for the beam parameters determined with the 200-μm aperture. If time permits, repeat this calculation for a different aperture (e.g., 400 μm).*

Aperture	Beam current	Beam size	Convergence	Brightness
200 μm	0.1 nA			
200 μm	10 nA			
400 μm	nA			

Is the brightness a constant for all these conditions? Why is the brightness measured at the specimen only an estimate of the gun brightness? Why does image resolution usually degrade at lower voltages, particularly for thermionic sources?

2.6 Depth-of-Field

Large depth-of-field (or depth-of-focus) is one of the most important characteristics of the SEM. The sharpness of images recorded at low magnifications depends more on the depth-of-field available than on a small beam size. The depth-of-field D depends on both the beam convergence α and the magnification M:

$$D = \frac{2r}{\alpha} = \frac{0.2 \text{ mm}}{\alpha M} \tag{2.2}$$

where r is the effective beam radius at the distance from the plane of focus where blurring begins (see Figure 2.4). The 0.2 mm factor arises from the limited resolving power of the eye when viewing an SEM screen or micrograph.

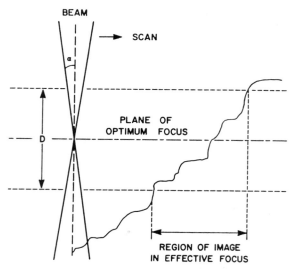

Figure 2.4. Schematic diagram showing how the depth-of-field may be calculated in an SEM image.

Experiment 2.8: Effect of Aperture Size. Insert the fracture surface specimen. For a small working distance (e.g., 5 mm), calculate the depth-of-field for the smallest and largest available apertures at magnifications of 50x and 5000x. Hint: Estimate the convergence for 200-μm and 400-μm apertures at W = 5 mm from your data for the 200-μm aperture at W = mm. Look for areas of greatly differing height and obtain images of the fracture surface under these conditions.

Working distance = 5 mm			
Aperture	Convergence	D at 50x	D at 5000x
200 μm			
400 μm			

Is the calculated depth-of-field reasonable when compared with the experimental images? How would you measure the depth-of-field obtained in your images?

Experiment 2.9: Effect of Working Distance. For the largest aperture (e.g., 400 μm), increase the working distance from 5 mm to 30 mm (or more) and record another image of the same area after refocusing. Which image has the greater depth-of-field? Calculate the expected improvement in depth-of-field for the longer working distance.

Aperture diameter = 400 µm			
W	Convergence	D at 50x	D at 5000x
5 mm			
30 mm			

Experiment 2.10: Image Rotation. *At long working distance, observe the effect of image rotation as the objective lens is oscillated. Why does this occur?*

2.7 Small Beam Size versus High Beam Current

You should now be able to set up the two extremes in secondary electron imaging: small beam size and high beam current. What SEM settings are necessary for each?

Experiment 2.11: Small Beam Size. *Image the fracture surface specimen. Take a high-magnification (10,000x) photograph of a stigmated image using conditions expected to produce the highest possible resolution for a 100-µm aperture at W = 5 mm. If time permits, measure the beam current for Experiments 2.11 and 2.12 after taking the required photos.*

Condenser setting for high resolution =

Beam current for high resolution =

Experiment 2.12: High Beam Current. *Take the same micrograph as above (10,000x) but with about 100 times more beam current. Without changing any beam parameters, take another photograph at 500x or 1000x.*

Condenser setting for high beam current =

Beam current for this situation =

How does the 10,000x photo of Experiment 2.12 differ from that of Experiment 2.11? Comment on the sharpness of the high-current image taken at low magnification. What situations require high beam current?

Laboratory 3

Image Contrast and Quality

Purpose

This laboratory demonstrates: (1) the two major types of contrast in SEM images, known as atomic number contrast and topographic contrast, (2) the factors affecting the quality of the image and how they ultimately limit the image resolution, and (3) the effects of electronic signal processing on the visibility of features in the image. More details and references may be found in *SEMXM*, Chapters 3 and 4.

Equipment

1. SEM with a backscattered electron detector (scintillator or solid-state variety), an Everhart-Thornley (E-T) detector, a specimen current detector, and some analog signal processing functions. Switch off all automatic features.
2. Stereo viewer.
3. Optional: a digital framestore to allow rapid evaluation of images without photography.

Specimens

1. Metallographically polished multiphase sample (e.g., a fine-grained multiphase basalt, Raney nickel, or a multiphase tool steel) containing phases of different average atomic numbers.
2. Fracture surface of a metal exhibiting strong topographic (relief) effects. The sample used in Laboratory 2 is suitable.

Time for this lab session

Three hours but may be shortened if a digital framestore rather than photographic recording is used to compare images.

3.1 Atomic Number Contrast

A flat, metallographically polished surface will have little or no topography, so most of the contrast can be attributed to variations in atomic number Z. Variations in Z must come from differences in chemical composition. Thus, Z contrast is a method to discern the chemistry of the sample in a qualitative manner. The elastic scattering of electrons is strongly dependent upon atomic number, and this behavior results in a relatively strong dependence of the backscattered electron coefficient on the atomic number. In order to observe atomic number contrast one usually utilizes either the backscattered electron (BSE) signal or the specimen current (SC) signal, which is also sensitive to electron backscattering.

Experiment 3.1: Z Contrast Using a Dedicated BSE Detector. Insert the basalt or other multiphase specimen. Using the dedicated "overhead" BSE detector located under the final pole piece, observe the contrast in the image. How many phases (areas of different intensity level) can be detected? *Examine the edge of a phase boundary at high magnification (10,000x or more).* How abruptly does the signal change at the boundary?

Note: No sample can ever be polished perfectly flat, particularly a multiphase sample. So topographic contrast (see Section 3.2) may be present at the interface between two phases, or in regions containing porosity or inclusions.

Experiment 3.2: Z Contrast Using the E-T Detector. Observe the same field of view with a positively and then a negatively biased E-T detector. Compare these images to that obtained with the dedicated BSE detector used in Experiment 3.1. Which detector provides better Z contrast? The positively biased E-T detector is usually thought of as a secondary electron detector. Why then is atomic number contrast visible in an image formed with this detector when secondary electrons do not carry significant atomic number information?

Experiment 3.3 (time permitting): Z Contrast Using the SC Detector. Using the specimen current detector, observe the same field of view and compare it to the other two images. Describe the main difference between the emissive (BSE) and absorbed (SC) signals. *Examine the images for fine-scale differences.* SC images are only sensitive to the number of electrons absorbed and not to trajectory effects in electron emission. Thus, directionality in BSE images due to trajectory effects should be absent in SC images. Is there any evidence in the SC image for the directionality of electron backscattering at minor topographic features such as scratches? Examine the interfaces between different phases where topography is evident in the BSE image. How does the interface appear in SC imaging?

3.2 Topographic Contrast

The fracture surface specimen is from a single-phase sample with a constant average atomic number. Most of the image contrast arises from the topography (shape) of the fracture surface. Topographic contrast is a complex phenomenon which may be carried by both the backscattered and the secondary electron signals. The appearance of a specimen exhibiting topographic contrast depends greatly on the nature and position of the electron detector. To examine these effects, we shall image the same field of view with all available detectors.

Experiment 3.4: Topographic Contrast from Secondary Electrons. Insert the fracture surface specimen. Select a magnification of about 500x or 1000x and a portion of the image which shows a significant amount of detail of the fracture surface and in which a large change in topography is evident (e.g., near the edge of the fracture or around a deep hole in the middle of the surface). Record an image of the area with the positively biased E-T detector. Can you tell the sense of topography of the specimen (i.e., which regions are "hills" and which are "valleys"?)

Note: The contrast effects that we want to demonstrate will be best observed if the edge of the fracture surface observed is closest to the E-T detector, and not tilted too much towards the detector. In this orientation part of the sample will be shadowed from direct line of sight to the E-T detector.

Experiment 3.5: Topographic Contrast from Backscattered Electrons. (a) *Obtain an image of the same area as in Experiment 3.4, but with the E-T detector biased negatively or turned off to exclude the secondary electrons.* Can this image be recorded successfully at the same beam current as that used for Experiment 3.4? How much must the beam current from a thermionic source be increased to obtain a signal similar to that obtained i Experiment 3.4?

(b) *Compare the positively biased and negatively biased E-T detector images. Note the location of the highlights in both images.* What is the origin of the highlights in the positively biased and negatively biased E-T detector images? *Obtain an image of the same area with a dedicated BSE detector.* How do the two images compare? Which facets of the fracture surface are not directly in view of the E-T detector?

Note: Field emission sources provide enough current in the probe so that an increase in current is not necessary.

Experiment 3.6 (time permitting): Topographic Contrast from the SC Signal. Record an image from the same area with the SC signal, both direct and inverted. How do the two images compare with the image from the E-T detector? Is the sense of the topography as easy to understand?

Experiment 3.7: Stereo Images. Prepare a secondary-electron stereo pair of the same area with the positively biased E-T detector. Take two photos of the same area with about 4°-7° tilt between them. Refocus the second image with the z control to maintain the same magnification. View the pictures with the tilt axis parallel to the vertical edge of the prints:

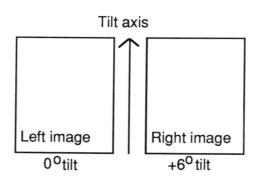

Does the stereo pair confirm the sense of topography observed in the single image obtained in Experiment 3.4? What happens if the photographs are reversed left-for-right?

Note: Some modern computer-based SEMs and/or multichannel analyzers have the capability of storing a stereo series which can be used to present a stereo image directly on the cathode ray tube (CRT). The stereo effect may be generated by high-speed playback of the images on the screen. Alternatively, some instruments can present a stereo image by storing the stereo images as described above and then displaying them in color-coded overlay format. Appropriately colored viewing glasses are required to see the stereo effect on the CRT screen. *Form a stereo image with this technique if it is possible on your machine.* More details about stereo images may be found in Laboratory 4.

3.3 Image Quality

The electron source brightness ultimately controls image quality. The beam diameter d and the beam current i are related through the brightness equation:

$$\beta = \frac{\text{Current}}{(\text{Area})(\text{Solid angle})} = \frac{4i}{\pi^2 d^2 \alpha^2} \quad \text{A/(cm}^2\text{sr)}$$

where α is the beam convergence and β is the electron optical brightness.

One of the most important concepts in scanning electron microscopy is the relationship of the visibility of an object in the final image to the characteristics of the signals which are generated in the sample and subsequently detected. This relationship is stated mathematically by the threshold equation:

$$i_{th} > \frac{4 \times 10^{-12}}{\varepsilon C^2 t_f} \frac{C}{\text{sec}} \quad \text{or (A)} \qquad \text{(threshold equation)}$$

where i_{th} is the threshold current needed to image a contrast level C; ε is the collection efficiency factor, combining the signal generation and collection characteristics (ε = the number of collected signal carriers/incident electron); and t_f is the frame time for a 1000 x 1000 picture point scan. The practical consequences of the threshold equation will be illustrated in the next three experiments.

Field-Emission Gun (FEG) Sources: With an FEG, a very small beam size can be obtained with 1-nA beam current; therefore, it may not be possible to reduce the beam current below threshold. If the threshold current is exceeded even with the fastest frame times, the images for each beam current and frame time will be very similar. Look at Figures A3.3 and A3.4 to see the effects of inadequate beam current.

Experiment 3.8: Effect of Beam Current on Image Quality. (a) Observe the fracture specimen at TV scan rates with beam currents of 10 nA, 1 nA, and 100 pA. Look for the finest detail which can be observed for each beam current. Use a magnification of 1000x or 5000x. By decreasing the beam current over two orders of magnitude in this experiment, we should obtain a significant reduction in beam diameter, and presumably an increase in the fine structure which we can observe. Does this actually happen? How does the visibility vary with the beam current? You may expect different answers for different electron sources, e.g., tungsten, lanthanum hexaboride, or an FEG. What finally limits the detail that can be seen, the beam diameter or the beam current?

(b) Prepare a series of photographic images of the fracture surface at 1 nA, 100 pA, and <10 pA with a constant frame time of 100 sec. At which current is there a loss of detail in the nearly smooth regions? Calculate i_{th}, using the threshold equation, for 100-sec frame time assuming the E-T detector is 25% efficient ($\varepsilon = 0.25$) and the minimum detectable contrast is 10% ($C = 0.1$). What beam size must you accept to detect this contrast?

Calculated threshold current (nA)	Consequent beam size (nm)

(c) Observe and sketch the signal waveform for 1 nA, 100 pA, and <10 pA. Note the relative change in the signal-to-noise ratio S/N.

Experiment 3.9: Effect of Scan Rate on Image Quality. When we examine an unknown specimen in the SEM, we generally search at a rapid scan rate, 1 sec/frame or faster, up to TV scan rate (0.03 sec/frame), but we photographically document the image at much slower scan rates, typically 60-100 sec/frame. If the SEM scan generator permits, obtain a series of photographic images of the fracture surface at 10 sec/frame, 1 sec/frame, and 0.03 sec/frame with a beam current of 100 pA. Compare the fine detail in the images. What strategy does this experiment suggest when seeking a very small feature of low contrast on the specimen?

Using the threshold equation, calculate i_{th} for frame times of 100 sec, 10 sec, 1 sec, and 0.03 sec, assuming that the E-T detector is 25% efficient ($\varepsilon = 0.25$) and that the minimum detectable contrast is 10% ($C = 0.10$).

Frame time	Calculated threshold current	Consequent beam size
100 sec		
10 sec		
1 sec		
0.03 sec		

Thus, for a given beam current (e.g., 100 pA) it is possible to determine when the resolution of the image (i.e., beam size) is limited by threshold current considerations. What is the minimum frame time to just see 10% contrast? How does the value of i_{th} for the TV rate image compare with your experience at the beginning of Experiment 8?

FEG Sources: For a beam current of 0.1 nA, what is the minimum contrast that can be detected at TV rates ($t_f = 0.03$ sec) and at $t_f = 100$ sec? Will specimen visibility degrade in the search mode using TV rates?

3.4 Signal Processing

A major strength of the SEM is derived from the variety of processing techniques which can be readily applied to the measured signals since they are presented in time serial fashion. Signal processing can be used to overcome some fundamental limitations of the human eye in recognizing intensity differences on the CRT screen or on a photographic recording of the image. It is important to realize that while we can manipulate the image display contrast, the modified signals do not overcome the limitations imposed by the threshold equations, i.e., we cannot create contrast or resolution that was not present in the generated image.

The basic analog signal-processing functions which can be found on virtually all SEMs include black level suppression and differential amplification, gamma processing (nonlinear amplification), and signal differentiation (time-derivative signal processing).

Experiment 3.10: Analog Image Processing. Locate a low-contrast feature on the fracture surface. Prepare the following series of images of this area:

(a) Normal image, no processing applied
(b) Black level suppression, two different levels
(c) Gamma processing (nonlinear amplification)
(d) Signal differentiation

Compare the images. How does the visibility of the fracture surface vary with processing? Does any one image provide all the possible information about the specimen? What special information is obtained in each image?

3.5 Digital Images (time and equipment permitting)

Digital image collection and digital signal processing are features of the electronics packages of some recent SEMs and energy-dispersive x-ray spectrometers. In digital image collection, the beam is brought to rest at the center of a discrete region of the image called a pixel (for example, one of 256x256 = 65,536 pixels). The computer directs the beam to dwell on that pixel for a specified length of time while the signal is collected, and then the computer moves the beam to the next pixel. The signal at each pixel may be collected using an analog intensity scale (the traditional method) or in digital form (discrete intensity levels). Digital SEM images offer many additional ways to process the image (see Laboratory 17).

Experiment 3.11: Digital Image Collection. At a pixel density of 256x256, what is a comfortable dwell time to view the image on the viewing screen? For photography? Will these conditions be different with different electron sources? At what screen magnification does a 10-nm diameter beam just fill an image pixel at 256x256? Under high-resolution conditions, photograph a 40,000x image at a very short dwell time (10 µsec/pixel) and at a much longer dwell time (1 msec/pixel)? *Compare the images.*

Experiment 3.12: Digital Image Processing. If facilities are available, apply the digital forms of the image signal-processing techniques mentioned in Section 3.4.

Laboratory 4

Stereo Microscopy

Purpose

This laboratory will introduce the basic concepts of stereo microscopy, which comprises the techniques of perceiving and measuring the "third" dimension of rough, topographic specimens. More details and references can be found in *SEMXM*, Section 4.3.

Equipment
 1. SEM equipped with a conventional Everhart-Thornley (E-T) detector and a stage which permits tilting of the specimen by at least 5°-10°.
 2. Optical stereo viewer.

Specimens
 1. Rough, irregular specimen such as a fracture surface or an insect.
 2. Specimen with regular topography such as a coin with raised lettering or a crystal, such as galena (lead sulfide), which has surface steps.

Time for this lab session

One hour with some advance preparation. *Note:* The stereo images can be prepared in advance and duplicated for distribution to each student.

4.1 Qualitative Stereo Microscopy

The most general class of specimens examined in the SEM is that comprised of rough, irregular objects which have changes in elevation along the optic axis (z axis) of the microscope. Although such objects obviously have "three-dimensional" surfaces and the electron beam intersects the specimen at different elevations during the scan, this z axis information is lost as the image is simultaneously generated on the cathode ray tube (CRT) because the image is effectively constructed by projecting that three-dimensional surface onto a two-dimensional plane. Stereo microscopy provides a means of reconstructing the lost third dimension.

Qualitative stereo microscopy refers to the process of *visualizing* the third dimension of a rough specimen. The technique involves the use of two SEM images (the so-called *stereo pair*) of the same field of view prepared at slightly different angles and an optical stereo viewer to trick an observer into seeing a "three-dimensional" stereo image. The stereo technique operates by mimicking the natural stereo viewing capability of the observer's eye/brain system. To understand the principles of the "recipe" given below for preparing the SEM stereo pairs, it is necessary to understand in at least a cursory fashion the working of the human stereo vision system. The human vision system generates two separate views of a scene which are slightly different because of the different positions of our two eyes. If we focus on an object located within about a meter of our face and alternately open and close each eye, we will see the object shift right and left relative to a vertical axis. This shift in position of the object in the two views is called parallax. Our subconscious image processor in the brain fuses the two distinct views, making use of the parallax to create the sense of depth or three-dimensionality in the final single image.

To mimic this process in the SEM, we need two different views of the specimen *relative to the incident beam*. These views are commonly obtained by recording the first image at a low angle of tilt (e.g., 0°) (the LEFT image of the stereo pair) and then tilting the specimen to a higher angle of tilt (the RIGHT image of the stereo pair). The convention is that tilt *toward* the E-T detector constitutes increasing the tilt angle. The difference in the tilt angles is typically chosen in the range from a minimum of 4° to a maximum of 10°. For qualitative use of stereo images, the choice of the tilt angle depends on the nature of the specimen topography. If the specimen has extremely rough topography, the difference in tilt angle between the members of the stereo pair should be chosen to be a low value to aid the observer in fusing the images with the stereo viewer. Conversely, slight topography will produce a better stereo effect with a larger tilt angle difference. Most observers find differences in tilt angle greater than 10° to be difficult to fuse.

Because the tilt axis in most SEMs is horizontal relative to the CRT, and the human visual system expects parallax perpendicular to a vertical axis, both the LEFT and RIGHT SEM images must be rotated by 90° counterclockwise. (Note that as a result of this rotation, the apparent lighting comes from the left side of the image rather than the top edge which is the normal convention to establish the light-optical analogy.) The images are then placed in the stereo viewer, Low-tilt-angle image in the LEFT ocular of the stereo viewer and the High-tilt-angle image in the RIGHT ocular. The separation of the images is adjusted manually in small increments to allow the observer's visual system to fuse the two views. It should be noted that a significant percentage, 10% or more, of observers cannot observe the stereo effect even after following carefully every step of this procedure. An additional fraction of observers have difficulty achieving the subconscious fusing. While it is difficult to give more specific guidance in achieving this subjective effect, difficulties in achieving the stereo effect can sometimes be overcome by relaxing the eyes. Once the stereo effect has been observed for the first time, subsequent efforts are usually more successful.

Stereo microscopy is not confined to the SEM. There also exists the well-known stereo optical microscope, which operates on the same principle of presenting one of a pair of images, each slightly different in angle of view, to each of the observer's eyes. Optical stereo microscopy with conventional optics is limited to magnifications of about 100x due to the limited depth-of-field of the optical microscope. The SEM is capable of creating useful stereo images throughout its magnification range because the depth-of-field of the SEM is so much greater than that of the optical microscope. For preparing stereo views in the SEM, the microscopist should seek to maximize the depth-of-field of the image by choosing the smallest beam convergence which is practical in terms of useful beam current. With careful choice of the beam convergence, it is generally possible to achieve a depth-of-field at least equal to the field width. If the beam-defining aperture is placed in the plane of the final lens, the beam convergence, α, is defined as

$$\alpha = R/W \tag{4.1}$$

where R is the radius of the final aperture (mm) and W is the working distance (mm) from the plane of the final lens to the crossover of the focused beam below the lens. The depth-of-field, D, can be expressed as:

$$D = 0.2 \text{ mm}/\alpha M \tag{4.2}$$

where M is the magnification and 0.2 mm corresponds approximately to the minimum object which the unaided human eye can resolve in the final SEM image.

4.2 Stereo Photography

The following procedure assumes that the tilt axis is perpendicular to the line from the specimen to the E-T detector. The alternative case is discussed at the end of the procedure.

1. Locating the tilt axis. Determine if the mechanical stage tilt action is arranged to tilt the specimen toward the E-T detector (increasing value of the tilt angle dial corresponds to specimen motion toward the E-T detector). Observers have a strong preference for viewing images with apparent top illumination and for having the tilt axis parallel to an edge of the stereo pair. Note that the scan control, called "scan rotation," may be applied so that the apparent location of the tilt axis in the recorded images is arbitrary. Tilt the stage and observe the SEM image to determine where the tilt axis is located. It may be necessary to adjust the position of the scan rotation control to establish that the apparent location of the tilt axis is horizontal in the recorded image.

2. Centering the eucentric point. If the stage in use is of the eucentric type, adjust the stage position to place the eucentric point at the middle of the image coincident with the optic axis of the SEM, which can be found by centering all shift controls and then progressively increasing the magnification to collapse the scan on the optic axis.

3. Left image. Record an image of the field of interest at an appropriate magnification, and record the specimen tilt. This is the *low-tilt-angle image*. To aid the relocation of the exact field of view, mark several significant features of the specimen directly on the CRT screen with a grease pencil.

4. Tilting. Tilt the specimen by a known amount between 4° and 10° toward the E-T detector.

5. Refocus and recenter. Note that in preparing the images, it is important that the scan field be oriented identically in the two images. When the specimen tilt is changed, the image usually goes slightly out of focus. *Do not refocus with the objective lens control!* Such refocusing changes the lens strength and will alter the relative rotation and/or magnification of the image! Instead, bring the image back into focus by using the motion of the z-axis drive on the stage. Carefully center the image to bring the reference points into coincidence relative to the grease pencil marks on the CRT. Note that because of the parallax, only motion parallel to the tilt axis can be corrected by translation.

6. Right image. Record a second image. This is the *high-tilt-angle image*.

7. Viewing stereo images. For viewing in a stereoscope, place the low-tilt-angle image under the left optical path of the viewer and the high-tilt-angle image under the right optical path. Rotate both images of the stereo pair by 90° counterclockwise to bring the tilt axis vertical in the stereo viewer. Note that this rotation places the apparent illumination at the left-hand side of the final stereo image as perceived by the observer. Most observers find it necessary to adjust the position of the images slightly to fuse the stereo pair. This is a subjective process, and not all observers are successful. A useful procedure is to locate a small, millimeter-sized feature which is easily recognizable in both images. Fix one image (arbitrarily the left-hand image) in the stereo viewer and move the right-hand image to bring the feature into coincidence in the viewer. Only slight additional adjustments (rotation, translation) are generally needed to fuse the images.

Special Case: SEM with stage tilt axis pointing to E-T detector. Some SEM stages are arranged such that the tilt axis is in fact pointed toward the E-T detector so that top illumination is established on the CRT and simultaneously the tilt axis is vertical on the display screen. For such instruments, it is not necessary to rotate the images when placing them in the stereo viewer, since the parallax is created about the vertical axis. The same convention applies to placing the left-hand (lower-tilt-angle) and right-hand (higher-tilt-angle) images. *This is the ideal situation since apparent top illumination is achieved in the stereo pair as perceived by the observer.*

Experiment 4.1: The Stereo Effect. Select a specimen with a large amount of topography. Prepare a stereo pair for a tilt angle difference of 4° according to the instructions above. Observe the stereo effect. Repeat the procedure on the same field of view for a tilt angle difference of 10°. Which stereo pair produces the more satisfying result? Repeat the procedure for a sample with a small amount of topography. At higher magnifications should the degree of tilt be increased or decreased?

4.3 Quantitative Stereo Microscopy

Quantitative stereo microscopy refers to *measuring* the third dimension of a rough specimen. Qualitative stereo microscopy is restricted to those observers who are able to subconsciously fuse the images of a stereo pair into a single image, giving the stereo effect. However, quantitative stereo microscopy is available to all observers, since it only involves positional measurements on recognizable structures common to the two images. These positional measurements are used to calculate the parallax, which is then converted into a relative height difference along the axis of the beam.

There are several approaches to the problem of calculating the parallax and converting it into a height difference. For this laboratory, we will assume that a condition of parallel projection exists, that is, the images of the stereo pair are effectively generated by parallel electron beams at all points in the scan frame of both images. This approximation yields simple formulas which provide useful estimates of the height differences. For magnifications above 1000x, the condition of parallel projection is well satisfied, and at lower magnifications, the approximation becomes progressively poorer. At a magnification of 100x (e.g., scanned field on the specimen 1 mm square at a working distance of 10 mm), the distortion due to the deviation from parallel projection results in an error of only 0.3% for an incremental distance measured near the edge of the field relative to the same measurement at the center of the field. At a magnification of 10x, the error due to projection distortion increases to a value of 25% for edge measurements relative to center measurements, which can clearly lead to unacceptable errors in the calculated heights. For more rigorous equations for the low-magnification stereometry regime, see the review by Howell and Boyde [1].

The critical parallax measurement is determined by first establishing a coordinate system relative to an arbitrarily chosen point common to both images, generally taken as an easily recognized feature. The x coordinates in the left (x_L) and right (x_R) images are then determined relative to this point. The scale distance printed on the image can be employed directly for this measurement, or the distance can be stated in the linear dimensional units of the photograph (e.g., millimeters or centimeters) and then divided by the magnification, M, to give the correct distance.

It is especially convenient if the coordinate system can be written directly on the recorded image with a function of the scan generator. Otherwise, the measurements must be made with a straight edge directly on the photographs, and extra care must be taken to ensure that the photographs are mounted without rotation relative to each other. Using the conventions for recording the stereo pairs given above, the displacement due to parallax is entirely along the x axis. The y coordinates of a feature of interest should be the same in both images if they have been recorded correctly with stage z axis adjustment for focusing and careful repositioning after tilting has been performed. The parallax, P, is calculated as:

$$P = x_L - x_R \tag{4.3}$$

By letting $\Delta\theta$ = the difference in tilt angle, the z coordinate of the feature of interest relative to the reference point, which is arbitrarily designated the origin of coordinates (0,0,0), is given by:

$$z = P/[2M \sin(\Delta\theta/2)] \tag{4.4}$$

The difference in height, D_n, between the reference point and the feature of interest, i, is

$$D_n = z_i - 0 \text{ relative to the reference point } (0,0,0) \qquad (4.5)$$

and

$$D_n = z_2 - z_1 \text{ relative to any other point} \qquad (4.6)$$

In keeping with the proper sense of a right-handed coordinate system, left-going features from the reference point have negative coordinates and right-going features have positive coordinates.

Note: If the parallax calculated with equation (4.3) is positive, the feature of interest is above the reference point, while if the parallax is negative, the feature of interest is below the reference point. The sense of the topography can be checked with the aid of the qualitative stereo effect.

Experiment 4.2: Height Measurements. Carry out quantitative measurements on the height differences of the features marked in the stereo pair in Figure A4.2 using point 1 as the reference point. Calculate the relative heights for features 2 to 7. How well do the corresponding y coordinates of these points match?

Location	x_L	x_R	P	z_i
1				
2				
3				
4				
5				
6				
7				

Reference
[1] P. G. T. Howell and A. Boyde, "Comparison of Various Methods for Reducing Measurements from Stereo-Pair Scanning Electron Micrographs to 'Real 3-D Data,'" *Scanning Electron Microscopy*, Chicago, IITRI (1972) 233-240.

Laboratory 5

Energy-Dispersive X-Ray Spectrometry

Purpose

The modern energy-dispersive x-ray spectrometer (EDS) coupled with a computer-based multichannel analyzer (MCA) provides a powerful analytical facility in the SEM lab. The purpose of this laboratory is to introduce the student to the basic concepts of energy-dispersive x-ray spectrometry and to examine some of the spectral artifacts inherent in the technique. More details and references can be found in *SEMXM*, Chapter 5.

Equipment
1. SEM with an EDS system.
2. Picoammeter (electrometer with high impedence) to measure beam current.

Specimens
1. Flat polished sections of copper, iron pyrite (FeS_2), magnesium, and titanium.
2. Faraday cup (platinum or molybdenum aperture over blind hole in brass, aluminum, or titanium block).

Time for this lab session
Three hours.

5.1 Spectrometer Setup

The solid-state EDS utilizes the fact that x-rays create electron-hole pairs in an intrinsic semiconductor. These electron-hole pairs are swept by the bias voltage to collection electrodes for measurement as a charge pulse--the higher the x-ray energy, the larger the charge pulse. The charge pulse is then amplified and registered as a count in a channel of the appropriate energy in the MCA.

The modern EDS-MCA has evolved into an integrated system in which many of the functions which were previously determined by the settings of potentiometers on electronic modules are now determined by software which controls circuits incorporated into the MCA. Operational parameters such as gain, fine energy calibration, channel width, and time constant are now often set in software rather than hardware. These concepts have thus become somewhat divorced from the apparent reality of hardware. Nevertheless, it is important for analysts to be aware of their responsibility to know the values of these and other parameters, particularly when stored spectra recorded at an earlier time are to be compared with spectra obtained under current operating conditions.

A conventional Be window Si(Li) detector with a 1024-channel MCA, set to 10 eV per channel, gives an energy range of 0.5-10 keV. An MCA with more memory would be preferable, allowing inspection of the full x-ray range 0.5-20 keV, at 10 eV per channel.

Experiment 5.1: Spectrum Acquisition. *Choose a specimen of polished copper. Select an electron beam energy E = 20 keV to ensure an overvoltage of at least U = 2 for analytical lines in the range 1-10 keV. Set the following parameters on the MCA:*

(a) *Energy range - at least 0-10 keV, preferably 0-20 keV*
(b) *Channel width - preferably 10 eV/channel*
(c) *Resolution - lowest count rate range (longest time constant on main amplifier). The lowest count rate range corresponds to the best resolution.*

Adjust the beam current to collect a spectrum with 30% dead time. What does dead time mean?

Experiment 5.2: Energy Calibration Check. *Using the cursor determine the energy of the CuL_α and the CuK_α lines. If these lines are not located at 0.93 and 8.04 keV, respectively, the spectrometer must be calibrated. Consult the manufacturer's manual for the calibration procedure.*

5.2 Beam Current and Dead Time

For a fixed beam energy and specimen type, the x-ray flux which is produced is proportional to the rate at which electrons reach the specimen. The portion of that flux which reaches the detector is determined by the detector area, distance from specimen, and take-off angle. The analyst must be aware of the rate at which the detector is processing x-ray pulses, which is usually indicated by a parameter called dead time. Although the spectrum appears to an observer to build up in parallel in all channels, only one x-ray pulse is actually processed at a time. While the analyzer is busy processing this pulse it is not available to process another pulse which may enter the detector. The second pulse, and sometimes the first pulse, must be discarded. Dead time can conceptually be thought of as

$$\text{Dead time} = \left(1 - \frac{R_{out}}{R_{in}}\right) \times 100\%$$

where R_{in} is the input count rate to the main amplifier and R_{out} is the output count rate from the main amplifier. In practice, manufacturers use electronic circuits to provide a measure of dead time. Many display techniques are used to show the dead time. The first task of the analyst using a new system is to discover what method has been chosen by the manufacturer to display this important parameter. In order to place all x-ray spectrum measurements on a time-equivalent basis, all spectrometers make use of the concept of "live time." The analyst selects a live time for the spectrum measurement, e.g., 100 sec, and the dead time correction circuit automatically guarantees that the system will be available for the specified live time. Thus, the live time and the clock time (real time) will diverge significantly for high dead times (high count rates).

Experiment 5.3: Comparing Live Time and Real Time. *Place the beam on pure copper. MCAs may display both real and live times or live time only. For this exercise, a reading of real time is necessary. If clock time is not provided on your MCA, a stopwatch will be suitable. Adjust the beam current to produce an indicated dead time of 10%. (If a Faraday cup and a specimen current meter are available, measure and note the beam current.) Record a spectrum for a live time of 30 sec, and simultaneously determine the clock time with the stopwatch or MCA function. Increase the beam current to give a dead time of 40% and repeat the measurement. Finally, increase the beam current to give a dead time of 80% and repeat the* *measurement. Report the results of live time and clock time in the table below. How do the values compare?*

Dead time	Live time	Real time	Beam current
10%			
40%			
80%			

While modern MCAs do a remarkably reliable job of correcting for dead time, occasional failures will occur. These failures may not be equipment failures but rather environmental effects such as unexpected and transient ground loops. The prudent analyst will periodically check the performance of the dead time correction. This can be readily accomplished with a specimen current meter (picoammeter). Since the x-ray count rate reaching the detector must be proportional to the current reaching the specimen, a series of spectra recorded for constant live time but with progressively increasing beam current should produce a linear plot of integrated x-ray counts (over either the entire spectrum or any energy window within the spectrum) versus beam current.

Experiment 5.4: Testing the Dead Time Correction Circuit. *For a live time of 50 sec, record spectra from the copper specimen for a series of beam currents, such as 0.1 nA, 0.5 nA, 1 nA, 2 nA, 4 nA, and so on, so as to span the dead time range up to 80%. Read Experiment 5.5 and measure counts after 10 sec real time as you perform Experiment 5.4. This will save considerable time. Plot the integrated intensity (e.g., 1-10 keV window) as a function of beam current. Any significant deviation from linearity, up to a dead time of 60%, should be considered possible evidence of a system problem requiring consultation with the manufacturer.*

Beam current	Counts for 50 sec live time	Dead time	Counts for 10 sec real time
0.1 nA			
0.5			
1.0			
2.0			
4.0			
6.0			
8.0			
10.0			

Because of dead time, there exists a maximum input x-ray count rate, determined by the beam current, beyond which the output count rate actually decreases with further increases in the beam current. This maximum output count rate can be determined by recording a series of spectra for the same beam current choices as used in Experiment 5.4, but with the accumulation time determined by real time.

Experiment 5.5: Maximum Output Count Rate. Select a real time of 10 sec. If the MCA does not permit the use of preset real time to terminate the accumulation, this experiment may be performed manually with a stopwatch. Determine the integrated counts in a wide energy window (e.g., 1-10 keV) over 10 sec real time. Plot the integrated count as a function of beam current. The integrated count should increase initially, rise to a peak, and then decrease with further increases in the beam current. If necessary, to find the maximum, record additional spectra with finer changes in the beam current near the peak in the counts.

In those situations where the analyst requires the best possible precision (i.e., maximum count) in the shortest time, it is necessary to do this experiment to determine the optimum beam current to use. Note that the total x-ray production depends on the atomic number of the specimen and the operating voltage. Consequently, the optimum beam current only applies to a given specimen type. Specimens with a higher average atomic number will produce a higher flux of x-rays and will therefore require a lower beam current to stay at or below the optimum value.

The detector resolution is intimately associated with the selection of the amplifier characteristics which affect the dead time. The best resolution is obtained with the longest amplifier time constants, and therefore the lowest limiting count rate. Since both high resolution and high limiting count rate capability are desirable, the analyst is faced with an unpleasant choice. There are usually three choices of resolution, typically selected by a hardware switch on the main amplifier. The choices, high, medium, and low, refer to the count rate range, and the resolution varies in the opposite sense, i.e., high count rate range gives the lowest resolution.

Experiment 5.6: Detector Resolution. Record spectra of iron pyrite for each selection of count rate range so that the height of the sulfur K_α peak is approximately 5000 counts. Using the sulfur K_α peak, estimate the EDS resolution by measuring the full-width-at-half-maximum (FWHM) for each spectrum at 2.31 keV.

Is this the guaranteed resolution stamped on the detector dewar? Why not? At what resolution setting is the SK_β sufficiently resolved from the SK_α to measure the ratio of peak heights? These spectra illustrate a classic example of the trade-off between the need for resolution and the need for high count rates.

5.3 Spectral Artifacts

Now that we have observed the general characteristics of EDS operation, it is appropriate to consider the artifacts of the detection process. Although these artifacts are, in general, second-order effects, their presence in the spectrum will influence both qualitative analysis of minor and trace constituents, possibly leading to incorrect identification of the elements present in a specimen, and quantitative analysis, possibly leading to serious errors in the concentrations calculated for some minor and trace constituents.

Experiment 5.7: Stray Radiation. Using the composite Faraday cup, which consists of an electron microscope aperture of platinum pressed into a blind hole in a titanium block, obtain 50-sec spectra with the beam placed

(a) On the block
(b) On the aperture metal
(c) In the aperture hole

Examine the "in-hole" spectrum. Is there any characteristic radiation from the elements of the aperture or block? *Intentionally misalign the final aperture of the microscope and repeat the three spectra.* Is there any increase in the stray radiation reaching the specimen plane?

Experiment 5.8: Escape Peaks, Sum Peaks, and System Peaks. Record a 100-sec spectrum on pure titanium with a system dead time of 60%. Examine the spectrum for the presence of titanium escape peaks, sum peaks, and peaks from elements in the specimen chamber.

Experiment 5.9: False Peaks. Obtain a 20-sec spectrum on titanium at a dead time of 20%. Observe the background of the raw spectrum and compare it with the same spectrum after smoothing. Can small peaks be seen after smoothing? Determine if these are true peaks by repeating the spectrum and smoothing again. Do the peaks appear in the same locations?

Experiment 5.10: Pulse Pile-up (time permitting). Record a 100-sec spectrum of magnesium with a system dead time of 60%. Examine the spectrum for a magnesium sum peak and a pulse pile-up continuum. Why is the pulse pile-up continuum present?

Experiment 5.11: Pile-up Correction Failure (time permitting). Plot the integrated intensity for the magnesium peak as a function of beam current. Why does the intensity roll over after a certain point?

The phenomenon of incomplete charge is an important second-order distortion of EDS spectra. Incomplete charge is a result of a number of individual effects, and it is therefore difficult to design a single experiment to illustrate adequately the overall impact on spectra. Incomplete charge has a negligible effect on qualitative analysis, but becomes significant for quantitative analysis of minor constituents below major constituent peaks.

Experiment 5.12: Incomplete Charge Collection (time permitting). *Examine the low-energy side of the major peak for both a titanium and a copper spectrum. How would incomplete charge collection affect the quantitative analysis of an element?*

Laboratory 6

Energy-Dispersive X-Ray Microanalysis

Purpose

The modern energy-dispersive x-ray spectrometer (EDS) coupled with a computer-based multichannel analyzer (MCA) provides a powerful analytical facility in the SEM lab. The purpose of this laboratory is to introduce the student to the wide range of analytical capabilities of the EDS/MCA system and to illustrate the basic appearance and characteristics of electron-excited x-ray spectra. Procedures for both qualitative and quantitative analysis will be examined. More detail on these subjects can be found in *SEMXM*, Chapters 6, 7, and 8.

Equipment
1. SEM with an EDS system. A conventional Be window Si(Li) detector with a 1024-channel MCA, set to 10 eV per channel, gives an energy range of 0.5-10 keV. An MCA with more memory would be preferable, allowing inspection of the full x-ray range of 0.5-20 keV at 10 eV per channel.
2. Picoammeter (electrometer with high impedance) to measure beam current.

Specimens

All specimens should be metallographically polished (to 0.3 µm grit) and unetched.
1. Aluminum, copper, gold, and uranium oxide are suitable for Experiment 6.1. If uranium oxide is not available, uranyl acetate, used for staining biological specimens, may be used. If no uranium compound is available, use lead. Aluminum, silicon, iron pyrite (FeS_2), titanium, copper, zirconium, silver, barium oxide, gold, tantalum, lead, and uranium oxide are used in Experiments 6.2-6.4.
2. A metal alloy and a glass sample, both of unknown composition.
3. Al-Ni alloy and Al and Ni pure element standards.

Time for this lab session

Three hours for Experiments 6.1 and 6.5 to 6.11.
Four hours for Experiments 6.1 through 6.11.

6.1 Families of X-Ray Spectra

The analyst must be familiar with the appearance of typical x-ray spectra. In this section, we will record x-ray spectra from a variety of elements and make use of the x-ray data display of the MCA to identify each x-ray peak. The EDS/MCA system is an ideal tool to study spectral energies and relative peak intensities.

Experiment 6.1: K, L, and M Spectra. With 20 kV electron beam record separate spectra for 50 sec live time and at 30% dead time from the elements aluminum, copper, gold, and uranium. Store the copper spectrum for Experiment 6.8. Identify all the peaks by looking up their energies in Figure 6.1 [1], in a reference table [2], or on an "energy slide rule."

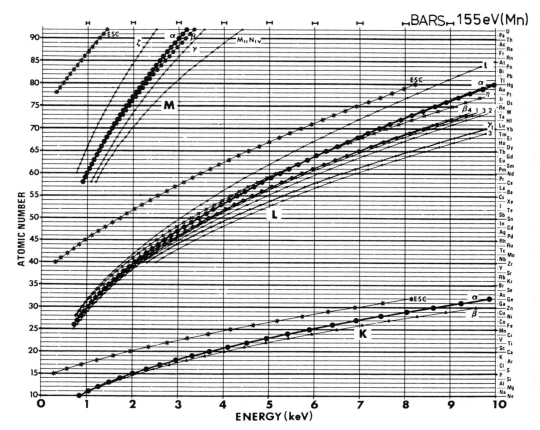

Figure 6.1. Plot of the x-ray emission lines observed in the range 0.75-10 keV by energy-dispersive spectrometry [1].

Record your data in Table 6.1 and compare the measured and tabulated energy values. How can you recognize the K, L, and M families of spectra? Use the KLM markers on the x-ray system to identify the peaks. Do the software markers identify all the peaks that you found yourself? Is the spectrometer energy calibration correct to within one channel?

Experiment 6.2: K Family X-Rays (time permitting). Record spectra for 50 sec and at 30% dead time from the following elements and compounds: Al, Si, pyrite (FeS$_2$), Ti, and Cu. With the MCA, locate the K-family x-ray lines in each spectrum. How does the appearance of K spectra change as a function of atomic number? *If possible, prepare a hard copy output of each spectrum.*

Table 6.1. X-Ray Spectra from Pure Elements

Element	Peak No.	Measured energy	Tabulated energy	Peak identification
Al				
Cu				
Au				
U				

Experiment 6.3: L Family X-Rays (time permitting). Record spectra for 50 sec and at 30% dead time from Cu, Zr, Ag, Ba (as a compound, e.g., BaO), and Au. Locate the L-family x-ray lines for each element. Prepare hardcopy output. Compare the appearance of the L family spectra as a function of atomic number.

Experiment 6.4: M Family X-Rays (time permitting). Record spectra for 50 sec and at 30% dead time from Ti, Au, B6, and UO_2. Locate the M-family lines for each element. Prepare hardcopy output. Compare the appearance of each M family. For uranium several minor M-family peaks, such as $M_{II}N_{IV}$, may not be included in the MCA data table. To identify such peaks, it may be necessary to resort to a more complete x-ray data compilation such as that of Bearden [2].

6.2 Qualitative Analysis

Qualitative analysis is defined as the identification of the elemental constituents of the specimen by recognition of the characteristic x-ray peaks associated with those elements. With a conventional beryllium-window EDS, elements with $Z \geq 11$ (sodium) can be directly measured. With the windowless or ultra-thin-window (UTW) detector, the range is extended to $Z \geq 5$ (boron). Qualitative analysis with the EDS involves some knowledge of basic x-ray physics, intuition, and common sense. From x-ray physics, we require such information as the energies of the x-ray lines for each element, the approximate relative intensities of the lines, and the artifacts which arise from the detection process. Intuition can be developed as a result of experience by starting with qualitative analysis of pure element spectra (Experiments 6.1-6.4), progressing to binary mixtures, and eventually attacking complex, multicomponent spectra. If this laboratory has been followed in sequence to this point, the novice has examined EDS x-ray spectra from pure elements and a binary compound (FeS_2) and has also become familiar with the artifact peaks (escape peaks and sum peaks) which are associated with the high-intensity peaks (Laboratory 5). The analyst must always be ready to apply common sense to a proposed identification, whether it is done manually or is suggested by an automatic peak identification program. For example, scandium is an exceedingly rare element, and if it is identified in a spectrum, the analyst should be cautious and ascertain if the apparent scandium peak is really from that element. One should be even more cautious with technetium and promethium!

As a general rule, it is dangerous to identify an element on the basis of a single peak. Sometimes one peak is all that is available, as in the case of a light element, e.g., Na or Al, even if the element is present as a major constituent. When trace or minor constituents are considered, only one peak may be observed from even the heavier elements. If it is important to identify the presence of a constituent, then multiple peaks should be identified whenever possible, and the "Golden Dictum" should be applied.

Golden Dictum: Always accumulate an adequate number of counts in the spectrum to give statistically significant peaks. If it is difficult to decide if a peak exists against the continuum background because of noise (which can be estimated as the "thickness" of the MCA line plotted for the background), then either accumulate more counts to "develop" the peak or else forget about identifying it.

Corollary to the Golden Dictum: EDS systems become increasingly prone to the introduction of spectral artifacts as the count rate increases. Remember that the count rate of interest is the whole spectrum count rate, including the continuum up to the beam energy.

Qualitative Analysis Guidelines

The following guidelines are provided to give some direction to the novice who wishes to solve an EDS spectrum. Most of the suggestions, excepting the Golden Dictum, are simply guidelines and not iron-clad rules. There is, unfortunately, no rote procedure which guarantees a perfect solution without careful thinking on the part of the analyst, who must be prepared to deal with unexpected situations.

Good bookkeeping must be practiced throughout the qualitative analysis procedure. As each elemental assignment is made, all peaks and artifacts associated with that element must be marked, particularly the low-intensity peaks. If these low-intensity peaks are overlooked, they are likely to be misidentified later in the procedure as belonging to constituents at the minor or trace level. Because of computer memory limitations, some analyzers do not allow enough markers to denote all peaks associated with a particular element (check this with a specimen of high atomic number), including all artifacts. In such cases, the analyst must note these additional peaks manually with a written list as a function of energy (as in Experiment 6.1).

1. **Start at High Energy.** Begin with characteristic peaks at the *high*-energy end of the spectrum. Reason: At high energy the peaks of a given family, e.g., K_α-K_β, Ll-L_α-L_η-L_β-L_γ, and M_ζ-M_α-M_β-M_γ-$M_{II}M_{IV}$, are separated by the largest differences in energy and are most likely to be resolved as separate peaks by the EDS.

2. **Try the K Family.** Choose a large peak. Using the KLM markers of the MCA, check to see if it might be a K_α peak by immediately looking for the corresponding K_β peak at approximately 10% of the K_α peak height. If the energy calibration of the EDS system was confirmed in Experiment 6.1, then the peak locations should be within one channel of the correct value. For elements starting with sulfur and increasing in atomic number, the K_α-K_β peaks will be resolved with an ordinary 145-eV resolution detector.

3. **Try the L Family.** If a K_α-K_β pair does not fit, try various L lines with the KLM markers. If an L_α candidate is found, look immediately for the complete L family, L_α-L_β-L_γ (three peaks of decreasing intensity). Look also for the low-intensity Ll and L_η lines. Be aware that some "energy slide rules" and even some MCA-generated KLM markers may not include a complete listing of minor lines (Ll, L_η, and $M_{II}M_{IV}$).

4. **Related Families of Elements.** As soon as a tentative elemental identification is assigned to a set of peaks, the analyst should seek all other lines associated with that element. Thus, the presence of a K_α-K_β pair above 6 keV (e.g., iron and above) *requires* the presence of the L family for that element at a lower energy in the spectrum. Thus, if the CuK_α (8.04 keV) and CuK_β (8.90 keV) are located, the analyst should immediately look for the L family, which appears as a single peak at 0.93 keV. Similarly, the existence of L-family lines above 5 keV (neodymium and above) *requires* the presence of the corresponding M family at 1 keV or above. For example, locating WL_α (8.40 keV) should cause the analyst to also look for the M family, which appears as a single line W M_α (1.78 keV). Locating all of the possible lines at the correct energies establishes a high level of confidence that a tentative elemental assignment is correct. To help in identifying families of x-ray lines for various elements, Table 6.2 gives the approximate relative intensities of lines in each family.

Table 6.2. Relative Intensities of X-Ray Lines (Approximate)

X-Ray Family	X-Ray Lines (relative intensity within family)
K Family	K_α (1), K_β (0.1)
L Family	L_α (1), $L_{\beta 1}$ (0.7), $L\gamma$ (0.08), $L_{\beta 2}$ (0.2) Ll (0.04), L_η (0.01)
M Family	M_α (1), M_β (0.6), M_ζ (0.06), $M\gamma$ (0.05), transition $M_{II}N_{IV}$ (0.01)

5. **Artifact Peaks.** Once all of the possible x-ray families associated with a candidate element have been located, the analyst should also find and mark the artifact peaks associated with the high-intensity members of each family. The analyst should examine the positions of Si escape peaks (at 1.74 keV below the parent peak energy) and sum peaks using appropriate vertical scale expansion. Remember that the relative strength of the artifact peaks decrease as the parent peak energy increases. It is generally difficult to detect artifact peaks for parent peaks above 7 keV in energy. When sum peaks are present, particularly for low-energy (< 3 keV) parent peaks, be aware of the various possible combinations. Thus, if three large peaks E_A, E_B, E_C exist in close energy proximity, the sum peaks can include $2E_A$, $2E_B$, $2E_C$ and $(E_A+ E_B)$, $(E_A+ E_C)$, $(E_B +E_C)$. Some peaks will disappear if the spectrum is recollected for a longer time at a lower count rate.

6. **Repeat for the Next Large Peak.** Steps 1-5 are repeated for progressively lower peak energies until all high-intensity peaks in the spectrum have been identified. Only after all large peaks are identified should the analyst seek to identify minor and trace elements.

7. **Minor Peaks.** The same strategy is followed for minor and trace constituents, i.e. the analyst begins at the high-energy end of the spectrum and works downward in energy. It is often the case that only one line can be recognized for a minor or trace element. If it is necessary to improve the confidence with which an assignment can be made, it may be necessary to accumulate the spectrum for a much longer time to develop adequate statistics for detection of additional x-ray family members for the element of interest.

8. **Peak Overlaps.** When finished with the above procedure, the analyst should always ask the following question: What other elemental constituents might be present which cannot be detected because of severe interference situations? That is, what minor or trace constituents will be missed because they are "buried" beneath major peaks of other elements? An excellent example is the mutual interference among SK_α, MoL_α, and PbM_α. Such interference situations can be conveniently assessed by means of Figure 6.1 which depicts all elements which can produce an x-ray line measurable by EDS in the range 0.75-10 keV. By means of this figure, the analyst can rapidly determine what hypothetical interferences exist. If these hypothetical interferences are important, the analyst should consider what steps could be taken to find alternate x-ray lines with which to detect the elements in question. It may be necessary to increase the beam energy to efficiently excite higher-energy lines. For example, in the S-Mo-Pb situation the presence or absence of Mo may be confirmed by sufficiently exceeding the K absorption edge energy for Mo (20.0 keV) to excite the Mo K series. A beam energy of 30 keV would provide an overvoltage of 1.5 to excite this radiation.

Experiment 6.5: Qualitative Analysis of a Simple Spectrum. *Record a spectrum from the metal alloy sample for 100 live sec at $\leq 30\%$ dead time. Identify all the peaks according to the procedures described above. If possible store the spectrum. Prepare the following report table noting the energies of all peaks:*

Specimen identification:	Beam energy = keV
Major constituents:	
Minor constituents:	
Trace constituents:	
Possible interferences:	

Experiment 6.6: Qualitative Analysis of a Complex Spectrum. Record a spectrum from the glass sample for 100 live sec at ≤30% dead time. Identify the peaks according to the procedures described above. If possible store the spectrum. Prepare the following report table noting the energies of all peaks:

Specimen identification:	Beam energy = keV
Major constituents:	
Minor constituents:	
Trace constituents:	
Possible interferences:	

Experiment 6.7: Automatic Qualitative Analysis. If the MCA in use is equipped with software for automatic peak identification, use it to determine the elements present in the stored spectra of Experiments 6.5 and 6.6. How do the automatic analyses compare with your manual analyses? Were any additional elements detected, especially minor or trace elements? If so, examine these peaks closely. Do the assignments appear justified? *Obtain a spectrum of the glass sample for 10% of the time which was used previously, and apply a smoothing function to the spectrum. Repeat the* automatic qualitative analysis. How do the elemental assignments compare? Are any new minor or trace elements identified? Are these identifications valid?

6.3 Quantitative Analysis

This section lays the groundwork for quantitative analysis in the SEM. On-line processing of EDS spectra from unknowns and simple standards can yield quantitative analysis with relative errors of the order of ±3% and detection limits of the order of 0.1 wt%. These extraordinary capabilities sometimes blind us to the limitations of these systems.

The basis for quantitative x-ray microanalysis is that, to a first approximation:

$$\frac{C_{spec}}{C_{std}} \propto \frac{I_{spec}}{I_{std}} = K \text{ ratio} \tag{6.1}$$

where C is the mass (weight) concentration of an element, and I is the measured characteristic x-ray intensity for that element (corrected for background, peak overlap, and dead time). The subscript "spec" denotes the concentration or intensity of the element in the specimen, and "std" refers to the standard of known concentration. The ratio of characteristic intensities measured on the specimen and the standard is known as the "K ratio." The proportionality sign in Equation 6.1 indicates that the relationship between concentration and characteristic x-ray intensity is not exact. There exist "matrix effects," arising from the nature of the electron and x-ray interactions with matter, which modify the measured intensities, and which depend on the unknown composition of the specimen. A variety of approaches are used ("ZAF," "phi(rho-z)," "empirical") to calculate correction factors for these matrix effects. Descriptions of these correction procedures are given in *SEMXM*, Chapter 7. This laboratory will examine two extremes: fully rigorous ZAF analysis using standards and the so-called "standardless" technique. In addition, this laboratory will illustrate the errors which can arise when user-selected parameters are incorrectly established.

When careful attention is paid to operating conditions and analysis procedures, fully rigorous analysis with standards can yield a relative accuracy of $\pm 3\%$ for most elements. For the EDS case, optimal conditions will be considered to be the following:

1. Element atomic numbers of 11 (sodium) and greater.
2. Concentrations greater than 5%.
3. No significant interelement peak overlaps.

The precision of the concentration determination depends on the measurement statistics for the characteristic and bremsstrahlung (background) x-ray intensities.
There are two key assumptions, often assumed as being obvious, which form the foundation of the x-ray microanalysis technique:

1. Homogeneous Specimen. The specimen is assumed to be homogeneous in the volume sampled, which typically has dimensions of micrometers. Note that this assumption precludes "overscanning" an inhomogeneous, multiphase specimen during spectral accumulation to obtain an "average" EDS spectrum which is mistakenly thought to yield an "average" composition for the specimen after matrix correction. Such a procedure can produce relative errors of several hundred percent, rendering the results meaningless!

2. Unknown and Standard Polished Flat. The only reason the x-ray intensity measured on the specimen differs from the x-ray intensity measured on the standard is that the concentration is different between the specimen and the standard. This situation is only achieved when both specimen and standard are polished flat (metallographic polish *but no etching*) and set at known and identical angles relative to the electron beam and the x-ray spectrometer. *Note that this requirement precludes the direct analysis of rough surface, particle, thin foil, and thin film on substrate specimens.* For these specimens, geometrical effects (size, shape) come into play and affect the measured x-ray intensities (see Laboratory 22 and *SEMXM*, Section 7.5). If we treat one of these special samples in the same way as an ideal flat, polished sample, relative errors of several hundred percent can occur, rendering the results meaningless!

Fully Rigorous ZAF Analysis. The most effective strategy for EDS analysis is to record and archive spectra from standards (pure elements, simple compounds such as gallium phosphide, alloys, and glasses) under standardized electron beam and x-ray spectrometer conditions. Provided data collection conditions can be reproduced from day to day, these archived spectra from standards can be used indefinitely.

Archived standard spectra may be used to process x-ray spectra from the specimen to extract characteristic x-ray intensities, automatically correcting for peak overlap and background. These spectra may even be used to calculate K ratios. The ZAF matrix correction method may then be applied to the K ratios to produce the final measured concentrations.

"Standardless" Analysis. There exists some ambiguity in the field over the definition of the "standardless" concept. For this laboratory, we will assume the most extreme case: only the spectrum of the unknown is measured. The standardless analysis technique then calculates the required standard intensities for the K ratios from first principles using mathematical descriptions of x-ray generation. The analytical total is normalized to 100%. The standardless technique is extremely sensitive to changes in take-off angle, changes in kV, and to the presence of unmeasured elements.

Experiment 6.8: Establishing Proper Working Conditions. A good procedure at the beginning of any quantitative analysis session is to record a spectrum from a pure element and compare that spectrum to the equivalent archived spectrum. In this way, possible errors can be evaluated which might arise from incorrect spectrometer settings (calibration, shaping time or count rate range, and dead time correction) or beam parameters (energy, which affects ionization efficiency, and beam current). An excellent choice for the test specimen is copper, which provides CuL, CuK_α, and CuK_β, bracketing most of the energy range of analytical interest. Collect a 50 sec copper spectrum at 20 kV and use the MCA to compare the peak eights and shapes with those recorded previously in Experiment 6.1 (or other library Cu spectrum recorded at the same take-off angle and at 20 kV).

Experiment 6.9: ZAF versus Standardless. *For an aluminum-nickel alloy and pure element Al and Ni standards, measure and record spectra at 20 kV and at a known specimen tilt and working distance. Choose the optimum detector resolution (longest shaping time), a beam current which gives a dead time on pure Al of approximately 30%, and a live time of 100 sec. Measure the beam current in a Faraday cup. Process the spectra through both the ZAF and the standardless software according to the manufacturer's instructions. Specify the correct values for the beam energy, the angle of beam incidence (or tilt angle), and the take-off angle. Examine the final report table for both techniques.*

1. Are the K ratios reported?
2. How do the measured concentrations compare for the two techniques?
3. Are the ZAF matrix corrections reported? What is the magnitude of the total matrix correction for each element? What are the magnitudes of the individual corrections? Which element requires the most significant correction *in this particular situation?*
4. What is the ZAF analytical total for the two elements measured? Are the final reported concentrations normalized to 100%? Is normalization a good idea? If the specimen were oxidized, and oxygen was not measured, does the software offer the option of calculating oxygen concentration by assumed stoichiometry? Would such a calculation be better than calculating the oxygen concentration by difference?
5. Is the precision of the measurement (of a single point) reported? How is the precision estimated? What statistical measure is used? What is the utility of this estimate of the precision?

Experiment 6.10: Effect of Take-Off Angle. *Change the specimen tilt by 20° and record a new spectrum on the Al-Ni alloy at the same beam energy. Calculate the composition of this unknown using the same values of the take-off angle and incidence angle as used for the previous ZAF and standardless measurements. How do the calculated concentrations compare between the two methods? Which method is more sensitive to operator error? Which element suffers the largest error? Note: if the electron beam instrument does not permit changes in the tilt, this effect can be simulated by altering the input take-off angle in the software.*

Experiment 6.11: Effect of Beam Energy. *Change the beam energy to 19 keV. This simulates a possible error which can easily arise in practice. Record a spectrum for the Ni-Al alloy at the same beam current and live time as previously. Repeat the calculations for the ZAF and the standardless techniques using 20 keV instead of the correct 19 keV in the beam energy input value to the software. How do the calculated concentrations compare between the two techniques? Which method is more sensitive to operator error? Which element suffers the largest error? If the electron beam instrument does not permit small changes in beam energy, this effect can be simulated by altering the input energy to the software.*

References

[1] C. E. Fiori and D. E. Newbury, *Scanning Electron Microscopy/1978*, vol. I, SEM Inc., AMF O'Hare, IL, p. 401.
[2] J. A. Bearden, "X-Ray Wavelengths and X-Ray Atomic Energy Levels," NSRDS-NBS 14, National Bureau of Standards, Washington (1967). Also published in recent editions of the *CRC Handbook of Chemistry and Physics*, The Chemical Rubber Company, Cleveland, Ohio.

Laboratory 7

Wavelength-Dispersive X-Ray Spectrometry and Microanalysis

Purpose

This laboratory demonstrates the operation of a wavelength-dispersive x-ray spectrometer (WDS) fitted to a scanning electron microscope (SEM) or electron probe microanalyzer (EPMA). The WDS has important advantages over the energy-dispersive spectrometer (EDS) in terms of the peak-to-background ratio, improved elemental sensitivity, and better energy resolution of characteristic x-ray peaks to avoid peak overlaps. Comparisons of the major characteristics of the WDS and EDS detectors will be made. More detailed background information may be found in *SEMXM*, Chapters 5-8.

Equipment

SEM or EPMA with both a wavelength-dispersive spectrometer and an energy-dispersive spectrometer.

Specimens

1. Zirconium dioxide (ZrO_2) cathodoluminescent specimen.
2. Polished but unetched 316 stainless steel (nominal composition in wt%: 1 carbon, 1 silicon, 2 manganese, 2-3 molybdenum, 10-14 nickel, 16-18 cromium, balance iron).
3. Polished, unetched pure manganese standard, preferably mounted in the same mount as 2.

Time for this lab session

Two to three hours.

7.1 WDS Operating Conditions

The WDS system usually requires a higher beam current than an EDS system. The specimen must be at the focal point of the WDS whereas specimen position is generally less critical for the EDS. Another difference is that the angle of the WDS analyzing crystal must be set to the correct value in order for the x-ray line of interest to be detected.

Experiment 7.1. Setting Beam Current and the Focusing Beam. Insert the 316 stainless steel specimen. Set the condenser C1 lens for high-beam-current operation (30 nA specimen current). Such a high beam current is necessary to obtain optimum x-ray intensity (e.g., 10,000-50,000 counts/sec on pure iron) because of the small collection angle of the WDS. *Focus the electron beam by scanning the sample at $\geq 5000x$ and focusing the secondary electron (SE), backscattered electron (BSE), or specimen current image.* If light optics are

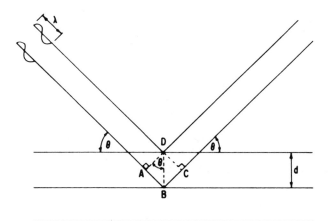

Figure 7.1. Schematic diagram showing the variables of Bragg's Law. To detect a particular wavelength λ the analyzing crystal of known atomic spacing d must be set to the appropriate Bragg angle θ for diffraction.

available, the focusing of the electron beam may be observed directly on the cathodoluminescent ZrO_2 specimen. Estimate the electron beam size for the 30-nA beam. What are the functions of the condenser and objective lenses?

Experiment 7.2: X-Ray Spectrometer Setup. Select the lithium fluoride (LiF) (200) analyzing crystal ($2d = 4.028$ Å). Calculate the 2θ angle on the spectrometer to detect the FeK_α x-ray peak ($\lambda = 1.936$ Å) using Bragg's law, $n\lambda = 2d \sin\theta$ (see Figure 7.1). Check that the detector voltage of the gas proportional counter is set correctly (~1500 to 2000 V). For the electron probe microanalyzer, the working distance and x-ray take-off angle are fixed, but both of these can be widely varied in the SEM. Set the working distance to a reproducible value consistent with obtaining a high x-ray signal. The manufacturer may provide a recommended value of the z-motion. What is the take-off angle? Why is an accurate and reproducible take-off angle important?

Experiment 7.3: X-Ray Intensity versus Sample Current. Measure the FeK_α peak x-ray intensity for specimen currents of 1 nA, 10 nA, 50 nA, and 100 nA. What is the relationship between x-ray intensity and specimen current? Why is it so critical to keep the specimen current constant (+ 1% relative) for quantitative x-ray analysis?

Sample current	FeK_α intensity
1 nA	
10 nA	
50 nA	
100 nA	

Is there a dead time associated with WDS measurements? Compare with EDS dead time.

7.2 Characteristics of the WDS

The WDS used on the SEM or EPMA is usually of the fully focusing type, maintaining a constant take-off angle with the specimen surface. Figure 7.2 shows the geometrical relationship among the spectrometer, specimen, and electron beam. The unique Johansson focusing optics allow excellent energy resolution for the spectrometer. However, large changes in measured x-ray intensity may occur if the WDS is misaligned, for example, if the specimen is not reproducibly placed at the same working distance.

Experiment 7.4: Energy Resolution of the WDS. *For a 20-kV operating potential, 30-nA specimen current on the 316 stainless steel specimen, and a 10-sec counting time per*

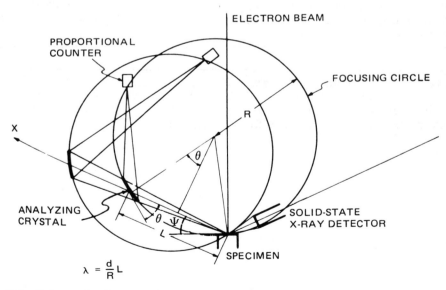

Figure 7.2. Schematic diagram showing the relative positions of the specimen, analyzing crystal, and proportional counter for a fully focusing WDS system.

point, measure the x-ray intensity as a function of 2θ at discrete 2θ values across the FeK_α peak (0.5 sec/10 eV). Note that the position of the spectrometer may be measured in terms of 2θ, wavelength (λ), energy, or distance of the specimen from the crystal in millimeters. Measure the full-width-at-half-maximum (FWHM) of FeK_α and determine the energy resolution of the spectrometer in electron volts. Compare the WDS energy resolution with that obtained from EDS for the FeK_α line from the stainless steel specimen.

WDS	EDS
Width of FeK_α	Width of FeK_α
FWHM = _____ deg 2θ	FWHM = _____ eV
= _____ Å	
= _____ eV	

Why is the resolution of the WDS so much better than that of the EDS? What is the trade-off between measured x-ray intensity and WDS energy resolution if the size of the focusing circle is different?

Experiment 7.5: Peak-to-Background Ratio for the WDS. *For a constant specimen current of 30 nA on the 316 stainless steel specimen and a 10-sec counting time, measure the FeK_α peak x-ray intensity and the off-peak background for operating voltages of 30, 20, and 10 kV. The off-peak background is obtained by measuring the average continuum background ~100 eV above and ~100 eV below the position of the characteristic FeK_α peak. What is the relationship between peak-to-background (P/B) ratio and operating voltage (E_c for FeK_α is 7.11 keV)? What happens to the penetration depth and x-ray spatial resolution at higher voltages? Last, compare the P/B ratio for the WDS with that of the EDS. For the EDS the background intensity is generally measured over the same energy interval as the peak intensity.*

WDS				EDS			
kV	P	B	(P - B)/B	kV	P	B	(P - B)/B
10				10			
20				20			
30				30			

Experiment 7.6: X-Ray Intensity versus Working Distance. For a 20-kV operating potential, 30 nA specimen current on the 316 stainless steel specimen, and a 10-sec counting time per point, measure the FeK_α peak intensity as a function of working distance for both the WDS and EDS detectors. Start with variations of 1 mm; smaller changes in working distance may be needed, especially for a vertical WDS, which is far more sensitive to height changes.

WDS		EDS	
Working distance	Fe intensity	Working distance	Fe intensity
mm		mm	

Plot the data. Does the peak intensity for both the WDS and the EDS change with sample height? Was the working distance set correctly initially? Does this result suggest why flat polished samples should be used for WDS analysis?

7.3 Typical WDS Analysis Situations

Experiment 7.7: Characteristic X-Ray Peak Overlaps. A serious peak overlap occurs in the measurement of MnK_α (7.894 keV) with CrK_β (7.946 KeV) using the EDS. For a 20-kV operating potential and a 30-nA specimen current on a flat polished 316 stainless steel specimen, set the WDS to detect MnK_α. A pure manganese standard may be useful for this since most of the manganese in the stainless steel will be tied up in manganese-sulfide inclusions. Scan 2θ through the MnK_α peak with enough room on the low-wavelength (high-energy) side to detect CrK_β. Compare this measurement of MnK_α with that shown on an EDS spectrum taken for the same length of time as the WDS scan. If possible, compare the WDS result with the deconvolution of these two peaks using the EDS x-ray microanalysis software. Even if MnK_α in the stainless steel did not overlap with CrK_β, why is analysis of small concentrations of an element more easily accomplished using the WDS than the EDS?

Experiment 7.8 (time and equipment permitting): **Quantitative Analysis.** *Form the K ratio of background-subtracted MnK_α from the 316 stainless steel matrix (avoid precipitates) to that from the pure manganese using both WDS and EDS. For the EDS spectrum a mathematical deconvolution will be necessary to obtain the MnK_α intensity separate from the CrK_β intensity. Use a ZAF-type quantitative correction program to correct the K ratios obtained. Compare the results.*

Experiment 7.9 (time and equipment permitting): **X-Ray Dot Maps.** *Take a 20-kV, 2000x BSE or SE image of a manganese-sulfide inclusion in the 316 stainless steel. Set the WDS spectrometer to detect sulfur K_α x-rays and record this signal as an analog dot map (frame time = 120 sec). K_α dot map of the same area for the same time. Record an EDS dot map for manganese. Explain the difference in contrast between the WDS and EDS dot maps of manganese. Could the sulfur dot map be obtained by EDS? What happens when this experiment is done at 100x?*

Part II

ADVANCED SCANNING ELECTRON MICROSCOPY

Laboratory 8
Backscattered Electron Imaging

Purpose
This laboratory will demonstrate the influence of the electron detector and its characteristics upon the final appearance of the SEM image. The experiments are principally centered around the use of the backscattered electron signal to obtain topographic and compositional information. More details may be found in *SEMXM*, Section 4.4.

Equipment
1. SEM equipped with a conventional E-T detector and a dedicated under-the-pole piece backscattered electron detector.
2. Additional detectors such as specimen current and other types of backscattered electron detectors will permit a more complete comparison of detectors to be made.

Specimens
Multiphase specimen with slight topography, e.g., metallographically polished specimen of Raney nickel, a multiphase tool, or basalt.

Time for this lab session
Two hours with some advance preparation.

8.1 Backscattered Electron Detectors

Several different types of detectors which are specifically sensitive to backscattered electrons (BSEs) are now commonly available. Most are based on either of two detector types: the solid state type, which converts BSE energy to charge in a silicon diode, or the scintillator/photomultiplier type, which converts BSE energy to light with subsequent reconversion to electrical charge in a photomultiplier. With these two basic detection schemes, further detector variations are possible through the use of arrays of discrete detectors, which provide the facility for combining signals in additive, subtractive, or ratio fashion. Most dedicated BSM detectors are located directly above the sample to provide the greatest possible collection efficiency; however, effects of BSEs can also be detected by the Everhart-Thornley detector and the specimen current meter. In this laboratory, images of the same specimen taken with a variety of BSE detectors will be compared.

8.2 E-T Detector as a BSE Detector

The conventional Everhart-Thornley E-T detector, which is often mistakenly thought of exclusively as a secondary electron detector, actually functions to a considerable degree as a BSE detector. In its normal mode of operation, the metal screen which surrounds the scintillator is biased positively a few hundred volts to establish a strong collection field for efficient capture of secondary electrons. While the positively biased E-T detector is highly efficient for secondaries, it must be realized that it also is simultaneously detecting backscattered electrons. There are two distinct ways that the E-T detector senses BSEs:

1. The BSEs emitted from the specimen directly into the solid angle of the E-T detector are detected directly. The E-T detector ordinarily subtends a relatively small angle around the specimen, and hence its geometric efficiency for direct collection of BSEs is poor. This BSE component can be selectively viewed by biasing the E-T detector negatively to reject all secondary electrons, no matter what their origin. Then only the BSEs directly emitted into the solid angle of the detector are collected.

2. BSEs may be detected indirectly through the formation of secondary electrons where the BSEs strike the polepiece and chamber walls. Although these secondary electrons have similar properties to the true secondary electrons emitted from the sample, their detection is actually equivalent to detecting the backscattered electrons striking the polepiece. This BSE component is difficult to measure, but estimates have shown it can form 50% or more of the total signal received by the positively biased E-T detector from targets of intermediate and high atomic number where backscattering is significant.

Experiment 8.1: E-T Detector Collection Efficiency. Estimate the solid angle of the E-T detector (negatively-biased) for direct collection of BSEs.
(a) Area, A, of scintillator (cm^2) = _____
(b) Distance, r, from specimen to scintillator (cm) = _____
(c) Solid angle, $\Omega = A/r^2$ = _____
(d) Relative geometric efficiency (assuming the maximum solid angle above specimen is a hemisphere of 2π steradians solid angle) $\varepsilon = \Omega/2\pi$ = _____
(e) For a specimen set normal to the beam (0° tilt), what is the approximate take-off angle (angle between the specimen surface and the detector axis) θ = _____

Experiment 8.2: Imaging with the E-T Detector. Use the Al-Ni (Raney nickel) polished specimen or a similar specimen which will display both atomic number contrast and topographic contrast. Set the beam energy to 20 keV and choose a condenser setting which provides enough beam current to produce a reasonably noise-free image on the CRT.
(a) Record an image at a magnification selected in the range 100x to 1000x with the positively biased E-T detector. The image of Raney nickel should show the three phases which comprise the microstructure as well as some surface topography.
(b) At the same magnification, record an image of the same field of view with the E-T detector biased negatively.
(c) Repeat steps above for a low beam energy, e.g., selected in the range 2-5 V keV.

Questions:
1. Compare these images. Can atomic number (compositional contrast) effects be observed in the +E-T image? In the -E-T image?
2. Can topographic effects be observed in the +E-T image? In the -E-T image?
3. Which image, +E-T or -E-T, gives the stronger impression of topography? Why?
4. How do images taken at 2 keV compare to their counterparts at 20 keV?

8.3 Dedicated BSE Detector in Sum Mode

The solid angle of collection and the take-off angle of dedicated BSE detectors are important parameters for the interpretation of BSE images. Knowledge of the placement of these detectors and their solid angle is critical for selecting the proper tilt of the specimen.

Segmented BSE detectors can be operated in two different modes, sum and difference. In the sum mode, the signal contributions of all segments are added to increase the effective solid angle of the detector. In the sum mode, number contrast effects, such as atomic number contrast, are emphasized while trajectory contrast effects which are highly directional in nature, such as topographic contrast, are suppressed.

Experiment 8.3: Dedicated BSE Detector Collection Efficiency. Estimate the solid angle for direct collection of the dedicated BSE detector:
  (a) Area, A, of scintillator or solid state diode (cm^2) = _____
  (b) Distance, r, from specimen to detector (cm) = _____
  (c) Solid angle, $\Omega = A/r^2$ = _____
  (d) Relative geometric efficiency of a single segment of detector $\varepsilon = \Omega/2\pi$. Relative geometric efficiency of all segments of detector: Ω_{total} = _____ x Ω. $\varepsilon = \Omega_{total}/2\pi$ = _____

  (e) For a specimen set normal to the beam (0° tilt), what is the approximate take-off angle (angle between the specimen surface and the axis of one of the BSE detectors)?
  θ = _____

Experiment 8.4: Imaging with a Dedicated BSE Detector in Sum Mode. On the same area of the Raney nickel sample set at 0° tilt and at the same magnification as that used for Experiment 8.1,
  (a) Record images at 20 keV using the BSE detector in the sum mode (all segments added).
  (b) Repeat the 20-keV image with a specimen tilt of 60°.
  (c) Repeat the 0° tilt procedure with a low beam energy selected in the range 2 to 5 keV.
Questions:
  1. Which contrast mechanism is more easily visible in the BSE sum image, atomic number contrast or topographic contrast?
  2. How do the images at 0° and 60° compare? Give two reasons why the atomic number contrast is reduced at 60° compared to 0°.
  3. Does the BSE detector function at low beam energy?

8.4 Dedicated BSE Detector in Difference Mode

The segmented BSE detector provides the possibility of displaying the signal difference between any two of the segments. The difference mode tends to emphasize topographic effects and suppress atomic number effects. To interpret difference mode images, two critical pieces of information are necessary:
  1. The position of the detector segments relative to the scan field.
  2. The order of subtraction, e.g., detector 1 - detector 2, or vice versa.

Experiment 8.5: Imaging with a Dedicated BSE Detector in Difference Mode. Identify the position of the detector segments relative to the scan field. On the same area of the Raney nickel sample set at 0° tilt and at the same magnification as that used for Experiments 8.1 and 8.2, record images at 20 keV using the BSE detector in the difference mode:
  (a) If the BSE detector has two segments, prepare images for both possible arrangements of the subtraction, e.g., segment 1 - segment 2 and segment 2 - segment 1.
  (b) If the BSE detector has four segments, prepare difference images for both pairs of opposing detectors, e.g., 1 - 2, 2 - 1, 3 - 4, and 4 - 3.

Questions:
1. Which contrast mechanism, atomic number or topography, is more prominently displayed in difference mode images? Why?
2. Does the order of subtraction affect the sense of the topography in the image? Which is the correct order for normal operation?
3. Locate fine scale, linear detail in an E-T detector image. Compare with images of the fine scale detail in the dedicated, difference-mode BSE images. Is there any evidence that information has been lost in the difference-mode images?
4. Rotate the specimen on its axis by 90°. How does the image change?

8.5 Specimen Current Signal as a BSE Detector

The specimen current (SC) signal provides another means to detect BSE signals and contrast mechanisms. The specimen current signal responds in the opposite sense to changes in the total emitted electron signal (backscattered plus secondary electron). The important property of the SC signal is that contrast forms strictly because electrons have left the specimen. As long as those exiting electrons do not return to the specimen, the SC signal is completely independent of the trajectory of the exiting electrons. The SC signal is also equivalent to a detector with 2π steradians of collection angle, since any electron emitted into the unit hemisphere above the sample surface contributes to the information in the SC signal.

Experiment 8.6: BSE Imaging Using Specimen Current. On the same area of the Raney nickel sample set at 0° tilt and at the same magnification as that used for Experiments 8.2, 8.3, and 8.5,
  (a) Record a specimen current image at 20 keV.
  (b) Using the "INVERT" function of the signal processing system, record a SC image under the same conditions so that the brightness increases with increasing atomic number as is the case for the BSE signal.
  (c) Observe the specimen current image as the visual scan speed is increased from the photographic scan speed through the slow visual scan speeds to TV scan.

Questions:
1. Compare the direct and inverted SC images with the E-T and dedicated BSE detector images. Is atomic number contrast visible? Is the atomic number contrast readily interpretable in the inverted SC image?
2. How does topographic contrast in the inverted SC image compare with that seen in the E-T and dedicated BSE images?
3. How does the image quality change as the scan speed is increased?

Laboratory 9

Scanning Transmission Imaging in the SEM

Purpose
To examine scanning transmission imaging as an alternative imaging mode to conventional scanning electron microscopy for the examination of small particles.

Equipment
SEM equipped with a conventional Everhart-Thornley (E-T) detector (above the specimen plane) and a scanning transmission detector (below the specimen plane). The specimen stage is typically modified to accommodate one or more specimens in the form of transmission electron microscope grids.

Specimen
Mineral particles dispersed on a thin (20 nm) carbon film carried on a copper electron microscope grid. The particles are obtained by crushing a mineral such as antigorite and dispersing the particle shards in freon. A drop of freon containing shards is placed on the grid and allowed to evaporate. Because the mineral is an insulator, a thin (~ 20-nm) carbon film is evaporated on the particles for charge dissipation and mechanical stabilty.

Time for this lab session
Approximately 30 minutes.

9.1 Scanning Transmission Electron Detectors

Scanning transmission imaging, referred to as scanning transmission electron microscopy (STEM), can be performed in an SEM for the special class of specimens which are sufficiently thin to permit penetration of beam electrons. Thin foils and particles often satisfy the requirements for beam penetration. Small particles may be dispersed on thin carbon support films for mounting. Depending on the atomic number, the specimen thickness must generally be less than 500 nm to obtain useful penetration at typical SEM beam energies 20-30 keV. For visualization of the internal structure of objects, the thickness should be less than 100 nm. Even if a particle is too thick for imaging internal structure in the STEM mode, it is often useful to view a particle as a dark silhouette against a bright field for high-resolution imaging of the edge morphology.

To perform STEM in an SEM, a detector for energetic electrons must be placed below the specimen. This detector is typically a scintillator connected by a light pipe to a photomultiplier, or alternatively a solid state detector. The scintillator detector usually offers advantages of bandwidth for rapid scan imaging, and so is often preferred as a STEM detector. Contrast is formed in a STEM image by the scattering of beam electrons outside the detector solid angle of collection. Elastic scattering produces an average deflection of approximately 5° for a single event, although any scattering angle from 0° to 180° is possible in a single event. By placing a detector with a small solid angle below the specimen, even a specimen with a thickness of the order of the elastic mean free path (variable with beam energy and specimen atomic number, but on the order of 20 nm) will produce sufficient scattering to generate detectable contrast.

Experiment 9.1: Characteristics of the STEM Detector. Estimate the solid angle of collection of the STEM detector:

    Detector area = _____ cm^2
    Specimen to detector distance = _____ *cm*
    Solid angle Ω *= area/distance^2 =* _____ *steradians*

9.2 STEM Images

It is useful to compare the transmission image of the mineral particles with the surface images of the same particles.

Experiment 9.2: Comparison of STEM and SEM Images. Image the mineral specimen with the conventional E-T detector placed above the specimen in its normal position and with a STEM detector placed below the specimen.

  (a) Compare the E-T and STEM images at rapid visual scan speeds, including TV scan speed.
  (b) Record E-T and STEM images at fast and slow scan speeds (if possible) and at a range of magnifications, e.g., 1000x, 10,000x, and 50,000x on suitable fine structure.

Questions:

    1. Which imaging mode, E-T or STEM, permits finer scale particles to be detected in a rapid searching mode (fast scan speed) of operation?
    2. Under photographic conditions, which detector produces the best results?
    3. Does the combination of images prepared with the E-T and STEM detectors offer special advantages?

Laboratory 10

Low-Voltage SEM

Purpose

This laboratory explores some of the principal phenomena observed at low beam energies. Images prepared at "conventional" beam energies, e.g., 15 keV and above, are compared with low-beam-energy images, e.g., 5 keV and below. The possibility of examining uncoated insulators by taking advantage of enhanced emission of secondary electrons at low energy is also examined.

Equipment

SEM capable of low-beam-energy operation.

Specimens

Three types of samples, comounted on a carbon planchet, will be used for this laboratory:
1. Metal fracture surface or surface of steel abraded by sawing.
2. Uncoated minerals, e.g., mica, calcite, and quartz.
3. Polystyrene latex spheres.

Time for this lab session

Two hours.

10.1 Topographic Contrast versus Beam Energy

Low-voltage scanning electron microscopy (LVSEM) is a rapidly developing field which has been made possible by the evolution of SEM instrumentation to permit easy operation at low beam energies, which we shall arbitrarily define as $E_0 < 5$ keV. State-of-the-art SEMs can typically operate in the low-voltage range from 0.5 to 5 keV with relative ease, so that comparison of images taken at energies throughout this range becomes possible. Such comparisons will form the basis of this laboratory.

Experiment 10.1: Estimating the Electron Range. Estimate the primary electron range in steel for 30-, 15-, 5-, 2.5-, 1-, and 0.5-keV electron beams using the Kanaya-Okayama range equation:

$$R_{K\text{-}O} = \frac{0.0276 A E_0^{1.67}}{\left(Z^{0.89}\rho\right)} \ (\mu m) \qquad (10.1)$$

where A is the atomic weight (g/mole), E_0 is the beam energy (keV), Z is the atomic number, ρ is the density (g/cm³), and R is given in micrometers. For iron, $A = 55.85$, $Z = 26$, and $\rho =$

7.87 g/cm³. Enter calculated values in the range table below. From the range, estimate the lateral spread of the area sampled in forming an image. Calculate the pixel size for 1000x magnification (typical for a thermionic gun at low voltage). How does this spread compare with the pixel size (referred to the specimen)?

Target:	
Magnification:	
Pixel size:	
Beam energy (keV)	Primary electron range
30.0	
15.0	
5.0	
2.5	
1.0	
0.5	

Experiment 10.2: Contrast Effects at Low Beam Voltage. Using the electrically conducting steel specimen prepare a series of images of the same field of view at a magnification which is sufficient to show fine scale surface detail (e.g., select one magnification value in the range 200-2000x) and over a wide range of energies: e.g., a suitable series would be 30, 15, 5, 2.5, 1, and 0.5 keV. Use a positively biased Everhart-Thornley (E-T) detector. Compare these images and answer the following questions:

1. How does the general appearance of the specimen topography behave as a function of decreasing beam energy?
2. Are there any features visible at high beam energy that are lost at low beam energy?
3. Do any features become visible at low beam energy?
4. Are there any unexpected changes in contrast?
5. At low energies, do any of the other available detectors (e.g., dedicated BSE) produce a satisfactory image?

10.2 Minimizing Charging Effects on Insulators

The total emitted electron coefficient (which is a sum of the BSE coefficient and the secondary electron coefficient, $\eta + \delta$) changes dramatically as a function of beam energy at low energies. Above 5 keV, $\eta + \delta$ is typically less than 0.5. At energies below 5 keV, $\eta + \delta$ increases and becomes greater than 1 for a certain range of energies defined by the first, E_I, and second, E_{II}, crossover points where $\eta + \delta = 1$. Crossover points vary with specific materials, but, in general, are in the range 0.5-2.0 keV. Operation above E_{II} results in an unstable charging situation: fewer electrons leave the surface than enter it since $\eta + \delta < 1$, so that the sample charges negatively. The surface potential of the specimen becomes more negative, and in extreme cases, if surface breakdown does not discharge the surface, the specimen may actually become a mirror and reflect the scanning beam, resulting in a scan (and image) of the inside of the specimen chamber! Note that this mirror effect can occur with a conducting specimen which is not properly grounded. In more typical cases of charging, the negative surface charge tends to disrupt the collection of secondary electrons, producing areas of anomalous brightness and darkness in the field of view. For operation below E_I, $\eta + \delta < 1$, and the surface again charges negatively since more electrons are injected into the specimen than leave. Even modest charging of the specimen in this low-beam-energy regime may be sufficient to deflect the beam and distort the image. Operation in the beam energy range $E_I < E < E_{II}$ produces a stable charging situation. Since $\eta + \delta > 1$, the surface tends to charge positively, which increases the difference in potential between the electron source and the specimen surface. The effective beam energy thus increases, and $\eta + \delta$ decreases. Eventually a dynamic equilibrium is reached where $\eta + \delta = 1$.

Experiment 10.3: Charging Effects. We will investigate insulating samples which in the uncoated state show strong charging effects at high beam energy (> 10 keV). *Place the nonconducting calcite specimen (black crystals on surface) normal to the beam (0° tilt). Prepare a series of images of the same field of view at a magnification which is sufficient to show fine scale surface detail (e.g., select one magnification value in the range 200-2000x). Use a positively biased E-T detector.*

(a) *If an image can be recorded at high beam energy (> 20 keV), record two images of the same area in rapid succession and identify examples of charging artifacts in the first image. Compare the first and second images. Are the images identical? Are the charging artifacts stable with time?*

(b) *Take images of the same area over a wide range of energies. A suitable series would be 15, 5, 3, 2, 1.5, 1, and 0.5 keV. In recording the beam energy sequence, note if any time is needed as the energy is lowered to establish stable imaging conditions at the new operating voltage. After determining the best images in the above sequence change the beam energy in smaller steps to find a beam energy for which the image has no apparent charging artifacts. Take two images in rapid succession at that beam energy. Compare the two images. Are the images identical?*

(c) *Having established a suitable operating beam energy for calcite, move to the mica (flat, flakelike crystal) and, using the same beam, detector, magnification, and scan rate conditions, record two images of the mica in quick succession. Are these two images identical? Is the same beam energy appropriate to both? Examine the image recorded at 0.5 keV. Is the image satisfactory? What does this suggest about the location of the lower limit, E_I?*

(d) *Having established a suitable operating beam energy for calcite, tilt the specimen to 60° and record another pair of images in quick succession. Is the same operating voltage suitable for the highly tilted specimen? If the desirable beam energy has changed, what is one possibility for the origin of this change?*

10.3 Electron Beam Damage

The problem of minimizing beam damage and avoiding damage-induced artifacts is complicated by the enormous range of damage behavior seen in various materials. Beam damage may be a function of several different factors: beam energy; radiation dose; dose rate; specimen temperature; specimen dimensions, i.e., thick versus thin (electron transparent); and the presence of secondary materials. Although we might intuitively expect that, all other factors being constant, lowering the beam energy should lower or even eliminate beam damage, this is not necessarily the result in all cases. As long as the threshold energy for beam damage is still exceeded, lowering the beam energy may actually increase the rate of damage, since the electron range (see the Kanaya-Okayama range equation above) decreases as the 1.67 power of the beam energy. Thus, reducing the beam energy by a factor of 2 actually decreases the depth of interaction volume of the primary electrons by a factor of approximately 3, and the volume by a factor of approximately 30. Although for constant current, the power ($P = V \times i$) injected into the specimen is decreased by a factor of 2, the power per unit volume actually increases by a factor of 15.

Experiment 10.4: Electron Beam Damage of Polymers. *Using the uncoated nonconducting polymer specimen, record a series of images of the sample at the same magnification (e.g., select a magnification in the range 200x-2000x which reveals detail) at 20 10, 5, 2, and 1 kV. Because of the damage effects, select a new field of view for each energy.*

(a) *Find a beam energy which does not cause gross artifacts, such as changes in the surface during SEM viewing.*

(b) *If there exists a suitable beam energy for which the damage is minimized or eliminated, increase the magnification in steps and record an image at each magnification. Does the damage effect reappear when the magnification, and therefore the current density of the dose, is increased?*

(c) *Choose a higher magnification for which some detail still remains in the field of view e.g., 2000x-10,000x. Now increase the beam current by a factor of 2 and record an image. Repeat. Does the damage effect reappear when the beam current, and therefore the dose rate, is increased?*

Laboratory 11

High-Resolution SEM Imaging

Purpose

Secondary electron (SE) images of various specimens will be taken at different magnifications and accelerating voltages and will be compared to the BSE image. High-resolution SE-I image features will be demonstrated and the SE-I imaging conditions defined. This technique has been developed since the publication of *SEMXM* and so additional explanatory material is given in Section 11.1. More details can be found in references [1-3].

Equipment
1. An SEM with a high-brightness electron gun (LaB_6 or field-emission), Everhart-Thornley (E-T) and dedicated BSE detectors, and low voltage capability (1-2 kV).
2. SEMs with below-the-lens position of the specimen should be equipped with a BSE-absorption device and specimen bias capability (see Laboratory 12).
3. Measuring magnifying glass (10x with 1/10 mm divisions).

Specimens
1. Homogeneous bulk gold-on-carbon specimen: thick gold crystals (50 nm high, 100 nm wide) on carbon (SEM Resolution Standard available from several EM supply companies).
2. Microtopography specimen: small gold particles (10-20 nm) decorated by small Pt particles (3-6 nm) on bulk silicon.
3. Inhomogeneous surface film specimen: 2-nm-thick platinum film deposited onto a silicon substrate through a TEM grid with narrow grid bars followed by a 2-nm carbon film deposited through a grid with wide grid bars.
4. Topographic plus Z contrast specimen: specimen composed of ~10-nm low-Z particles (e.g., ferritin) adsorbed to a silicon substrate and shadowed with 2 nm of platinum or 2 nm of carbon.

Time for this lab session

Two to three hours provided that the SEM is already set up for high resolution.

11.1 Contrast Mechanisms from Secondary Electrons

Secondary electrons have both very low energies (about 1-10 eV) and very small escape depths (1-5 nm). These properties favor high-resolution surface imaging. However, interpretation of SE images is difficult because there are several components of the SE signal that contribute to image resolution and contrast in different ways.

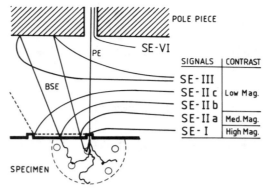

Figure 11.1. Signal components collected in a conventional SEM.

SE signal components originate from the specimen and from parts of the microscope. Their nomenclature refers to the distance from the probe site at which they originate (Figure 11.1). The specimen-specific secondary electron components are labeled SE-I and SE-II. High-resolution SE-Is are produced by primary electrons (PEs) of the probe when they enter the specimen and cause SEs within the immediate vicinity (~1 nm) of the electron probe site. SE-IIs are produced by primary electrons in the electron diffusion zone beneath the specimen surface. The primaries that escape from this relatively large region are called BSE-IIs. Although the escape depth of the SE-IIs is the same as that of the SE-Is their exit distance from the probe entrance site depends on the scattering of the PEs and may be over 100 nm away from the probe. SE-IIa electrons are generated by PEs that are scattered only a few times (< 25 times), mostly in the forward direction and occasionally in the backwards direction so as to allow escape without large energy losses. SE-IIb electrons are generated by multiply scattered (> 25 times) PEs such as the typical BSE that might emerge at a large distance from the probe

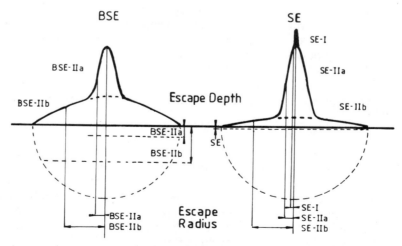

Figure 11.2. Lateral emission distribution of BSEs [4] and the consequent expected SE components.

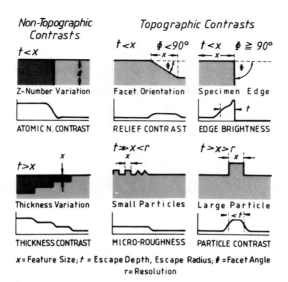

Figure 11.3. Types of surface contrast that may be imaged on bulk samples. Nontopographic contrast types may be called atomic number contrast and mass-thickness contrast. Topographic contrast types are relief contrast, edge brightness contrast, microroughness contrast, and particle contrast. t = escape depth, x = feature size, r = resolution, and Φ = facet angle.

site. A third component (SE-IIc) is generated by BSEs which reenter the specimen at some distance from the probe.

Unfortunately, the high-resolution SE-I electrons comprise only 1-10% of the total signal (Figure 11.2). If a high brightness electron gun is used, a large current may be placed in a small probe which enriches the SE-I signal versus the SE-II signal and thus increases the signal-to-background or signal/noise ratio. Non-specimen-specific SEs are produced by BSEs outside the specimen at the pole piece or the walls of the specimen chamber (SE-III) and by the probe electrons in the electron optical column (SE-IV). The latter two SE types contribute to the background which usually swamps the high-resolution SE-I signal.

In high magnification surface imaging, certain contrast mechanisms dominate [1] depending upon the internal structure of the specimen or its topography (Figure 11.3). If the specimen is composed of regions of various materials on smooth surfaces, "atomic number contrast" is observed when the region thickness (x) exceeds the escape depth (t) of the SE signal electrons. However, in the case when the thickness (x) of material of different atomic number is thinner than the SE escape depth, a "mass-thickness contrast" will be generated.

On uneven surfaces additional topographic contrasts are generated. They fall into two groups depending upon the ratio of SE escape depth t to feature size x. If the features are larger than t, variations in facet orientation produce "relief contrast" and "edge brightness contrast." If the features x are smaller than the escape depth t and are resolved ($t > x > r$ = spatial resolution), signal electrons are emitted from the entire surface of the particles and produce "particle contrast." However, if the features are so small that they are not resolved, they contribute a signal which creates "micro-roughness contrast." The topographic contrast types which are independent of escape depth (relief and micro-roughness contrast) contribute to low magnification image contrast, and those which are sensitive to escape depth (edge brightness and particle contrast) contribute to the high magnification image contrast. The latter can be used to measure the escape depth of signal-generating electrons. Escape-depth sensitive

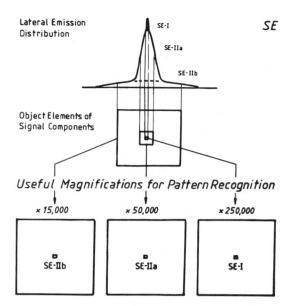

Figure 11.4. Magnifications required for imaging escape-depth sensitive contrast mechanisms of various signal components.

contrast mechanisms are only observed when the contrast-generating object element is imaged by a sufficient number of pixels to sufficiently define that object element (Figure 11.4). In practice a structural detail should be displayed to the unaided eye on a photograph about 5-10x larger than the eye's resolution, i.e., as a 1-2 mm feature. Considering the exit surface areas of the different signal components for a hypothetical material (Figure 11.4), it is evident that Type II image contrast generated by BSEs (Type II signals) is observed at medium magnifications (1000x-50,000x), while Type I contrast mechanisms, sensitive only to the SE escape depth, are identifiable only at very high magnifications (>50,000x).

11.2 Topographic SE-I Contrast

Since SE-Is and SE-IIs have the same escape depth, 1-5 nm, these specific signal components cannot be separated on bulk specimens. However, because SE-I and SE-II electrons have different exit distances from the probe site, these signal components may be identified by their characteristic escape-depth-dependent topographic contrast. On homogeneous specimens, the SE-II component contrast is similar to that of the BSE signal. Therefore, the BSE signal will be used to demonstrate a SE-II contrast contribution to the image since the width of edge brightness contrast should be similar for each. However, this procedure provides only a rough approach, which may not be accurate on complex specimens, since high magnification imaging (> 50,000x) requires very specific conditions of instrument performance, instrument operation, and specimen preparation.

Experiment 11.1: *Relief Contrast and Edge Brightness Contrast. (a) Image the "homogeneous bulk gold-on-carbon specimen" at high accelerating voltage (15-30 kV) with different magnifications (5,000x, 50,000x, 100,000x, and 200,000x) in SE mode (E-T* detector positively biased) and BSE mode (high-take-off angle BSE detector). *Identify relief contrast and edge brightness contrast produced by SEs. Compare SE and BSE images.*

1. Which contrast features are identical and which are different? *Measure the width of edge brightness contrast and estimate the exit radius of the signal-generating electrons at 50,000x and 200,000x.*
2. At which magnification are SE-I contrasts observed? At higher magnification, does the SE-I S/N increase or decrease?

(b) *Take the same image sequence at low voltage (1 kV).* Produce low-voltage images at high magnification even though a decrease in the S/N and an increase of probe diameter will deteriorate the image quality. If no commercial low-voltage BSE detector is available, use the converted BSE signal (SE-IIIs generated by BSEs striking the polepiece and collected with the E-T detector) which will provide a high-quality BSE image (converter plate and specimen biasing required, as described in Laboratory 12).

1. At low voltage is there a difference in SE and BSE contrasts? *Measure the width of the edge brightness contrast and estimate the exit radius of the signal-generating electrons accounting for the electron probe diameter.*
2. At low voltages is the SE-I yield so increased that SE-I contrast features are identifiable? Is the resolution increased when compared to high-voltage imaging?

Experiment 11.2: Particle Contrast. (a) *In SE-I image mode, image the "microtopography specimen" at high accelerating voltage (30 kV) and at high magnification (50,000x, 100,000x, 200,000x, and higher if possible). Measure the width of contrast features (edge brightness or particle contrast) and calculate the exit radius of the contrast-generating signal electrons.*

1. Identify particle contrast on the gold crystals.
2. Identify edge brightness on the gold crystals.

(b) Repeat the same imaging procedure as described above at 1 kV.

11.3 Nontopographic SE Contrast

On inhomogeneous specimens containing materials of different average Z, subsurface structure may vary in two ways. Different materials can be joined in bulk or in layers (or particles) which are thinner than the escape depth of the signal-generating electrons. The SE signal is especially modified by thin surface films of a thickness similar to the escape depth of the SE, i.e., 1-5 nm. However, such layers affect all types of SE. A thin-film specimen will be used for the evaluation of contrast-generating SE signal components.

Experiment 11.3: Atomic Number Contrast and Mass-Thickness Contrast. Image the "inhomogeneous surface film specimen" at low (50x and 500x) magnifications in SE imaging mode using high (30-kV) and low (1-kV) accelerating voltage. Produce BSE images from the same areas.

1. Which areas of the specimen are imaged in the SE and BSE images?
2. Which components of the SE signal image carbon as dark islands and which components image the metal as light areas?
3. Do the contrast features vary with voltage or magnification?

11.4 SE-I Imaging Requirements

Both the SE-I and SE-II signal yields depend on the average atomic number of the outermost specimen surface layer. Since SE-I contrast features are only identifiable on small particles at high magnification, a small-particle specimen will be used to establish the SE-I imaging conditions on high- and low-Z surfaces.

Six conditions must be met to establish imaging conditions for the type of topographic imaging called the "SE-I imaging mode":

A. Increase signal-to-noise of SE-I image by

1. Employing high-brightness electron sources with sufficient probe current
2. Increasing magnification

B. Reduce background signal level by

3. Eliminating SE-III
4. Reducing probe diameter
5. Increasing voltage

C. Enhance contrast and reduce the signal excitation volume by

6. Coating with thin (<1 nm), continuous, low-Z metal films (see Laboratory 30).

The specimen used to establish some of these parameters consists of small, low-Z particles (10-12 nm in diameter), which are mounted on a homogeneous bulk substrate (silicon) to reduce background signal contrast and retain background signal level. Some particles are shadowed with a 2-nm (average mass-thickness as measured by a quartz crystal monitor), discontinuous platinum film which is composed of ~3-nm-high metal crystals and produces a strong BSE signal. The platinum decorates the low-Z particles and accumulates at the tops. Other particles are shadowed with a 2-nm continuous carbon film which coats the particle surface and does not generate recognizable BSE signals.

Experiment 11.4: SE-I Yields from Thin Film Coatings. (a) Image the "topographic plus Z contrast specimen" at low magnifications (500x) and observe the contrast in SE mode at 30 kV.

(b) Image at medium (10,000x) magnification selecting an area which displays all three surfaces. Does the atomic number contrast increase or decrease? Are fine structural details identifiable?

(c) Image the platinum shadowed area (2 nm platinum on the support) at high magnification (200,000x) and look for the small low-Z particles ~10 nm in diameter, shadowed by platinum. Can shadows on the substrate be observed in areas behind the particles? Are fine structures of the discontinuous metal film identifiable (gap resolution)?

(d) Image the carbon shadowed area (2 nm carbon on the substrate) under the same conditions as (c). Are particles identifiable? How are particles imaged? Is the shadowed area behind particles identifiable? What are your conclusions about the best conditions for the SE-I imaging mode?

References
[1] K.-R. Peters, *Microbeam Analysis*, A. D. Romig and J. I. Goldstein (eds.), San Francisco Press (1984) 77-80.
[2] D. C. Joy, *Journal of Microscopy,* **136** (1984) 241-258.
[3] K.-R. Peters, *Scanning Microscopy* (1985) 1519-1544.
[4] K. Murata, *J. Appl. Phys.,* **45** (1974) 4110.

Laboratory 12

SE Signal Components

Purpose
This laboratory will demonstrate the contrast produced by the different SE signal components known as SE-I, SE-II, SE-III, and SE-IV. It will also describe methods to measure the proportion of the components and to enhance or suppress certain components. This laboratory is a continuation of Laboratory 11 in which the signal components and contrast mechanisms were described.

Equipment
1. An SEM provided with a high brightness electron gun (LaB_6 or field-emission), E-T and BSE detectors, low voltage capability (1-2 kV), and specimen current measurement via a Faraday cup.
2. SEMs with the specimen position below the final pole piece must be equipped with BSE-absorption or enhancement devices such as a converter plate made from an insulated board smoked with MgO fitted on the specimen side with a metal screen that is grounded. Specimen biasing provided by two connected Eveready 45-V "B" No. 487 batteries.
3. Carbon- and gold-coated aluminum plates to be attached to the retractable BSE detector.
4. Stereo prisms and measuring magnifying glass (10x with 1/10 nm divisions) will be helpful.

Specimens
1. Silver TEM grid with bulk crystals on the surface.
2. Inhomogeneous surface film specimen: 2-nm-thick platinum film deposited onto a silicon substrate through a TEM grid with narrow grid bars followed by a 2-nm carbon film deposited through a grid with wide grid bars.
3. Clean glass cover slip mounted to stub with double-stick tape (no electrical connection).
4. A conventional Faraday cup made from a platinum TEM aperture and a low-Z support provided with a deep drilled hole. The aperture is mounted over the hole with carbon "DAG."

Time for this lab session
Three hours with some advance preparation.

12.1 Enhancing SE Signal Collection Efficiency

The specific imaging of contrast generated by the different signal components can be enhanced through a number of procedures which include change of magnification, reduction or addition of background signal, and certain specimen preparation techniques. Since some of the high magnification signal components are very weak, increase of signal-to-noise ratio through efficient signal collection is as important as the reduction of background signal.

The E-T detector has a low SE signal collection efficiency (~50%) on smooth surfaces which may be further reduced on topographically rough surfaces. Collection of the specimen-specific and the non-specimen-specific signal components can be altered by modification of the E-T detector collection field. For this purpose the specimen may be biased or deflection electrodes may be mounted between the specimen and the detector. The first possibility will be demonstrated in the following experiments. (See Figure 12.1 but ignore the converter plate and specimen grid for Experiments 12.1 and 12.2.)

Experiment 12.1: Biasing of Rough Specimens. *With a positively biased E-T detector, image the silver TEM grid specimen at 30 kV and medium magnification (10,000x). Locate and, with the electron beam off, detach the specimen current cable. Connect two 45-V batteries in series (connect the positive pole of one battery to the negative pole of the other battery) and attach the common connection to the microscope ground. Use the specimen current feedthrough to connect the specimen sequentially to the positive, negative, and common poles of the batteries. Observe the changes in signal intensity. Increase or decrease the signal level to the same level using the photomultiplier only. Observe the contrast and document the changes through line scans.* Does the topographic contrast change? Which signal components are changed (SE-I, SE-II, or SE-III)? Measure the edge brightness and determine the escape depth of the dominant component.

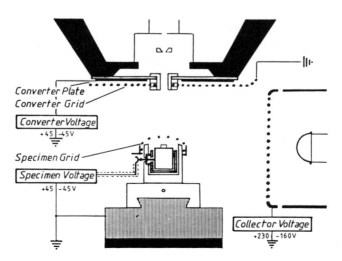

Figure 12.1. Modifications to the SEM specimen chamber for Experiments 12.1, 12.2, and 12.4.

Experiment 12.2: Biasing of Flat Specimens. Repeat Experiment 12.1 at 5 kV, but use the flat "thin film on Si" specimen. Does the atomic number contrast change? Does the signal intensity change? Can specimen biassing of +50 V eliminate SE-I+II (read as SE-I and SE-II) electrons?

12.2 SE-III Signal Contributions

Contrast is inversely proportional to the strength of the background signal. Since each microscope has its specific signal collection properties, measurement of the signal components becomes important for the improvement of imaging quality. These SE-III background components seriously degrade Type IIa and Type I imaging. However, the SE-III component can be easily reduced or enhanced by simple devices which will be described below. One way to determine the origins of SE-III electrons is to simulate the backscattering process by reflecting primary electrons with a specimen that is charged up to the potential of the primary electrons--mirror imaging.

Experiment 12.3: Mirror Imaging of the Specimen Chamber. (a) Use a freon-cleaned glass cover slip mounted without electrical conductive paste on a stub (use double-sided tape). Image the surface at low voltage (1 kV) and electrically charge the surface to form an electrostatic lens. In order to establish a symmetrical mirror, without changing the optical alignment, increase the accelerating voltage in steps within 1-2 min, as the magnification is increased, starting at 1,000x and 1 kV and ending at 100,000x and 30 kV. Finally, return to the initial low voltage and focus the deflected probe on the surface of the specimen chamber. Move the lens (the stage) and change the magnification to image different parts of the chamber, i.e., the pole piece and the SE detector. Which electrons collected by the E-T detector produce the image contrast? What does this contrast mean in terms of signal generation? (b) (Optional) Produce an image with the BSE detector using all or only some part of the available sectors. What contrast mechanism produces the image of the BSE detector?

Two devices can be used for increasing or decreasing the SE-III signal contribution: A converter plate and coated aluminum plates. These are easily made at small cost and with little effort. The converter plate is made from copper-coated phenolic printed circuit board (Radio Shack), cut to size and provided with holes for the probe and for nylon screws used to insulate an aluminum grid (aluminum window screen) from the copper board (Figure 12.2). The copper surface should be smoked with MgO to improve its SE emission (SE yield is very high on rough insulators).

Experiment 12.4: The Converter Plate. The BSE converter plate can be easily made from simple parts such as a blank printed circuit board and an aluminum screen (see above). The converter is mounted underneath the pole piece. The converter grid (the aluminum screen) is grounded, and the smoked plate is connected to an electrical feedthrough which can be biased using the batteries (see Figures 12.1 and 12.2). Image the plate fastened beneath the pole piece with an electrostatic mirror (see Experiment 12.3). Apply a negative (-45 V) or positive (+45 V) bias to the plate and image the converter. How does the contrast change? What signal electrons cause that change?

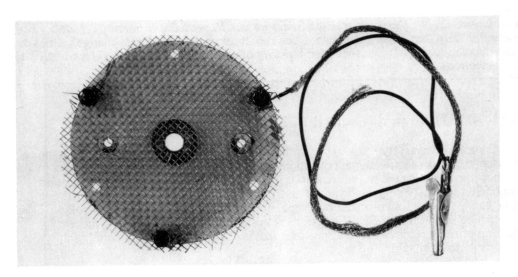

Figure 12.2. Photograph of the converter plate shown schematically in Figure 12.1. The wire is to be connected to the electrical feedthrough for the converter plate voltage.

Figure 12.3. Photos of three coated aluminum plates for mounting beneath the BSE detectors of various SEMs. The plate coated with gold enhances SE-III signals while the carbon-coated plates reduce SE-IIIs.

Experiment 12.5: BSE Absorption and Amplification Plates. If possible, mount on a retractable BSE detector two aluminum discs, one coated with carbon and the other coated with gold (see Figure 12.3). Both plates should be provided with central holes for the electron beam to pass through. Image the silver specimen at medium magnification (10,000x) using each of the plates directly under the pole piece. Adjust the photomultiplier to obtain similar signal levels. Are differences in signal levels and contrasts recognizable? Why does the contrast change?

12.3 Quantitative Measurement of SE Signal Components

Quantitative measurement of the different signal components is only possible under simplified imaging conditions. However, such measurements provide for a good assessment of the imaging capability of the instrument. Using a Faraday cup, and additional provisions to alter the SE-III signal contributions, measurement may be made of the relative proportion of some of the SE signal components.

The basic procedure uses Faraday cup with a 20-µm aperture. The aperture is imaged at medium magnification so that the hole covers the right part of the image. Then, a line scan is used to quantitate the total signal collected. On the aperture SE-I+II+III+IV components are collected. However, over the hole only SE-IVs are collected because the probe's electrons are trapped within the cup. As base line, a line scan is produced with unchanged amplifier setting but with the beam shut off (turn off the filament current or deflect beam with deflection coils or final aperture holder).

The following modifications will allow a more detailed analysis. Specimen biasing may be used but does not provide reliable data since the distortion of the collection field can not be reproduced for normal specimen imaging. Ideally, a grounded grid surrounding the specimen (a "specimen grid") should be mounted so that it shields the specimen from the detector. Then positive specimen grid biasing can be used to eliminate the SE-I+II components. Such a grid must be mounted on the specimen stage without touching the electrically isolated specimen holder (see Figure 12.1).

The Faraday cup required for these experiments can be made from a platinum TEM aperture and a low-Z support provided with a deep drilled hole. The aperture is mounted over the hole with carbon "DAG." Several types of measurements can be made depending on the alterations that have been made to the instrument.

Experiment 12.6: Imaging of the Faraday Cup. Produce an SE image at the aperture edge of the hole in the Faraday cup at medium magnification (3,000x) and high accelerating voltage (20 kV). Write two line scans over the image (triple exposure, make sure to set the brightness for line scan exposures). The first line scan is produced with unchanged amplifier settings using the SE signal; the second line scan is produced with unchanged amplifier settings but without a beam. The latter is used as a baseline for the signal detection: no beam = no signal.

(a) Observe the signal over the platinum surface and over the aperture hole. Measure the proportion of the signal over the hole. Which components are collected on the Pt and over the hole? If the specimen were biased negatively (-45 V), how would these signals change?

(b) Increase the magification to about 10,000x and record the signal near the rim of the cup with a large final lens aperture. Adjust the SE signal for a smaller aperture to the same signal level on the Pt that was obtained with the larger aperture. Measure the increase in signal over the hole in the Faraday cup.

Experiment 12.7: Modifying the SE Signal with a Converter Plate. Produce line scans from the platinum rim of the Faraday cup as described above. However, the SE signal can now be modified, since the converter plate can be biased negatively or positively. Alter the converter bias from negative to positive 45 V while writing the signal scan on the aperture surface. Allow the end of the scan to go into the hole. What signal components can be identified from the drop of the signal?

Experiment 12.8: Modifying the SE Signal with a Grounded Specimen Grid. Install a grounded grid over the specimen. The grid may be mounted on any grounded part of the specimen stage (Figure 12.1). Make similar measurements as described above but introduce the following variations: (a) Produce a scan with the specimen biased negatively and the converter plate biased neutrally. (b) Bias the converter positively and superimpose a line scan. (c) Bias the specimen positively and write a third scan followed by the baseline. Identify the signal contribution of the specimen grid.

Laboratory 13

Electron Channeling Contrast

Purpose

The objective is to acquaint the microscopist with the crystallographic and contrast effects of electron channeling. The initial emphasis is on channeling experiments which can be performed on any conventional scanning electron microscope: large area channeling patterns of single crystals and channeling contrast images to reveal the crystalline microstructure of polycrystalline materials. The optional advanced experiments on covering area electron channeling patterns can only be carried out on an SEM which is equipped with special scanning and/or electron optical modifications. More detail on this topic may be found in *ASEMXM*, Chapter 3.

Equipment
1. For large area electron channeling patterns and channeling contrast microscopy of polycrystalline materials: any conventional SEM.
2. For selected area electron channeling patterns: an SEM equipped for deflection focusing or a similar modification to permit a substantial angle of beam rocking while the scanned rays are brought to a crossover at the specimen plane.

Specimens
1. Large area electron channeling patterns: 1 cm x 1 cm fragment of semiconductor grade silicon wafer (available prepolished with a high-quality surface from semiconductor suppliers).
2. Channeling contrast microscopy: annealed polycrystalline metal or metal alloy with a large grain size (e.g., Fe-3% Si transformer steel, nickel, gold, brass, etc.) prepared with a high-quality flat surface by electropolishing or mechanical polishing followed by electropolishing.
3. Selected area electron channeling patterns: single crystal silicon for setting up the proper conditions and a polished polycrystalline specimen with fine grain size (< 50 μm) to demonstrate local orientation differences.

Time for this lab session
    Two hours.

13.1 Electron Channeling Contrast

Electron channeling contrast provides information on the crystallinity of a specimen. In principle, any SEM can be utilized to observe electron channeling contrast, but because of the weak nature of the contrast mechanism (natural contrast level approximately 0.05 or 5%), careful attention must be paid to proper selection of the electron-optical conditions and to the signal processing. A beam current of several nanoamperes and a beam collimated to a convergence of the order of 5×10^{-3} radians is required. Signal processing is applied by means of differential amplification (typically designated on the SEM as "black level suppression," "dark level," or "contrast expansion").

A second important point in successfully carrying out this laboratory is the selection and preparation of the specimen. Electron channeling contrast originates close to the specimen surface in a shallow layer approximately 100 nm deep. The surface of the specimen must therefore be as free from contaminants and oxide coatings as possible. The contrast mechanism is also highly sensitive to the perfection of the crystal, and so the sample should not be strained. Because channeling contrast is weak, it is important to suppress competing contrast from surface topography, so the specimen should be as flat as possible. The ideal material to begin studying channeling contrast is a fully annealed crystalline material with a flat clean and perfect surface. Chemical or electrochemical polishing can produce such a surface on annealed metals and alloys. If mechanical polishing is used to produce a flat surface, chemical or electrochemical polishing must be applied as a final step to remove the thin surface damage layer which inevitably results from mechanical damage.

13.2 Wide-Area Channeling Patterns

An orientation pattern ("electron channeling pattern") can be obtained from a large single crystal (about 1 cm wide) angular scan available in any SEM by observing the specimen at low magnification (<20x). A change in the beam-to-surface angle of 10° or more can be obtained at low magnification while the convergence of the beam is kept small.

Experiment 13.1: Obtaining a Wide-Area Channeling Pattern. Perfom the following steps:

(a) Choose a large (e.g., 1 cm x 1 cm) piece of polished single-crystal silicon.
(b) Place the specimen at a tilt angle of 0° (normal incidence) and at a long working distance so that the lowest possible magnification may be obtained.
(c) Select a beam energy of 20 keV.
(d) A collimated beam can be obtained by choosing a small final aperture, typically 100 μm in diameter (convergence angle at 1 cm working distance = 5×10^{-3} rad). Alternatively, the beam can be intentionally defocused by adjusting the objective lens to make the conventional image more blurry; a large probe has a smaller convergence for a given working distance.
(e) Using a specimen current meter, choose the condenser lens setting so as to obtain a beam current of the order of 5-10 nA.
(f) Channeling contrast is principally carried by the backscattered electron signal. Choose a detector which maximizes the BSE signal: (a) large BSE detector placed above the specimen use the sum mode); (b) Everhart-Thornley detector positively-biassed to maximize the sensitivity to BSE effect through the collection of BSE-produced secondary electrons on the polepiece and chamber walls; or (c) specimen current.
(g) Apply differential amplification with a high value of dc suppression ("black level," "dark level," "contrast expansion"). Because of the lack of standardization in SEM
instrumentation, it is not possible to specify exact settings for the signal processing. The microscopist should be prepared to increase the differential amplification in stages until the desired result is obtained.

Experiment 13.2: Properties of Channeling Patterns. Once the ECP has been obtained, determine its properties and its relationship to the specimen:

(a) If a channeling map is available (see ASEMXM, pp. 137-141), identify the orientation of the crystal relative to the beam.
(b) Using the x-y translation controls on the mechanical stage, move the specimen so that an edge of the specimen is visible near the limit of the field of view to provide a point of reference. Record an image. Translate the specimen so that the specimen edge moves 1/3 of the field width. Record a second image. Does the ECP move? Why or why not?
(c) Translate the specimen so that the specimen edge is again near the limit of the field of view. Record an image. Using the mechanical stage controls, rotate the specimen and maintain the edge of the specimen in the field of view. How does the ECP behave?
(d) Again translate the specimen to place the edge near the limit of the field of view. Record an image. Using the mechanical stage controls, tilt the specimen by 10° and translate if necessary to maintain the same field of view. Record a second image. How does the ECP behave and why?

13.3 Measurements with Channeling Patterns

The spacing of lines and bands in electron channeling patterns is controlled by the Bragg law:

$$n\lambda = 2d \sin \theta_B \tag{13.1}$$

where n is an integer, λ is the electron wavelength, d is the spacing for a particular set of crystal planes, and θ_B is the Bragg angle for that particular set of planes. The relativistically corrected electron wavelength can be found from the expression:

$$\lambda_{rel} = \frac{3.87 \times 10^{-9}}{\left[E^{0.5}(1+9.79\times 10^{-4}E)^{0.5}\right]} \tag{13.2}$$

where λ_{rel} is the electron wavelength in centimeters and E is the beam energy in keV. The beam energy can be measured within ±10 eV by measuring the Duane-Hunt limit of the x-ray bremsstrahlung (expand the vertical scale on the EDS system to measure the highest x-ray energy at 20 kV).

An angular scale for a channeling pattern can be determined if a crystal with a known spacing is measured at a known beam energy. For a *cubic* crystal, the Bragg angle θ_B is given by:

$$\theta_B = \arcsin\left[\frac{\lambda(h^2 + k^2 + l^2)^{1/2}}{2a_0}\right] \tag{13.3}$$

where (h,k,l) are the integer Miller indices and a_0 is the lattice parameter.

Experiment 13.3: Calibrating the Angular Scale. Use the tilt and translation of the stage to obtain a channeling pattern for silicon at 25 kV with the (100) pole nearly parallel to the electron beam. Determine the appropriate angular scale marker for the channeling pattern (for silicon, $a_0 = 0.542$ nm) using the (022) band. Is the SEM-generated angular marker correct? Repeat the calculation for the (044) band, which is more sharply defined and more widely separated than the (022) band, which improves the precision and accuracy of the linear measurement.

Experiment 13.4: Sensitivity to Lattice Parameter and Beam Energy Changes.
It would be of considerable use if ECPs could be used to accurately measure lattice parameter. Unfortunately, the distortions inherent in CRT displays prevent measurements of the type used in Experiment 13.3 from producing lattice parameter results of sufficient accuracy and precision. Another possibility is to use the fine scale detail in the center of low-index, high-symmetry poles. This fine scale detail actually arises from channeling lines with very high indices (hkl), e.g., (771) and (373) in a silicon [111] pole. As a result of these high values of (hkl), the channeling lines are very sensitive to changes in the lattice parameter, or alternatively for constant lattice parameter, the lines are very sensitive to λ and hence to beam energy, E_0.

Obtain an ECP of the silicon crystal at a beam energy in the range from 20 to 30 keV. Use the tilt and rotation controls to bring the [111] pole into the center of the CRT image. Increase the magnification until the image of the pole fills the CRT image. Adjust the signal processing to optimize the presentation of the fine structure. It may be necessary to reduce the beam convergence by weakening the objective lens to better "focus" the fine channeling lines in the pole. Photograph the optimized image. Next, change the beam energy by a small amount of the order of 1 keV, either as an increase or decrease. Optimize the ECP of the [111] pole again and record an image. Compare the two images. Has the fine structure changed noticeably as a result of the small percentage change in beam energy?

13.4 Channeling Contrast Imaging of Microstructures

Electron channeling contrast can be employed in images of polycrystalline material to reveal the crystalline microstructure (grains, grain boundaries, subgrains, twins, and other crystallographic details) in images obtained with magnifications up to several thousand times. Limiting resolutions are of the order of 500 nm for typical operation and specimens, and 100 nm can be achieved under optimum electron-optical and specimen conditions.

Channeling contrast imaging seeks to simultaneously produce a highly focused probe (to maximize the spatial resolution and fine scale detail in the image), a small convergence probe (to maximize the channeling contrast), and a large beam current (to exceed the threshold current and make it statistically possible to observe channeling contrast). These conditions are mutually exclusive, so a compromise must be achieved.

Experiment 13.5: Channeling Contrast of Microstructures. The initial specimen to be investigated should have a relatively coarse grain size: 100 µm to 1 mm. Fe-3% Si transformer steel typically has large grain sizes. Place the specimen at 0° tilt (normal beam incidence).

(a) Select a beam energy in the range 20 keV to 30 keV. Higher beam energy will increase the beam brightness, and improve beam characteristics for channeling.
(b) Select a final beam-defining aperture of approximately 200 µm diameter (divergence 1 x 10^{-2} rad at 10 mm working distance)
(c) Adjust the condenser lens to give a beam current at the specimen of 2-5 nA.
(d) Focus the beam with the objective to give the finest diameter probe.
(e) The best choice for the detector is a large solid angle solid state BSE or scintillator/PMT BSE detector. The solid state detector in particular has the property that it produces a larger response from the high-energy BSE which carry channeling contrast, and is generally insensitive to BSE with energies below 5 keV, which contribute only to the background of the signal.
(f) Adjust the signal processing (differential amplification) to observe the contrast. The settings should be similar to those needed to obtain the wide-area ECP (Section 13.2).

Experiment 13.6: Properties of Channeling Contrast Images. Once the grain structure of the polycrystalline material can be observed, carry out the following tests:
(a) Record an image with the specimen at 0° tilt.
(b) Change the tilt to 5° and record a second image of the same area. Keep a recognizable surface defect in the field of view. Do any of the grains change their gray level? Why does the grain shading change? Do any grain boundaries appear or disappear between the two images? If you increase the tilt to 10° and record another image of the same area, are any new grains discovered?

Experiment 13.7: Resolution of Channeling Contrast Images.
(a) Find a crystalline boundary (grain boundary, low-angle boundary, annealing twin, etc.). Annealing twins are especially useful for this experiment.
(b) Record a series of channeling contrast images at progressively higher magnification. Assess the spatial resolution using line scan.

What is the limiting spatial detail of the crystallographic microstructure which can be discerned? How does this compare with the minimum probe size calculated from the brightness equation? [See Laboratory 2, Equation (2.1); assume $\beta = 5 \times 10^4$ A/cm^2 str for a conventional tungsten hairpin filament at 20 keV; $\alpha = 5 \times 10^{-3}$ rad; beam current = 0.8 nA.]

13.5 Selected Area Electron Channeling Patterns

If the SEM has the capability to produce selected area electron channeling patterns (SACP), determine from the instruction manual what electron-optical ray paths are used. The deflection-focusing method is one possible technique (*ASEMXM*, pp. 113-117).

Experiment 13.8: Taking an SACP.
(a) Obtain a conventionally scanned image of the microstructure showing channeling contrast. Center a large grain in the field of view.
(b) Switch to the SACP mode. How does the image change? Is a single ECP visible? If not, slightly translate the specimen mechanically. Can a single ECP be obtained?
(c) What ultimately controls the size of the selected area? *If the deflection focusing method is used, change the strength of the objective lens to change the size of the selected area and observe the image.* Can several ECPs be observed simultaneously? As the selected area size is decreased, can a single ECP be obtained? Can the grain from which this ECP is obtained be readily identified?
(c) Use the large silicon single crystal to obtain a high-quality pattern. Estimate the beam convergence from the angular width of the finest channeling line which is visible in the ECP. Using this angular feature, change the strength of the final lens. What lens controls the convergence in the SACP mode? Does the sharpness (beam convergence) of the feature change? Repeat this experiment with the condenser lens? Does the sharpness change?
(e) Estimate the smallest area from which an SACP can be obtained. If a range of grain sizes is present in the specimen, select progressively finer grains, noting their sizes in the conventional microscope mode, and then determine if a single crystal channeling pattern is obtained in the SACP mode. If the available grains are too large for this determination, locate a grain boundary in the conventional imaging mode, and place it near the center of the field of view. In the SACP mode, translate the specimen mechanically, quantitatively measuring the amount of motion. What is the smallest area from which the ECP can be obtained?

Laboratory 14

Magnetic Contrast

Purpose

The purpose of this laboratory is to demonstrate the two types of magnetic contrast, Type I and Type II, which can be detected in suitable specimens in any standard SEM. Because of the special characteristics of the magnetic contrast mechanisms, the steps required to observe magnetic microstructures provide a good illustration of the need to follow an appropriate strategy in order to satisfy the threshold current/contrast relationship and to apply proper signal processing. More detail on this topic may be found in *ASEMXM*, Chapter 4.

Equipment

Any standard SEM equipped with an Everhart-Thornley (E-T) detector.

Specimens

1. For Type I magnetic contrast, the specimen is a section of an ordinary magnetic recording tape upon which has been recorded a saturated square wave with a frequency such that stripe domains are formed with a spacing of approximately 100 µm.
2. For Type II magnetic contrast silicon steel, available from any scrapped transformer, can be used to prepare a suitable specimen. The specimen should be mechanically polished to eliminate surface relief and then chemically or electrochemically polished remove the surface damage layer left by mechanical polishing.

Time for this lab session

Two hours.

14.1 Type I Magnetic Contrast

For magnetic materials which have structures that permit the magnetic flux to leave the specimen, the leakage magnetic fields above the specimen surface can be viewed by their influence on the trajectories of the secondary electrons which are emitted. This is Type I magnetic contrast. Artificial magnetic structures such as recording tape and magnetic disk material have this type of magnetic microstructure, as well as certain natural crystalline materials such as cobalt.

Experiment 14.1: Obtaining Type I Magnetic Contrast. Prepare a test specimen which consists of ordinary magnetic tape upon which a square wave of known frequency has been recorded at a known rate of tape transport.

(a) Select a beam energy of 10 keV for initial work.
(b) Select a beam curent of 1 nA.
(c) Place the specimen at a tilt of 0°.
(d) Use the positively biased E-T detector.
(e) Select a magnification of 200x to 500x.
(f) Apply a large amount of contrast expansion (black level or differential amplification)
(g) If the magnetic domains (which should appear as stripes running across the width of the tape) are not visible, rotate the specimen about its normal by 45° and 90° and reexamine the image for the stripes.
(h) If the stripes of the recording are still not visible, try increasing the beam current and further increasing the black level processing.
(i) If the domain stripes are still not found, the E-T detector may be too efficient! Try reducing the collector voltage (Faraday cage) to reduce the collection efficiency. Try tilting the specimen to grazing incidence.
(j) When the domains have been found, prepare a positively biased E-T detector image.
(k) Repeat the image on the same field of view with the E-T detector biased negatively.

How do the these images compare?

Experiment 14.2: Effect of Specimen Rotation on Type I Magnetic Contrast. After the domains are found, try rotating the specimen about its normal to maximize the contrast. Note the stage rotation position. Choose a field of view which also has some recognizable topography, such as a scratch or other defect to serve as a point of reference, and record an image. Rotate the stage by 90° and record a second image with the reference defect in the field of view. Rotate the specimen a further 90 (180° from the initial maximum contrast position) and record a third image. How does the contrast change as a function of specimen rotation?

Experiment 14.3: Spatial Resolution of Type I Magnetic Contrast. With the domains showing maximum contrast, prepare a series of images at increasing magnification. Estimate the spatial resolution. Are the edges of the domains sharp at high magnification?

14.2 Type II Magnetic Contrast

For magnetic materials which have structures that permit the magnetic flux to be retained within the specimen so that no leakage fields exist outside the surface, the internal magnetic fields can be viewed through their influence on the beam electrons as they scatter within the specimen, causing changes in the backscattered electron coefficient. This is Type II magnetic contrast. Natural ferromagnetic materials of cubic symmetry, such as iron and nickel, have this type of magnetic microstructure.

Experiment 14.4: Obtaining Type II Magnetic Contrast: The specimen should be a polished section of polycrystalline iron or iron alloy, such as Fe-3.2%Si transformer steel.
  (a) Select the maximum beam energy available on the SEM, preferably 30 keV or higher. Below 20 keV the contrast is so weak as to be virtually impossible to detect.
  (b) Adjust the condenser lens to produce a beam current greater than 200 nA.
  (c) Place the specimen at a tilt of 55°.
  (d) Maximize the contrast expansion by the use of black level processing.
  (e) Search at a magnification in the range 100x-400x.

(f) *If domains are not in evidence, rotate the specimen about its normal by an angle of 45°; repeat by rotating another 45°.*
(g) *If domains are still not in evidence, increase the beam current to 500 nA and further increase the contrast expansion.*
(h) *After domains are found, rotate about the specimen normal to maximize the contrast.*
(i) *From the maximum contrast position, rotate the specimen by 45°, 90°, and 180° from the original maximum contrast position and record images.*

How does the contrast behave as a function of specimen rotation?

Experiment 14.5: Spatial Resolution of Type II Magnetic Contrast. Place the specimen at the maximum contrast position (keep at a tilt of 55°). Prepare a magnification series to the maximum useful magnification on a suitably fine-scale magnetic structure. For the highest magnification image, reduce the beam current by a factor of 5, refocus, and increase the record time by a factor of 5. What is the maximum effective spatial resolution which can be obtained for Type II magnetic contrast?

Laboratory 15

Voltage Contrast and EBIC

Purpose

The purpose of this laboratory is to understand voltage contrast (VC) and electron beam induced contrast (EBIC) as important tools for the examination of semiconductor materials which aid in the production of microcircuit devices. More details may be found in *ASEMXM*, Chapter 2.

Equipment

1. Almost any SEM, although to achieve the most effective demonstrations of the phenomena discussed here it is desirable that the instrument be able to operate down to at least 2 keV beam energy and up to 25 or 30 keV.
2. Separate amplifier for EBIC mode is required. Most currently available solid-state "specimen current amplifiers" are quite suitable for this purpose, but older SCAs which used nuvistors or tubes should be avoided as these have such a high input impedance that they are unable to respond to changes in the signal current at a high enough rate to permit acceptable imaging.
3. Small power supply (± 9 V dc) or some PP9 size batteries are needed to supply the specimen with operating voltages.
4. A function generator able to produce sine or square waves at frequencies up to 100 kHz with an amplitude of a few volts is desirable.
5. An electrometer, such as a Keithley 607, capable of reading potentials of the order of 100 mV.
6. Special specimen mounts. In order to examine finished devices (i.e., integrated circuits or "chips") in either voltage contrast or EBIC mode, it is convenient to have specimen mounts which are equipped with suitable chip bases and the necessary wiring harness to a connector on the outside of the microscope column so that devices can be readily plugged or unplugged from the stage, and monitored or activated by the operator, without the need to break the vacuum. Most SEM manufacturers can now supply such mounts for their instruments. When such an arrangement is not available, then the chip must be mounted firmly on the usual specimen stub, taking care that the electrical leadouts are not inadvertently shorted together or to ground. The connections to external power supplies or amplifiers must then be made through wiring connected to a vacuum electrical leadthrough. To avoid induced ac field pick-up, all of the leads from the device should go to the same leadthrough, and the leads should be dressed so as to be out of line of sight from the specimen. If this is not done then stray backscattered electrons may hit the wires causing the insulation on them to charge and subsequently deflect the beam or interfere with the collection of secondary electrons.

   For EBIC observations of materials only two connections are required, one of which is grounded the other of which goes to the amplifier. Again, so as to minimize the possibilities of ground loops and ac pick-up, the specimen should be floated above electrical ground inside the microscope. Both of these leads should go to the same feedthrough on the column, and the ground wire should only be earthed at the input of the amplifier.

Specimens

1. For voltage contrast and EBIC studies of an integrated circuit, an ideal choice of device is a Radio Shack type 741 operational amplifier (part number 276-007). This chip costs less than $1, and is packaged in a plastic header. To prepare the device for observation in the SEM gently support the chip in a benchtop vise and remove the top of the package by horizontal cutting with a jewelers' hacksaw to expose the chip surface (covered by a glass passivation layer). Remove any debris by gently blowing the surface with clean, dry, air . The pin-outs for this chip are shown in Figure 15.1.

2. For EBIC observations of a semiconductor material, a good specimen is a Radio Shack "Solar Cell" (part number 276-124A). As purchased this comes complete with surface tracks which allow easy solder connections to both the top (shiny side) and back surfaces. To prepare for use in the SEM, select one corner of the solar cell wafer with the connection track in place and mark a rectangle about 5 mm on each side with a diamond scriber. Place a straight edge or metal ruler along each edge in turn and cleave the silicon by pressing down on it. Solder a fine-gauge insulated wire (preferably solid rather than stranded) to the top and bottom surfaces.

15.1 Voltage Contrast

The voltage contrast mode permits us to visualize the potential distributions on the surface of an operating semiconductor device and even, under some conditions, make a quantitative determination of the potential itself. By applying suitable inputs to the device we can also study the behavior of the microcircuit as it processes real signals.

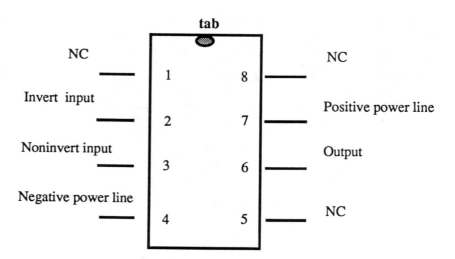

Figure 15.1. Type 741 operational amplifier viewed from underside of chip.

Experiment 15.1: Static Voltage Contrast. Prepare the chip and mount it in the SEM being careful not to short together, or to ground, any of the leads. Using the power supply or the battery apply +9 V to pin number 7, -9 V to pin number 4, and the return lead (ground) to pin number 5. Place the specimen normal to the beam, choose secondary electron mode, select an operating voltage of about 2 kV, a rapid scan rate, and a magnification which is low enough to allow the whole of the chip area to be observed (about 100x).

(a) Note that the lead from the header pin 7, connected to +9 V, is dark while the lead from pin 4, connected to -9 V, is bright. This is an example of "voltage contrast." Explain this contrast.

(b) *Look for signs of voltage contrast from features on the device area of the chip itself.* Most probably little or nothing will be visible because of the glass passivation layer on top of the chip. This layer will charge up and the electrostatic field produced by this charging will be sufficiently strong to swamp voltage contrast from the device itself. *Determine the polarity of the charge on the surface in the following manner: Increase the SEM magnification to 1000x for about 5 sec then quickly switch back to 100x. Look at the scan square in the center of the screen.* If this area is bright, then the passiviation is charging negatively (repelling electrons). If the area is dark, then the charge-up is positive (recollecting secondary electrons). If the charge-up is negative, then the specimen is receiving more electrons than it is emitting, that is to say that the incident beam energy is above E2--the energy at which the total emission coefficient (backscattered+secondary yields) becomes unity. *Try reducing the beam energy in steps of 100 eV until the charge-up becomes zero or positive. If the original charge-up was positive then the incident energy is below E2, so try increasing the beam energy in 100-eV steps until the charging just disappears.*

(c) *Determine the value of E2 for this passivation material. Repeat this experiment with the specimen tilted 10, 20, 30, 45 and 60 degrees.* How does E2 vary with tilt angle?

(d) *With the beam at the E2 energy look again for signs of voltage contrast from the device structures.* If none is visible, try disconnecting the power supplies or batteries momentarily from the chip and then reconnect them.

Experiment 15.2: Dynamic Voltage Contrast. *If a function generator is available, use it to apply a square or sinewave signal of about 1 or 2 V to the input of the amplifier. Use the fastest scan rate that the SEM can provide and choose a frequency of a few hundred hertz.* The screen will now be crossed with a diagonal pattern of bright and dark lines. These will be strong on some parts of the device area and weaker or absent on other regions. This "barber pole" effect is the voltage contrast from the ac input to the amplifier with bright regions corresponding to the negative excursions of the signal and dark regions to the positive part of the signal. We can see the variations in the ac signal because the SEM image is itself time-resolved since successive pixels on the image are recorded at slightly different times. *Try varying the frequency of the oscillator and note how the spacing and orientation of the bands vary. If the SEM has a framestore device, then record the image from the specimen and observe how the visibility of the barber pole stripes varies from one feature to another on the sample. Examine how the visibility of the contrast varies with the choice of beam energy.*

15.2 Electron Beam Induced Current (EBIC)

The EBIC technique allows us to examine unprocessed semiconductor materials and to look for such features as grain boundaries and crystallographic defects within them as well as giving us important quantitative information about the parameters (e.g., diffusion length, depletion depth) which characterize the semiconductor. EBIC also provides a means of examining finished devices and inspecting each of the active junctions in the device.

Essentially, the contrast in an EBIC image arise because an incident beam electron can promote an electron in the valence band of a semiconductor specimen to the conduction band creating an electron-hole pair. By collecting and amplifying the charge from the promoted electron and the resultant hole, a signal can be obtained to modulate the CRT brightness as a function of electron beam position on the specimen. Since many types of signals may be collected in a similar manner, this technique is often more appropriately called *charge collectio microscopy* rather than the more restrictive term electron beam induced current.

Experiment 15.3: EBIC Observation of Devices. Remove the power supplies and function generator connected to the 741 op-amp for the previous experiments and connect up the chip as shown in Figure 15.2. The specimen current amplifier (SCA) is connected across the positive and negative rails of the chip. Be careful to ensure that none of the leads from the chip is touching ground inside the microscope column since this will cause high contrast hum bars to be visible across the image. The lead connected to pin 4 of the amplifier should be attached to the grounded side of the SCA input. Set the microscope to 20 kV and, if possible set the spot size to produce an incident beam current of about 1 nA (10^{-9} A). The sensitivity o, the SCA should be chosen to be in the 10^{-8} to 10^{-7} A range, and one of the inputs to the SEM should be selected to be that output from the SCA. With a beam scan rate of several seconds per frame observe the image, adjust the contrast and gain controls on the SCA to produce a pleasing image. What is the significance of the bright and dark regions now visible on the

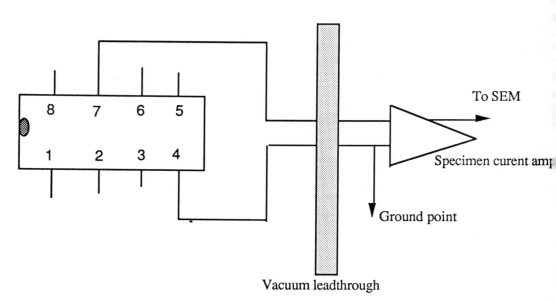

Figure 15.2 Connections for the EBIC experiments

image? What happens if the scan speed of the image is increased? By switching to secondary electron imaging mode see how these regions are related to visible structures on the device. If the SCA has a meter, note how the indicated current compares with the magnitude of the incident beam current.

Experiment 15.4: Effect of Electron Beam Energy. Record EBIC images of the same area of the device but at beam energies of 17, 15, 13, and 10 keV. Note that as the energy of the beam is reduced the signal becomes weaker, fewer features are visible in the image, and at the lowest voltage no contrast at all is visible in the EBIC mode. What is the origin of these changes in the signal and why does the contrast ultimately disappear? How does the sharpness of detail in these images vary with the energy?

Experiment 15.5: EBIC Observation of Materials. Replace the 741 device by the solar cell and connect it to the SCA once again ensuring that neither the leads nor the cell itself is accidentally grounded within the microscope. Select a beam energy of about 20 keV, a beam current of a few nanoamps, an SCA sensitivity of about 10^{-8} A, and choose a magnification which is low enough to permit the whole area to be viewed at the same time. Adjust the SCA brightness and contrast controls to form an acceptable image. Note that if the physical size of the solar cell fragment is large, it may be necessary to use a very slow scan rate before an acceptably sharp image is obtained. Record the image and note the average signal current received by the SCA. Vary the beam energy. How do the appearance of the image and the current gain of the diode vary with the incident beam energy?

Experiment 15.6: EBIV Measurements of a Schottky Diode. Replace the SCA with the Keithley 607 electrometer, and set this to measure voltages between 10 mV and 500 mV. In this mode we will be measuring the open circuit voltage of the Schottky diode, so this technique is called electron beam induced voltage (EBIV). Choose an incident beam energy of 20 keV and an incident beam current of about 1 nA. Scan the electron beam over the surface of the Schottky diode. Record the output voltage as measured by the electrometer. Increase the probe size to put more current into the specimen and note that the output voltage increases. For incident currents between about 0.5 nA and about 20 nA measure the open circuit output voltage of the diode. Measure the gain G of the diode (output current/input current) and also total area of the Schottky barrier.

Laboratory 16

Environmental SEM

Purpose

In this laboratory the new capabilities of environmental SEM will be analyzed and demonstrated. Comparison of BSE and SE imaging under different gas pressures will allow characterization of the various signal components available for imaging of insulators and wate Dynamic imaging of crystallization phenomena will be demonstrated using a cold stage.

Equipment

An environmental SEM operating with the specimen at >20 Torr equipped with SE and BSE detectors and a cold stage.

Specimens
1. Inhomogeneous surface film specimen: 2-nm-thick platinum film deposited onto a silicon substrate through a TEM grid with narrow grid bars followed by a 2-nm carbon film deposited through a grid with wide grid bars.
2. Metal stub with 2.5-mm-wide groove covered with 400 mesh nickel grid used as water condensation chamber.
3. Nonconductive, inhomogeneous surface film specimen: 2-nm-thick platinum and 2-nm-thick carbon films on bulk sintered polyvinylchloride.
4. Polymer foam or biological tissue (rat lung, fixed, dried but not metal coated) attache with double-stick tape to a support which will be used as a bulk inhomogeneous, lov mass-density sample of low electrical conductivity.
5. Salt solution (NaCl) saturated at room temperature which will be used for crystallization experiments.

Time for this lab session
Two hours.

16.1 Environmental SEM

The limitations encountered in conventional high-vacuum SEM when electrical insulator or wet specimens are examined can be overcome though the utilization of residual gas in the specimen chamber. Gas/electron interactions can be used to neutralize surface charge on insulators. Specific gas/specimen interactions can be used to amplify the emitted electron signal. Liquids can be maintained in a partial vacuum by raising the pressure to that of the saturated vapor pressure. Additionally, the gas composition may be altered for investigation c chemical reactions. Control of the gaseous specimen environment allows the extension of microscopy to fluid, wet, or irregularly shaped surfaces independent of their electrical conductivity. The gas pressure (> 20 Torr) has to be maintained only in the specimen chamber which is separated from the high vacuum of the electron optical column by several differential pressure zones formed with pressure-limiting apertures within the probe-forming lens. microscopy at high gas pressures is not different from conventional SEM. However, new procedures are required for generation and evaluation of the different imaging possibilities as well as for stabilization of water within the specimen chamber.

The interactions of the probe electrons, the signal electrons, and the specimen with the gas are closely connected with each other by gas pressure, working distance (specimen distance), signal collection field, and accelerating voltage. Gas molecules become ionized within the electrostatic field formed on insulating specimens as well as through interaction with probe electrons and signal electrons. The positive ions bombard regions of negative charge build-up on the specimen surface in a self-regulated fashion (i.e., positive charges are automatically attracted to negative charges) so that the specimen surface remains electrically neutral. Environmental secondary electrons (ESEs), generated through gas ionization, are collected by a positively biased gaseous detector. If water vapor is used to provide the

specimen environment, the saturated water vapor pressure at room temperature is about 20 Torr. Under this condition water does not evaporate and can be maintained indefinitely. A cooling/heating stage may be used to reduce or increase specimen temperature, and thus, to condense or evaporate water or other substances under microscopic observation. This new kind of high-gas-pressure microscopy allows imaging of specimens unobscured by metal coatings and without further preparation.

16.2 Signal Quality

The gaseous detector, positively biased, collects electrons emited from the specimen and from the gaseous environment between specimen and the detector. The proportion of electrons collected from the specimen surface and the environment depend on the working distance, the gas pressure, the accelerating voltage, and the collector bias. In order to analyze the signal components, a thin-film test specimen will be used which generates distinct specimen-specific SE and BSE signal electrons.

Experiment 16.1: Collection of SE and BSE Signals. Image the "inhomogeneous thin surface film specimen" at high accelerating voltage (10-20 kV) and low magnification (500x) in BSE mode. Identify the platinum metal patches and find an area where platinum covers only part of the substrate. Image the same area using the gaseous detector. Vary the gas pressure from <0.1 Torr to 20 Torr and find that gas pressure at which the carbon patches are imaged with strongest contrast. Which signal components image the metal and carbon patches, respectively? Which environmental SE components image the metal at high, medium, and low gas pressure?

16.3 Imaging of Liquid Water

If water vapor is used as the environmental gas, it is possible to stabilize liquid water at pressures equal to the vapor pressure of water. At a given vapor pressure, the specimen temperature may be adjusted to maintain, evaporate, or condense water. Water vapor partial pressures are given in Figure 16.1.

Thus, wet samples may be imaged at appropriate stabilizing water pressures. If the SE

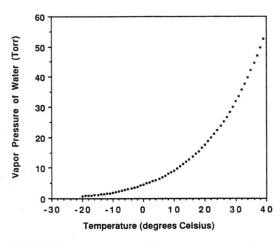

Figure 16.1. Water vapor partial pressure versus temperature.

signal is established at a certain pressure, the specimen temperature can be used to stabilize the water. This important aspect of environmental SEM will be demonstrated by producing and imaging thin water films.

Experiment 16.2: Control of Saturated Water Pressure. Use the "water condensation chamber" inserted into the cold stage and image the rim of the chamber at 20 kV and low magnification (500x) with the gaseous detector. Use a water pressure between 10 and 20 Torr.

(a) Reduce the specimen temperature until water condenses at the walls of the chamber and continue to accumulate water at that particular pressure until the chamber is partially filled. Then increase the specimen temperature slightly to evaporate some of the liquid. The retracting water meniscus will leave thin water films in the mesh of the grid cover. When a stable film is produced, measure the specimen temperature and the water vapor partial pressure.

(b) Go to medium magnification (2,500x) and increase or reduce the thickness of the films. Use topographic relief contrast to identify if the film is convex, concave, or evenly thick. Image the same films with the BSE signal and observe the image contrast.

16.4 Imaging of Insulators

Moderate gas pressure (>10^{-3} Torr) provides for self-regulating charge neutralization on electrical insulators. As surface charge builds up an electrostatic field, gas molecules are ionized and the appropriate counter ions are accelerated toward the charged surface and arrive with a low (~10 eV) kinetic energy. If the field strength increases, more ions are generated. This charge neutralization generally does not interfere with the SE signal emission and allows imaging of rough insulating surfaces without metal coating or use of low acceleration voltage.

Experiment 16.3: Surface Imaging of a Polymer.
(a) Image the "nonconductive inhomogeneous thin surface film specimen" composed of

tempered PVC covered with platinum and carbon patches. Use 20 kV and 500x magnification. Find the conditions for SE imaging of the platinum patches. Describe the contrast phenomena which image the platinum patches. Why is the metal not imaged with positive contrast? Use the BSE image to confirm signal type.

(b) Image the fine structure of the polymer seen in the granular surface depressions. Compare the BSE and the SE images. Define the surface topographic contrast and the non-topographic contrast (mass density variations). Identify the types of signal components and measure their escape depths.

Experiment 16.4: Imaging of Uncoated Polymer Foam or Biological Tissue.

(a) Image the dried and "uncoated biological tissue sample" at low and high accelerating voltage (6-20 kV) using the SE imaging mode. Change magnification, accelerating voltage, scan speed, specimen position, and gas pressure. Try to identify "specimen charging" and "beam damage."

(b) Compare environmental microscopy of this inhomogeneous bulk nonconductor with low-voltage microscopy performed in Laboratory 10 at high vacuum and 1 kV accelerating voltage. Compare these images for artifacts such as specimen charging, beam damage, deposition of contaminations.

16.5 Dynamic Environmental Experiments

The ability of environmental SEM to image specimens without any specimen preparation and at high gas pressures allows imaging of dynamic environmental processes. Crystallization can be performed either from aqueous solutions using a cold stage for the regulation of water condensation, or from solid phase using a hot stage for the melting.

Experiment 16.5: Crystallization of Salts.

(a) Use the "water condensation chamber" and place a droplet of room-temperature saturated salt (NaCl) solution onto the cover grid. Establish SE imaging conditions at TV rate. Warm up the specimen table until the water is evaporated and the salt forms large crystals. Reduce the stage temperature in order to hydrate the crystals and to condense enough water to dissolve the salt.

(b) Repeat the cycle. Observe the structural changes of the crystals when they form or dissolve. Compare the strength of topographic relief contrast of the crystal facets with that of the liquid droplet.

Laboratory 17

Computer-Aided Imaging

Purpose

This laboratory is concerned with use of the digital computer to aid in the acquisition, display, and interpretation of various digital images obtained with the scanning electron microscope, e.g., secondary electron, backscattered electron, x-ray, etc. We will be concerned with all aspects of SEM imaging from the generation of the signal, when the electron beam interacts with the specimen surface, to the perception and interpretation of this information in the mind of the observer. Indeed, we will view all steps involved as an information channel which has certain imperfections and nonlinearities that must be examined. This aspect of microscopy is a rapidly developing area and consequently it is difficult to condense all of the important concepts into a single laboratory exercise. Nevertheless, the topics covered in the following experiments should give a sense of the power of this new tool for the microscopist. For more details see *ASEMXM*, Chapter 5, and references [1-4] at the end of the laboratory.

Equipment

SEM equipped with digital imaging and processing capabilities. The computer system may be a composite which includes the energy dispersive x-ray spectrometer/multichannel analyzer, or it may be a stand-alone image acquisition and processing system. In addition, commercial and public domain programs exist for personal computers such as the IBM-PC and the Apple Macintosh which can carry out many computer-aided imaging functions.

Specimens
1. Fracture surface such as the one used in Laboratory 2.
2. Multiphase, flat-polished specimen such as the one used in Laboratory 3. Both basal and Raney nickel will work well.

Time for this lab session
Three hours.

17.1 Digital Image Acquisition

The application of an on-line interactive computer to the SEM results in a substantial improvement of several key capabilities. These capabilities derive mainly from the ability of the computer to record more of the information generated and detected when the primary electron beam interacts with a point on the specimen. The recorded information may be modified and combined in a great variety of ways before presentation to the operator-analyst. This permits structural and/or analytical information to be seen in a micrograph where none could be seen before.

In the conventional analog display system used on most SEMs, the electron beam moves over the face of the recording oscilloscope in synchronism with the beam inside the column moving over the specimen surface. When a detector in the vicinity of the specimen receives an increase in its signal (e.g., secondary electrons), the beam on the recording oscilloscope is increased in brightness. Consequently, an image is "painted" onto the face of the display tube. The synchronous beams are displaced by imparting horizontal and vertical velocity components V_x and V_y, where V_x is typically between 500-2000 times V_y. The direction in which the velocity component is greater is historically called the line direction and the other direction is called the frame direction. When the beam reaches the end of a line on the oscilloscope it is inhibited from producing light (it is blanked), moved to the beginning of the next line, and the scan is repeated. This conventional display system has been used since the earliest days of the SEM since only simple analog circuits are required.

An alternative to the continuous beam rastering method is discrete rastering accomplished with what is usually called a "digital scan generator." In this technique the x and y velocity components of the synchronous electron beams are not constant but remain zero for a finite period of time and then the beam is rapidly stepped to the next point. The displacements along the line direction are equal. When the end of the line is reached, the beam is moved back to the beginning of the line and displaced one step along the frame axis with a step size equal to the line step. Each point in the image at which the beam dwells is called a "pixel" (picture element). Digital scan generators typically use between 100 and 4000 pixels along the frame and line directions. The digital method of scan generation is very convenient for digital computer applications such as mathematical manipulations.

Experiment 17.1: Digital Scanning. The frame time of digitally scanned images, and their appearance, depends on the number of pixels and the dwell time per pixel.

(a) Acquire a digital image at low pixel density (e.g., 64 x 64 or 128 x 128) with 100-usec dwell time per pixel. Adjust contrast and brightness to produce a pleasing image. How many pixels are in the image? What was the frame time?

(b) Acquire a digitally scanned image with a high pixel density (e.g., 512 x 512 or greater) at 100 µsec/pixel. How many pixels are in the image? What was the frame time? Can you perceive any difference between a 512 x 512 digital image and a typical 1000-line analog image?

All images are eventually viewed in the form of analog signals such as phosphors excited on a cathode ray tube or photographic chemicals fixed on paper or transparency film. In the past, all images viewed in the SEM were also processed entirely in the analog signal domain where continuously varying electrical signals were manipulated with electronic circuits which involved various sorts of linear and nonlinear amplifiers to produce the final electrical signal controlling the intensity-modulation of the CRT. However, analog image processing of intensities is now often replaced by operations carried out in a digital computer. In digital imaging, the concept of an analog image is replaced by the concept of a mathematical matrix in the computer memory which contains all of the information that constitutes an "image." The continuous intensity range (gray-level scale) of the analog image is replaced by a range of discrete numerical values (usually the display is 0-255 levels or 8 bits) stored at the matrix addresses. Digital images may be collected and/or displayed with intensity levels of 1 bit (2 levels), 2 bits (4 levels), 4 bits (16 levels), 8 bits (256 levels), 16 bits (65,535 levels), etc.

Experiment 17.2: Digital Intensity Levels. To illustrate the concept of discrete digital intensity levels, we shall digitize the intensity scale of a typical SEM image to several different numbers of bits. Note that the analog signal from the amplifier chain must fill the input acceptance voltage range of the analog-to-digital converter (ADC). If this condition is not fulfilled, the digital output range of the ADC will be incomplete and an insufficient number of levels will be produced.

Choose a suitable field of view at 5000x to 10,000x on a topographic specimen such as a fracture surface. Obtain digitized images with 1-bit, 2-bit, 4-bit, and 8-bit digital intensity levels. Compare the images. Are there any unusual features?

Note: Local hardware may permit this experiment to be performed in either of two ways: (a) as described, with the digitization occurring at the specified number of bits, or (b) with the digitization occurring only at the largest number of bits for the particular system but with control over the number of bits displayed on the viewing screen.

17.2 Image Contrast

Two important concepts in computer-aided imaging are image processing and image analysis. *Image processing* constitutes those operations which are principally intended to alter the visibility of a feature of interest as it is presented to the viewer. Contrast enhancement is an example of image processing. *Image analysis* constitutes those operations which are principally intended to derive numerical information about features of interest in the image, e.g., number, size, areal fraction, etc. The two topics are linked because an image must have sufficient contrast in order to be properly analyzed. Unfortunately, owing to the wide variety of hardware and software encountered in the field, it may not be possible to carry out all of the following experiments in the form in which they are described.

A dominant problem throughout all of microscopy is the contrast of features of interest in the final micrograph. Often that contrast is inadequate, and enhancement is needed. Many different digital image processing functions are available under the general topic of contrast enhancement. The simplest method is simultaneous linear contrast expansion of all pixels.

Experiment 17.3: Single Pixel Contrast Enhancement. Collect a digital image of the fracture surface, typically a high-contrast image situation, with at least 8 bits of digital intensity levels. Increase the contrast by altering the slope and intercept of the image input/output function with the enhancement software that controls brightness and contrast. Observe the effects on the displayed image. Is contrast enhancement useful in this high contrast case?

Repeat this experiment for a low-contrast image, e.g., the compositional contrast between two phases of similar average atomic number. Is contrast enhancement useful in this low contrast case?

17.3 Quantitative Intensity Measurements from Images

It is extremely difficult to obtain quantitative intensity information from direct observation of a micrograph or display screen. Perceptual limitations and artifacts of human vision cause many distortions in the information we observe. The following are just a few of those effects which involve image intensities.

(a) **Perception of distinct intensity levels.** Experience would lead us to believe that we can distinguish far many more levels than the 16-20 levels shown in Figure 17.1. For example, on a brilliant day with the sun overhead a typical illuminance would be 100,000 lumens/m^2. On a clear moonless night the stars alone provide an illuminance of 0.0003 lumens/m^2. This is a dynamic range of some ten powers of ten. Clearly, then, we should be able to distinguish more than 20 or so levels of brightness! However, our vision does not simultaneously have a dynamic range of 10^{10}. It requires some 30 min for our eyes to adjust from one intensity to another. Under typical ambient conditions of illuminance, the hard fact remains that we can only usefully distinguish about 20 levels. We stress this point because it determines the number of levels of information we can "simultaneously distinguish" coming

Figure 17.1. Intensity wedges of distinct gray levels. 64 levels (top), 32 levels, (middle), 16 levels (bottom).

from the specimen. Thus, if we wish, for example, to display an image derived from some elemental signal such as characteristic x-rays from a bulk specimen over the full range of possible concentrations (0-100%), each detectable level would correspond to about 5 wt% at best.

  (b) Logarithmic perception of light intensity. Our eyes can detect differences in intensity better at low intensity levels than at high intensity levels. This is also shown in Figure 17.1 where it appears that the darker rectangles are further apart in intensity than the brighter rectangles even though all intensity differences are the same between each wedge.

  (c) Perception of nonuniform intensity from adjacent uniform objects. This edge enhancement effect can again be seen in Figure 17.1 as an apparent uneven intensity across each rectangle when, in fact, each rectangle is of uniform intensity. Even though the exact cause of this effect is not completely understood, the implications of this perceptual artifact are quite serious for the interpretation of micrographs, particularly those that convey elemental concentration information. If it is important to the investigator to know, even qualitatively, that homogeneity exists on either side of a sharp interface, he cannot trust his eyes to provide this information.

  For all of these reasons, quantitative information cannot be obtained unaided from the brightness information of a micrograph or a display screen. We will now see that quantitative information can be obtained from an image in other ways. Line scans and histograms are two methods which extract different but complementary information from images.

Experiment 17.4: Line Scans. Display the image of the fracture surface recorded for Experiment 17.3 or of the polished multiphase specimen. Select a locus of pixels across the image (line scan) using the appropriate computer function for your system (mouse-selection of endpoints, coordinate specification, etc.). Superimpose the intensity trace along this locus on the original image. How can numerical information be extracted from this line scan?

Experiment 17.5: Image-Intensity Histograms. An intensity histogram of an image displays on the horizontal axis the intensity value, and on the vertical axis the number of pixels having that particular intensity value. *Prepare a backscattered electron image of a polished, multiphase specimen. Produce the histogram of the image.* Which gray level is the most highly populated? Can you detect "phases" (related intensity regions) in the histogram? Where are these phases located in the original image? Can you estimate the relative amounts of the phases from the information in the histogram?

Experiment 17.6: Histogram Modification. Contrast modifications to the image are easily detectable in the image histogram. The original histogram and the modified histogram provide a direct basis for a quantitative understanding of what has been done to the image to modify the contrast.

    (a) Using the histogram prepared from the image of the fracture surface or the multiphase specimen, modify the histogram by means of the function called "histogram stretching." Observe the image generated from this modified histogram and compare it to the original. Did histogram stretching improve the visibility of features or phases? *Note:* This procedure will not necessarily lead to desirable results in all cases.

    (b) Repeat this procedure with the function denoted "histogram equalization" (often called normalization). Observe the image generated from this modified histogram and compare it to the original. Did this histogram equalization improve the visibility of features or phases? Again note that this procedure will not necessarily lead to desirable results in all cases.

    (c) Produce a density slice image to enhance one phase in the image and set all others to black.

17.4 Image-Processing Kernels

The real power of computer-aided imaging is derived from the plethora of mathematical operations which can be readily applied to the matrix representation of the digital image which would be difficult or impossible to perform in the analog domain. To operate on this set of discrete points (pixels) it is convenient to use "kernel operators" [3]. Kernels are mathematical operators that modify the intensity value of a particular pixel using information from surrounding pixels. This process is easy for a digital computer to perform and provides a wide range of useful image-processing functions. For the following experiments, determine which kernel operators are available in your computer system (consult the operator's manual), and test their effects on your images.

Experiment 17.7: Smoothing Noisy Images. Often SEM images or x-ray maps are taken under conditions in which very few counts are accumulated in each pixel. A class of processing kernels is available to smooth such images to improve the visibility of certain features of interest.

    Acquire an image of the fracture surface under conditions of very low beam current, e.g., 1-10 pA, and/or employ a very short pixel dwell time to produce a very noisy image. Capture and retain as separate image files two separate scans of the same area. Process this pair of images through a variety of smoothing kernels, e.g., block, tent, and Gaussian. Examine the effects of various sizes of smoothing kernels, e.g., 3x3 up to 27x27. Compare the image pair after processing through each kernel. *Ideally the same kernel should produce the same result.* Do they? In particular, look at the fine scale detail. Watch out for worms!!!!! Only the features which are the same in both images are significant and can be considered real. Features that differ are almost certainly artifacts.

Experiment 17.8: Edge Enhancement. Edges of features usually provide the most important perceptual information to an observer. A broad class of image processing operations act on images to selectively enhance edges.

Select the image of the multiphase alloy. Apply various edge enhancing kernels available in your particular system. Look up the matrix representation of each kernel that was used.

References
[1] K. R. Castleman, *Digital Image Processing*, Prentice-Hall, Englewood Cliffs, New Jersey (1979).
[2] R. C. Gonzalex and P. Wintz, *Digital Image Processing*, Addison-Wesley, Reading, Massachusetts (1977).
[3] W. K. Pratt, *Digital Image Processing*, Wiley-Interscience, New York (1978).
[4] A. Rose, (1970) "Quantum Limitations to Vision At Low Light Levels", *Image Technology* (June/July, 1970) 13-32.

Part III

ADVANCED X-RAY MICROANALYSIS

Laboratory 18

Quantitative Wavelength-Dispersive X-Ray Microanalysis

Purpose

The purpose of this laboratory is to demonstrate quantitative x-ray microanalysis as practiced on an electron column instrument equipped with a wavelength-dispersive spectrometer (WDS) and appropriate detector electronics. Although the calculation of composition is performed automatically in the computer, the analyst must be aware of the responsibility to select operating conditions that optimize accuracy and precision. The experimental measurements will show that increasing the beam voltage will increase x-ray count rate and the peak-to-background ratio and therefore the precision and sensitivity. But, at the same time, higher voltages dramatically increase the absorption of lower-energy (longer-wavelength) x-ray lines. More details may be found in *SEMXM*, Chapters 3, 5, and 7.

Equipment

Electron probe microanalyzer (EPMA) or SEM equipped with a wavelength-dispersive spectrometer.

Specimen

NiCrAl alloy containing 58.5 wt% nickel, 38.5 wt% chromium, and 3.0 wt% aluminum. The alloy was heat treated at 1135°C for 5 days to produce a single-phase homogeneous alloy. Pure element standards of nickel, chromium, and aluminum are included in the same specimen mount.

Time for this lab session

Two to three hours.

18.1 Operating Conditions for WDS Analysis

The operating voltage (kV) and electron beam current for microanalysis (as measured in a Faraday cup) are determined by the elements under investigation and the spatial resolution required. Ideally, we would like to select x-ray lines which are similar in energy for each of the analyzed elements in order to obtain similar excitation volumes from each element. The particular specimen chosen for study illustrates a frequently encountered dilemma which confronts the analyst in practical analysis situations. The analyst should always seek to optimize the selection of operating conditions to ensure adequate excitation for all of the analyzed elements. The excitation edge energies E_c of the analyzed elements are considered and a beam energy E_o is selected to provide an overvoltage, $U = E_o/E_c$, of approximately a factor of 2 for the highest edge energy. When this procedure is followed for the specimen under consideration, the edge energy for nickel, $E_c = 8.33$ keV, determines the minimum acceptable beam energy, 15 keV, which will adequately excite NiK_α radiation. This choice of

Table 18.1a WDS Standards Data for 10 kV

10 kV	Measurement time = _____ sec		Beam current = _____ nA		
X-ray line	Background N_{B-} (counts)	Peak N (counts)	Background N_{B+} (counts)	Average bkgnd. N_B (counts)	Net counts $N_P = N - N_B$
1 NiK_α					
2 NiK_α					
Average					
1 CrK_α					
2 CrK_α					
Average					
1 AlK_α					
2 AlK_α					
Average					

beam energy has an unfortunate consequence for aluminum, where $E_c = 1.559$ keV and $U = 9.6$. Such a high overvoltage results in a large absorption correction because the aluminum x-rays are produced very deep in the specimen. We will look specifically at the analytical errors as a function of beam energy.

An additional issue in WDS analysis is the choice of the diffracting crystal for each element. For each x-ray wavelength of interest, a specific crystal will provide optimum performance in terms of count rate, peak-to-background, and resolution. The wavelengths of the NiK_α, CrK_α, and AlK_α lines are 0.1658 nm, 0.2290 nm, and 0.8339 nm, respectively. The analyzing crystal will diffract x-rays with wavelengths about the same magnitude as the crystal interplanar spacing. Typical diffracting crystals for each of the lines above are LiF (lithium fluoride $2d_{002} = 0.4028$ nm), for nickel and chromium, and TAP (thallium acid phthalate $d_{1010} = 2.59$ nm) for aluminum, although other crystals are possible.

Experiment 18.1: Measurement of X-Ray Intensities. (a) Total intensity N. Using Tables 18.1 and 18.2, tabulate the measured x-ray intensities from the alloy (four or more locations) and from standards before and after the analysis at 3 operating voltages: 10, 15, and 30 kV. Choose a beam current at each energy which produces a count rate of not more than 20,000 counts per second on the aluminum standard. This choice ensures that deadtime correction will be less than 2% and can be safely ignored. For this experiment use a counting time of 100 sec in order to obtain a reasonable total count N on the peaks of interest.

(b) Background intensity N_B. You will also need to measure the background N_B. This can be done by measuring the average value of the total counts obtained at settings just above (N_{B+}) and below (N_{B-}) the peak position on your sample and your standards. Before you do

Table 18.1b WDS Standards Data for 15 kV

15 kV   Measurement time = _____ sec   Beam current = _____ nA

X-ray line	Background N_{B-} (counts)	Peak N (counts)	Background N_{B+} (counts)	Average bkgnd. N_B (counts)	Net counts $N_P = N - N_B$
1 NiK_α					
2 NiK_α					
Average					
1 CrK_α					
2 CrK_α					
Average					
1 AlK_α					
2 AlK_α					
Average					

Table 18.1c WDS Standards Data for 30 kV

30 kV   Measurement time = _____ sec   Beam current = _____ nA

X-ray line	Background N_{B-} (counts)	Peak N (counts)	Background N_{B+} (counts)	Average bkgnd. N_B (counts)	Net counts $N_P = N - N_B$
1 NiK_α					
2 NiK_α					
Average					
1 CrK_α					
2 CrK_α					
Average					
1 AlK_α					
2 AlK_α					
Average					

Table 18.2a WDS Specimen Data for 10 kV

10 kV X-ray Line	Measurement time = _____ sec		Beam current = _____ nA			
	Background N_{B-} (counts)	Peak N (counts)	Background N_{B+} (counts)	Average bkgnd. N_B (counts)	Net count $N_P = N - N_B$	RCE (%)
1 NiK_α						
CrK_α						
AlK_α						
2 NiK_α						
CrK_α						
AlK_α						
3 NiK_α						
CrK_α						
AlK_α						
4 NiK_α						
CrK_α						
AlK_α						

Table 18.2b WDS Specimen Data for 15 kV

15 kV X-ray Line	Measurement time = _____ sec Beam current = _____ nA					
	Background N_{B-} (counts)	Peak N (counts)	Background N_{B+} (counts)	Average bkgnd. N_B (counts)	Net count $N_P = N - N_B$	RCE (%)
1 NiK$_\alpha$						
CrK$_\alpha$						
AlK$_\alpha$						
2 NiK$_\alpha$						
CrK$_\alpha$						
AlK$_\alpha$						
3 NiK$_\alpha$						
CrK$_\alpha$						
AlK$_\alpha$						
4 NiK$_\alpha$						
CrK$_\alpha$						
AlK$_\alpha$						

Table 18.2c WDS Specimen Data for 30 kV

30 kV	Measurement time = _____ sec		Beam Current = _____ nA			
X-ray Line	Background N_{B-} (counts)	Peak N (counts)	Background N_{B+} (counts)	Average bkgnd. N_B (counts)	Net count $N_P = N - N_B$	RCE (%)
1 NiK_α						
CrK_α						
AlK_α						
2 NiK_α						
CrK_α						
AlK_α						
3 NiK_α						
CrK_α						
AlK_α						
4 NiK_α						
CrK_α						
AlK_α						

the background measurement check to make sure that there are no peak overlaps, for example from higher-order K_β, L_α or M_α lines. Use a table of characteristic x-ray lines to investigate these possible problems. The data for the nickel, chromium, and aluminum standards and the specimen can be entered in Tables 18.1 and 18.2. Use Table 18.1a for 10 kV, Table 18.1b for 15 kV, etc.

(c) Calculate peak intensity N_P. In the same tables used to record the x-ray intensities, calculate and report the net peak counts (N_P) where:

$$N_P = N - N_B \tag{18.1}$$

for each measured peak.

(d) Calculate RCE. It will then be possible to calculate the relative counting error RCE:

$$RCE = \sqrt{N_p}/N \times 100\% \tag{18.2}$$

(e) Calculate the K ratio. Finally, by dividing the specimen peak intensity (N_P) for each element by the peak intensity from the standard (N_{PO}) for that element, we obtain the intensity ratio or K ratio:

$$K \text{ ratio} = \frac{\text{specimen intensity}}{\text{standard intensity}} = \frac{N_P}{N_{PO}} \tag{18.3}$$

Report the K ratios in Table 18.3.

Table 18.3 WDS Intensity Ratios Calculated for Nickel, Chromium, and Aluminum from Data in Tables 18.1 and 18.2

10 kV	k ratio 1	k ratio 2	k ratio 3	k ratio 4
Ni				
Cr				
Al				
15 kV	k ratio 1	k ratio 2	k ratio 3	k ratio 4
Ni				
Cr				
Al				
30 kV	k ratio 1	k ratio 2	k ratio 3	k ratio 4
Ni				
Cr				
Al				

18.2 Calculation of Composition

The x-ray intensity ratios (K ratios) of Table 18.3 are converted to concentrations for each element through multiplication of the K ratio by the three correction factors for atomic number (Z), absorption (A), and fluorescence (F). Thus, this method became known as the "ZAF method" where

$$\text{Measured composition} = (K \text{ ratio})(Z)(A)(F). \tag{18.4}$$

Experiment 18.2: ZAF Correction of Intensities. (a) Use the ZAF quantitative analysis procedure resident in the software package included with the spectrometer control system (or with a separate computer) to reduce the spectral data for the alloy (K ratios) to elemental concentrations. Report the measured concentrations in Table 18.4.
(b) The accuracy of your measurements can be determined by comparing your calculated composition values with the known or true composition of the alloy given at the beginning of Laboratory 18. The accuracy can be described by the percent relative error:

$$\text{Percent relative error} = [(\text{measured} - \text{true})/\text{measured}] \times 100\% \tag{18.5}$$

Record the percent relative error in Table 18.4.
(c) Calculate the ratio of net peak intensity divided by the background (N_P/N_B or P/B) and record this value for each element and kV in Table 18.4. Plot (N_P/N_B) as a function of

Table 18.4 Summary of WDS Measurements for Point Analyses on the Alloy

10 kV	True wt%	Point 1	Point 2	Point 3	Point 4	Ave.	Percent relative error	N_P/N_B
Ni								
Cr								
Al								
15 kV	True wt%	Point 1	Point 2	Point 3	Point 4	Ave.	Percent relative error	N_P/N_B
Ni								
Cr								
Al								
30 kV	True wt%	Point 1	Point 2	Point 3	Point 4	Ave.	Percent relative error	N_P/N_B
Ni								
Cr								
Al								

voltage for each element of the pure element standards and for each element in the sample.
Inspect your report tables and your plots as you consider the following questions:
1. At which beam energy is the maximum percent relative error for aluminum encountered?
2. How do the analytical errors behave for chromium and nickel as a function of energy?
3. How does the relative counting error, RCE, vary for nickel and chromium as a function of beam energy?
4. How does the measured concentration vary as a function of position for the several measurements made on the NiCrAl sample? If the range (%) of the calculated concentrations is less than 3 x RCE, can the sample be considered homogeneous based on this small number of sample points? Is your alloy homogeneous?
5. Does *P/B* increase with voltage for all the elements? If not, explain the phenomenon involved.

Optional Experiment: If other methods of quantification are available, e.g., $\phi(\rho z)$, "standardless," etc., repeat the above calculations and compare those calculated concentrations with the values found by ZAF. Further comparisons can be made with the EDS results obtained in Laboratory 19.

Laboratory 19

Quantitative Energy-Dispersive X-Ray Microanalysis

Purpose
 The purpose of this laboratory is to demonstrate quantitative x-ray microanalysis as practiced on an electron column instrument equipped with an energy-dispersive spectrometer (EDS) and a computer-based multichannel analyzer (MCA). Although the difficult calculation of quantitative x-ray microanalysis are performed automatically in the MCA, the analyst must be aware of the responsibility to select operating conditions that optimize accuracy and precision. As with the WDS, increasing the accelerating voltage will improve the x-ray count rate and peak-to-background ratio and therefore the precision and sensitivity, respectively. However, increasing the beam voltage will also increase the absorption of lower-energy x-ray lines such as aluminum. More details may be found in *SEMXM*, Chapters 3, 5, and 7.

Equipment
 Electron probe microanalyzer (EPMA) or SEM equipped with an energy-dispersive spectrometer.

Specimens
 1. NiCrAl alloy containing 58.5 wt% Ni, 38.5 wt% Cr, and 3.0 wt% Al. The alloy was heat treated at 1135°C for 5 days to produce a single phase homogeneous alloy. Pure element standards of nickel, chromium, and aluminum are included in the same mount.
 2. MgO, SiO_2, and Al_2O_3 polished standards. $Mg_3Al_2Si_3O_{12}$, Mg_2SiO_4, $MgSiO_3$, $MgAl_2O_4$, and Al_2SiO_5 specimens polished for analysis in a manner identical to the standards.

Time for this lab session
 Two to three hours.

19.1 Operating Conditions for EDS Analysis

 As with WDS (Laboratory 18), the selection of operating voltage and electron beam current is determined by the elements under investigation and the spatial resolution required. Ideally, we would like to select x-ray lines which are similar in energy for each of the analyzed elements in order to obtain similar excitation volumes from each element. The particular specimen chosen for study illustrates a frequently encountered dilemma which confronts the analyst in practical analysis situations. The analyst should always seek to optimize the selection of operating conditions to ensure adequate excitation for all of the analyzed elements. The excitation edge energies E_c of the analyzed elements are considered, and a beam energy E_o is selected to provide an overvoltage, $U = E_o/E_c$, of approximately a factor of 2 for the highest edge energy. When this procedure is followed for the specimen under consideration, the edge

Table 19.1 EDS Standards Data for 10 kV, 15 kV, and 30 kV

10 kV	Measurement time = _____ (sec)	Beam current = _____ nA	
X-ray line	Peak, N (counts)	Background, N_B (counts)	Net Counts $N_P = N - N_B$
NiK_α			
CrK_α			
AlK_α			

15 kV	Measurement time = _____ (sec)	Beam current = _____ nA	
X-ray line	Peak, N (counts)	Background, N_B (counts)	Net Counts $N_P = N - N_B$
NiK_α			
CrK_α			
AlK_α			

30 kV	Measurement time = _____ (sec)	Beam current = _____ nA	
X-ray line	Peak, N (counts)	Background, N_B (counts)	Net Counts $N_P = N - N_B$
NiK_α			
CrK_α			
AlK_α			

energy for nickel, $E_c = 8.33$ keV, determines the minimum acceptable beam energy, 15 keV, which will adequately excite NiK_α radiation. This choice of beam energy has an unfortunate consequence for aluminum where $E_c = 1.559$ keV and $U = 9.6$. Such a high overvoltage results in a large absorption correction because the aluminum x-rays are produced very deep in the specimen. We will look specifically at the analytical errors as a function of beam energy.

With EDS systems all x-ray peaks are collected simultaneously, and there is no need to worry about selecting the appropriate analyzing crystal as in the case of WDS. However, there is an issue of equivalent importance in the selection of the window width for each element. Since x-ray peaks statistically broaden at higher energies there is a need to collect the same proportion of a peak at each energy. This is usually done by setting the window on each x-ray peak to 1.2 x FWHM (full-width-at-half-maximum), the window width that gives the best peak to-background ratio.

Experiment 19.1: Measurement of X-Ray Intensities. *(a) Total intensity N. Choose a beam current at each energy which produces a dead time on the aluminum standard of approximately 20%. This choice ensures that the spectral artifacts due to count rate effects will not be significant. For this experiment, the live time should be 100 sec to obtain a reasonable integrated count on the peaks of interest. Set windows on the K_α peaks for nickel, chromium, and aluminum. Make the width of each window about 1.2 times the full-width-at-half-maximum (FWHM) of the x-ray peak.*

Collect spectra and measure x-ray intensities from the alloy (3 to 4 or more locations) and standards at 10, 15, and 30 kV. Use Tables 19.1 and 19.2 to record your data.

(b) Background intensity N. Use the EDS background subtraction procedure which is resident in the software package of the computer to calculate the background under each peak.

(c) Calculate net peak intensity. Record the gross peak counts N the background under the peak N_B and calculate the net peak counts N_P as given by

$$N_P = N - N_B \qquad (19.1)$$

(d) Calculate RCE. In the same tables, report the measurement statistic called the relative counting error, RCE, defined as

$$RCE = \sqrt{N_p}/N \times 100\% \qquad (19.2)$$

(e) Calculate the K ratio. Finally, by dividing the specimen peak intensity N_P for each element by the peak intensity from the standard (N_{P0}) for that element, we obtain the intensity ratio or K ratio

$$K\ ratio = \frac{specimen\ intensity}{standard\ intensity} = \frac{N_P}{N_{P0}} \qquad (19.3)$$

Report your K ratios in Table 19.3.

Note: If your sample has to be tilted, you may find it necessary to vertically reposition the individual standards and the alloy to the same height to ensure that the collection efficiency does not vary between the standards and the sample. This can be done by using the largest objective aperture, locking the objective lens focus, and shifting the vertical sample position until a sharp image is obtained at 1000x or higher. To determine how sensitive your SEM/detector combination is to this effect, with the sample tilted to the desired angle, observe the change in count rate as the sample is displaced a few millimeters up and down from the position selected for the analysis. This change in count rate with specimen position is caused by a variation in detector geometrical collection efficiency. It is due in part to a change in shadowing of the detector crystal by the collimator.

Table 19.2a EDS Specimen Data for 10 kV

10 kV   Measurement time = _____ sec   Beam current = _____ nA

X-ray line	Peak N (counts)	Average background N_B (counts)	Net counts $N_P = N - N_B$	RCE (%)
1 NiK_α				
CrK_α				
AlK_α				
2 NiK_α				
CrK_α				
AlK_α				
3 NiK_α				
CrK_α				
AlK_α				
4 NiK_α				
CrK_α				
AlK_α				

Table 19.2b EDS Specimen Data for 15 kV

15 kV	Measurement time = _____ sec		Beam current = _____ nA	
X-ray line	Peak N (counts)	Average background N_B (counts)	Net counts $N_P = N - N_B$	RCE (%)
1 NiK_α				
CrK_α				
AlK_α				
2 NiK_α				
CrK_α				
AlK_α				
3 NiK_α				
CrK_α				
AlK_α				
4 NiK_α				
CrK_α				
AlK_α				

Table 19.2c EDS Specimen Data for 30 kV

30 kV	Measurement time = _____ sec		Beam current = _____ nA	
X-ray line	Peak N (counts)	Average background N_B (counts)	Net counts $N_P = N - N_B$	RCE (%)
1 NiK_α				
CrK_α				
AlK_α				
2 NiK_α				
CrK_α				
AlK_α				
3 NiK_α				
CrK_α				
AlK_α				
4 NiK_α				
CrK_α				
AlK_α				

Table 19.3 EDS Intensity Ratios for Ni, Cr, and Al Calculated from Data in Tables 19.1 and 19.2

10 kV	K ratio 1	K ratio 2	K ratio 3	K ratio 4
Ni				
Cr				
Al				

15 kV	K ratio 1	K ratio 2	K ratio 3	K ratio 4
Ni				
Cr				
Al				

30 kV	K ratio 1	K ratio 2	K ratio 3	K ratio 4
Ni				
Cr				
Al				

19.2 Calculation of Composition

The x-ray intensity data (K ratios) are converted to concentrations for each element by multiplying the K ratio by the correction factors for atomic number (Z), absorption (A), and fluorescence (F). Thus, this method became known as the "ZAF method" where

$$\text{Measured composition} = (K \text{ ratio})(Z)(A)(F) \tag{19.4}$$

Experiment 19.2: ZAF Correction of Intensities. (a) Use the ZAF quantitative analysis procedure resident in the computer to calculate concentration values from the standard and sample data in Tables 19.1 and 19.2. Report the measured concentrations in Table 19.4.
(b) The accuracy of your measurements can be determined by comparing your calculated composition values with the known or true composition of the alloy given above. The accuracy can be described by the percent relative error:

$$\text{Percent relative error} = [(\text{measured} - \text{true})/\text{measured}] \times 100\% \tag{19.5}$$

Record the percent relative error in Table 19.4.

(c) *Calculate the ratio of net peak intensity divided by the background (N_P/N_B or P/B) and record this value for each element and voltage in Table 19.4. Plot (N_P/N_B) as a function of voltage for each element in the pure element standards and the sample.*

Inspect your report tables and your plots as you consider the following questions:

1. At which beam energy is the maximum percent relative error for aluminum encountered?
2. How do the analytical errors behave for chromium and nickel as a function of energy?
3. How does the relative counting error, RCE, vary for nickel and chromium as a function of beam energy?
4. How sensitive were the measured x-ray intensities to the sample height? How would neglect of the effect of height affect your quantitative analysis?
5. In addition to sample position effects, what other factors can influence quantitative results?
6. How does the measured concentration vary as a function of position for the several measurements? If the range (%) of the calculated concentrations is less than 3xRCE, can the sample be considered homogeneous based on this small number of sample points? Is your alloy homogeneous?
7. Does P/B increase with voltage for all the elements? If not, explain the phenomenon involved.

Table 19.4 Summary of EDS Analysis Data Measurements for Point Analyses on the Alloy

10 kv	True wt%	Point 1	Point 2	Point 3	Point 4	Ave.	Percent relative error	N_P/N_B
Ni								
Cr								
Al								

15 kv	True wt%	Point 1	Point 2	Point 3	Point 4	Ave.	Percent relative error	N_P/N_B
Ni								
Cr								
Al								

30 kv	True wt%	Point 1	Point 2	Point 3	Point 4	Ave.	Percent relative error	N_P/N_B
Ni								
Cr								
Al								

Optional Experiment. If other methods of quantification are available, e.g., $\phi(\rho z)$, "standardless," etc., repeat the above calculations and compare those calculated concentrations with the values found by ZAF. Further comparisons can be made with the WDS results obtained in Laboratory 18.

19.3 Comparing "Standardless Analysis" with Quantitative Analysis Using Standards

With the extreme ease of data acquisition inherent in the EDS system, it is tempting to try to completely automate the analysis without having to measure data from standards. Many procedures of this type work well within a limited range of samples and at a single accelerating voltage; however, when flexibility is desired to change the voltage for optimizing the accuracy or the spatial resolution these "standardless" methods show their weaknesses.

Experiment 19.3: Silicate Analysis. "Standardless" EDS analysis procedures often work well on specimens like stainless steel, where (a) the analyzed x-ray lines are well separated and are of reasonably high energy, (b) the background under these peaks is reasonably flat and the peak-to-background ratio is reasonably high, and (c) the absorption of these x-rays by the detector window and Si dead layer is minimal. Standardless procedures work less well when there are severe overlaps and when one is analyzing with low-energy x-ray lines, such as Na, Mg, Al ,and SiK_α. In such cases, more accurate results are obtained by analyzing standards and using these for spectral deconvolution and data correction.

This experiment will test full quantitative versus standardless quantitative analysis procedures.

(a) Collect data from standards. Collect and store a 100-sec (live time) spectrum at 20%-30% dead time of a polished, carbon-coated specimen of SiO_2 (quartz). Use an accelerating potential of 15 or 20 keV and the other analytical conditions employed in Experiments 19.1 and 19.2 (making sure that you know the effective detector take-off angle). Taking care to keep the beam current constant, collect and store a 100-sec spectrum from polished, carbon-coated standards of MgO (periclase) and Al_2O_3 (corundum). Store these spectra as library standards.

(b) Check calibration. Use your EDS system's software to determine the energies of the centroids of the Mg, Al, and SiK_α peaks. Compare these with the theoretical values. If they differ, the standardless procedures will probably give larger errors than if the energy calibration were correct. *Note:* You should always check the energy calibration of your system before doing quantitative analysis.

(c) Collect and store two or three 100-sec spectra of $Mg_3Al_2Si_3O_{12}$ (pyrope). Note how the Mg, Al, and SiK_α peaks overlap and the complexity of the shape of the background in the vicinity of these peaks. If time permits and the materials are available, collect and store one or two 100 sec spectra of polished, carbon-coated specimens of Mg_2SiO_4 (forsterite), $MgSiO_3$ (enstatite), $MgAl_2O_4$ (spinel), and Al_2SiO_5 (kyanite).

(d) Process data from various silicates. Set up your system's standardless quantitative analysis program. If you need to input the elements to be analyzed, list MgK_α, AlK_α, SiK_α, CaK_α, and FeK_α. Note that calcium and iron should not be present in any of your samples; testing for them may provide information regarding the effective minimum detectability. Set your standardless program to calculate oxygen by difference or preferably by stoichiometry relative to the other elements. Read in each of the spectra for the silicates pyrope, forsterite, enstatite, spinel, and kyanite and process them. How do the results compare to the actual compositions?

(e) Do a full quantitative analysis using standards. Read in the MgO, Al_2O_3, and SiO_2 spectra to set up your system's full quantitative analysis program. Then read in and process the various silicates. How do these compare with the true compositions and the standardless results? Is true quantitative analysis possible in the SEM with an EDS detector?

Laboratory 20

Light Element Microanalysis

Purpose

The light element regime traditionally defined as the elements Be through F (Z = 4 to 9) has posed many difficulties for the analyst. Poor detectability due to low count rates caused by insufficient beam current at low voltage and small spot sizes has made *microanalysis* very difficult. Adequate matrix correction procedures are just being developed and quantification is difficult without very good standards. Contamination of the sample during analysis often makes the results questionable. These problems are slowly but surely being corrected in today's modern electron microprobes. New WDS crystals and recent ultra-thin-window (UTW) have contributed to improved detectability; modern electron columns have much better low-voltage performance; and finally, today's SEMs and microprobes are designed with anticontamination devices. However, with all these improvements, the measurement and subsequent quantification of light elements is far from routine. This laboratory will illustrate operational aspects of light element analysis background measurements, peak overlap problems, and quantitation techniques. More details may be found in *SEMXM*, Chapter 8.

Equipment

Electron probe microanalyzer (EPMA) or SEM equipped with a wavelength dispersive spectrometer (WDS) and/or energy-dispersive spectrometer (EDS).

Specimens

1. A sample of pure carbon such as graphite, colloidal graphite or diamond.
2. A steel sample containing carbides (Fe_3C) that are >10 μm in size.
3. A homogeneous steel sample such as 1080 carbon steel.
4. A set of Fe-C and Fe-Ni-C standards.
5. A sample of pure iron.

Time for this lab session

Two to three hours.

20.1 Data Collection for Light Element Analysis

The following aspects of light element analysis should be considered in detail before measurements are made:

Specimen preparation
  (a) Proper mounting media (epoxy, bakelite, etc.) to avoid contamination during polishing.
  (b) Sample cleaning to avoid contamination.

(c) Thin conductive coating material that will minimize x-ray absorption.
(d) Samples and standards must be coated to the same thickness.

Proper instrument conditions
(a) Low overvoltage—to minimize the absorption correction. Overvoltage must be high enough so that the analysis is not surface sensitive.
(b) High take-off angle (EDS) to minimize absorption correction.
(c) Pulse height analysis (WDS) to avoid peak overlaps.
(d) Beam current/counting time to minimize contamination and sample damage.
(e) Enlarged spot size (~1 µm) to minimize contamination and sample damage.
(f) Anticontamination devices (liquid nitrogen trap, air jet) to minimize surface contamination.

The WDS has very high x-ray energy resolution and as such is capable of measuring wavelength shifts due to chemical bonding effects in different compounds. To observe both the wavelength shift and peak shape change, run the following set of measurements.

Experiment 20.1: Carbon K Peak Shape.
Pure Carbon
(a) Optimize the instrument for CK_α measurement (for example: 10 voltage, lead stearate analyzing crystal, adjust detector voltage and amplifier gain).
(b) Run a slow wavelength scan (about 1 sec/eV) through the entire carbon peak with proper marking of the wavelength or energy scale. Be sure to collect plenty of background on each side of the peak, e.g., 0.22 to 0.35 keV.
(c) Measure the peak intensity with $\sqrt{N}/N \times 100 < 0.5\%$, where N is the mean value of the total number of counts in time t.
(d) Measure the average of the two backgrounds on each side of the peak for a time equal to the peak measurement (record data in the data Table 20.1 and save for Experiment 20.2).
(e) Collect a spectrum with a windowless or UTW EDS detector, if available.
Iron Carbide, Fe_3C
(a) Using the same instrument conditions as above, obtain a wavelength scan through the carbon peak in Fe_3C.
(b) Compare with the pure carbon scan. Identify the additional peaks.
(c) Measure the peak and background intensities with $\sqrt{N}/N \times 100 < 0.5\%$ as in (c) and (d) for pure carbon.
(d) Collect an EDS spectrum if a light element EDS detector is available.

Questions
1. Do you observe any differences in the wave length spectra corresponding to changes in the peak positions and/or any differences in the peak shapes? If so, what are the reasons for the wavelength shift and peak shape differences?
2. To analyze carbon in a steel sample, which standard should be chosen? Why?
3. Are there any differences between the two EDS spectra?

To do accurate light element analysis, the measurement of background intensity is very important. Average background can be obtained from measurements of the continuum background on either side of the peak or by using a background standard similar to the sample of interest but with none of the light element present in the sample.

Experiment 20.2: Background Intensity.
Method 1: Using WDS calculate the average of the two carbon backgrounds measured on either side of the carbon peak in the sample of Fe_3C or a time equal to the carbon peak measurement on Fe_3C (use data from Experiment 20.1 recorded in Table 20.1).

Table 20.1. Peak and Background Data for CK_α

kV = _____ nA Probe current =	Take-off angle = Counting time =	
Sample	Peak (counts)	Average background counts
Pure carbon		
Fe_3C		
Pure Fe or ferrite		

Method 2: Using WDS measure the background for carbon in pure iron at the carbon peak position determined in Experiment 20.1 for Fe_3C and save in the data Table 20.1.

Questions
1. What is the difference between background measurement Methods 1 and 2?
2. Are these methods more accurate than measuring the off-peak continuum background directly on the sample at the analysis position?
3. Is there any advantage in using WDS for background measurements rather than EDS? Consider $\sqrt{N}/N$ effects.

In WDS it is possible for peak overlaps to occur when $n\lambda$ for a heavier element is similar to the peak wavelength position for a light element. For example, overlap effects occur in light element analysis when the L lines of transition elements, present in the sample, are excited.

Experiment 20.3: Peak Overlaps.
   (a) Run a slow wavelength scan through the entire carbon peak on an Fe-10Ni-C standard. Compare this scan to that obtained from Fe_3C and see if a Ni L_α peak interferes. Calculate the position of the $n\lambda$ peak from the wavelength of the Ni L_α line.
   (b) Using the pulse height analyzer, remove the Ni L_α peak from the scan. Run a slow wavelength scan through the carbon peak on the Fe-10Ni-C standard to make sure the overlapped peak is removed.

20.2 Measurement of Light Element Concentrations

The ultimate objective of this laboratory is to make a quantitative analysis of carbon. Two methods will be used, ZAF (or $\phi(\rho z)$), and the calibration curve method. Either WDS or EDS may be used.

Experiment 20.4: Quantitation by the ZAF Method. *Use the same experimental conditions as for Experiments 20.1-20.3.*

Step 1. *Measure the carbon peak intensity*

$$I_{carbon}^{std}$$

and carbon background intensity

$$I_{carbon}^{std\ bg}$$

from the carbide and the chosen background standard (from Experiment 20.2). The measured standard peak intensity is I_{STD}:

$$I_{STD} = I_{carbon}^{std} - I_{carbon}^{std\ bg} \quad (20.1)$$

Enter data in Table 20.2 below.

Step 2. *Measure the carbon intensity, I_{carbon}, from the homogeneous steel sample. Measure background intensity directly on the sample. Record the data in Table 20.2. The measured sample peak intensity is I_{SAMPLE}:*

$$I_{SAMPLE} = I_{carbon} - I_{carbon}^{bg} \quad (20.2)$$

Step 3. *Determine the carbon content of the homogeneous steel sample with the ZAF or $\phi(\rho z)$ method using the following K-ratio:*

$$K_{carbon} = I_{SAMPLE}/I_{STD} \quad (20.3)$$

If time permits, make at least three independent measurements of the carbon concentration in the steel.

Table 20.2. Data for ZAF Analysis of Carbon in Steel (5 Independent Measurements)

Counting time = _____ sec						
Fe₃ C Standard		I_{STD}	Steel Sample		I_{SAMPLE}	K_{carbon}
CK_α Peak intensity	Background intensity		CK_α Peak intensity	Background intensity		
1						
2						
3						
4						
5						

Table 20.3. Data for Calibration Curve of Carbon in Iron

kV = Probe current = Take-off angle = Counting time =							
		Measurement 1		Measurement 2		Measurement 3		Average I_{STDi}
Sample		C Peak	C Backgr.	C Peak	C Backgr.	C Peak	C Backgr.	
Pure Fe								
Standard 1 (0.30 wt%C)								
Standard 2 (0.62 wt%C)								
Standard 3 (1.01 wt%C)								
Standard 4 (1.29 wt%C)								
Steel unknown								

Experiment 20.5: Quantitation by the Calibration Curve Method.

Step 1. *Measure I_{STDi} [equation (20.1)] for the four Fe-C standards, including the pure iron sample, taking the background directly from each standard. Be sure to measure all specimens at the same beam current, voltage, and take-off angle. A minimum of three measurements should be made on each sample.*

$$I_{STDi} = I_{carbon}^{std} - I_{carbon}^{std\ bg} \quad \text{for each standard } i \qquad (20.4)$$

Enter data in Table 20.3.

Step 2. *Plot I_{STDi} versus the carbon concentration of the standards. This is the calibration curve.*

Step 3. *Measure I_{SAMPLE} from the homogeneous steel sample. Enter data in Table 20.3.*

Step 4. *Using I_{SAMPLE} and the calibration plot, determine the carbon content in the steel sample.*

Questions
1. Which carbon quantitation method is more accurate? Why?
2. If the operating voltage were increased, would your answer to question (1) change?

Laboratory 21

Trace Element Microanalysis

Purpose
This laboratory is designed to give you a feel for the amount of care and time that must into any measurement involving trace elements. The trace element concentration range is ofte defined as concentrations below 0.5 wt% for most elements and below 1 wt% for the light elements. The example used here is the measurement of small amounts of phosphorus (P) in FeNi alloys. More details on these techniques may be found in *SEMXM*, Chapters 7 and 8.

Equipment
Electron probe microanalyzer (EPMA) or SEM equipped with a wavelength dispersive spectrometer (WDS) and/or an energy-dispersive spectrometer (EDS).

Specimens
1. (FeNi)$_3$P, a phosphide mineral known as Schreibersite in meteorites, containing 15.5wt% P (25 at%P).
2. An Fe-Ni-P alloy containing 14.6 wt%Ni, 0.35 wt% P, balance Fe.
3. An Fe-Ni binary alloy.
4. A meteorite containing 0.08wt% P in the low nickel metal phase.

Time for this lab session
Two to three hours.

21.1 Data Collection for Trace Element Analysis

The use of x-ray microanalysis to measure trace concentrations has been a standard procedure since the electron probe microanalyzer (EPMA) was developed in the early 1950s. Minimum detectable concentrations of <100 ppm for wavelength-dispersive spectrometers (WDS) and <1000 ppm for energy-dispersive spectromers are commonly quoted in textbook: and by manufacturers. The analyst must understand, however, that these numbers represent the "best case" situations and are not trivial to achieve. Furthermore, once achieved they can only be reported as having relative errors of 100% (that is, ±100 ppm for WDS and ±1000 ppm for EDS).

The following instrumental aspects of trace element analysis should be considered in detail before measurements are made:
1. Accelerating voltage: adjust to maximize the P/B ratio, the peak (N_P) over backgrou (N_B), for the element of interest.
2. Beam current: increase to maximize peak intensity.

Table 21.1. WDS Data from High-Phosphorus Standard

15 kV	Beam current = _____ nA					
Element	Counting time (sec)	Peak position (keV, λ, or mm)	Background (counts) N_{SB-}	N_{SB+}	N_S (counts)	N_{SB} (counts)
FeK_α						
NiK_α						
PK_α						

3. Counting times: increase to improve detection limits.
4. Spot size (x-ray source region): may increase as beam current is increased.
5. Pulse height analysis (PHA) settings (for WDS): to minimize peak overlaps.
6. Peak and off-peak positions (WDS): to obtain correct continuum background values.
7. Selection of analyzing crystal (WDS): to maximize peak intensity.
8. Region of interest settings (EDS): to minimize peak overlaps.

In the following sections N refers to the number of counts at the peak position for an element in a given sample and N_B is the average background intensity for that element in the same sample. For an element standard N_S and N_{SB} refer to the number of peak and background counts, respectively.

Experiment 21.1: Data from a High-Phosphorus Standard.

(a) Using the phosphide standard, measure with the WDS the peak counts N_S and background counts N_{SB} for phosphorus, iron, and nickel, respectively. Note the appropriate peak and background positions. Use a beam current ≥100 nA, 15 kV, and an analysis time of at least 10 sec for iron and nickel and 100 sec for phosphorus. Record your results in Table 21.1.

(b) Repeat the experiment using the EDS. Use a current which gives <20% dead time and an analysis time of 100 sec. Peak and background intensities can be measured by using windows defined by full-width-at-half-maximum (FWHM) regions. Record your results in Table 21.2.

Table 21.2. EDS Data from High-Phosphorus Standard

15 kV			Counting time = 100 live sec		Deadtime = _____ %		
Element	Peak position (keV)	FWHM (ev)	Background measurement			N_S (counts)	N_{SB} (counts)
			keV range	N_{SB-}	keV range	N_{SB+}	
FeK_α							
NiK_α							
PK_α							

Table 21.3. WDS Data from Fe-Ni-0.35 wt%P Alloy

WDS, Counting time = _____ sec					
Element	Phosphorus peak position (keV, λ, or mm)	N (counts)	Background positions and intensity		Ave N_B (counts)
			N_{B-} (counts)	N_{B+} (counts)	
Phosphorus					

Experiment 21.2: Establishing the Background at Low Phosphorus Concentration. Since we are looking for small peaks barely above the background, it is appropriate to more clearly delineate the method for measuring background. In EDS, the background fit or the background region of interest must be chosen carefully to avoid other peaks, absorption edges, etc. In WDS the shape of the background must be well understood. A choice must be made of the energy below and above the peak to obtain two background intensities which are usually averaged to determine the background.

(a) On the Fe-Ni-P alloy containing 0.35wt%P obtain a WDS wavelength scan 30 eV wide around the PK_α peak position at 2.014 keV. A counting time of 10 sec for each 1 eV step is suggested. Determine the continuum background positions to use for the analysis. Be sure to avoid the PK_β peak. Measure the phosphorus background, N_B, using these positions with a counting time of 100 sec and record your results in Table 21.3.

(b) Obtain an EDS spectrum of the Fe-Ni-P alloy containing 0.35wt% P for 100 sec. Is the phosphorus peak clearly distinguishable? If it is not, acquire the EDS spectrum for 400 sec. Measure the phosphorus background, N_B, using either a background fitting technique or an average of background regions above and below the phosphorus peak, as used for the standards measurement in Experiment 21.1. Record your results in Table 21.4.

(c) Measure the phorphorus background, N_B, on the pure FeNi binary standard. Compare the values obtained here to the measurements in (a) and (b). Record these values in Table 21.5.

Table 21.4. EDS Data from Fe-Ni-0.35wt%P alloy

EDS, Counting time = _____ sec							
Element	Phosphorus peak position (keV, λ, or mm)	N (counts)	Background regions and intensity (counts)				Ave N_B (counts)
			keV range	N_{B-}	keV range	N_{B+}	
Phosphorus							

Table 21.5. Comparison of Phosphorus Background Measurements

Background standard	N_B WDS	N_B EDS
Fe-Ni-0.35P		
Fe-Ni binary alloy		

21.2 Minimum Detectability Limits

Liebhafsky et al. [1] suggest that an element can be considered present if the value of N exceeds the background N_B by $3(N_B)^{1/2}$.

Experiment 21.3: Estimation of Minimum Detectability from Alloy Data. Calculate $3(N_B)^{1/2}$ from your WDS and EDS measurements of N_B obtained in Experiment 21.2. To estimate a minimum detectability limit, C_{DL}, assume $N-N_B$ just equals $3(N_B)^{1/2}$ and use the linear formula:

$$C_{DL} = \left(\frac{N-N_B}{N_S-N_{SB}}\right) C_S = \left[\frac{3(N_B)^{1/2}}{N_S-N_{SB}}\right] C_S \qquad (21.1)$$

where C_S is the concentration of phosphorus in the phosphide standard. Place your result for C_{DL} in Table 21.6 below. A more detailed formula for C_{DL} is given in SEMXM (p. 436).

Experiment 21.4: Estimation of Minimum Detectability from Pure Element Standards. One can also estimate the minimum detectability limit using the following equation due to Ziebold [2]:

$$C_{DL} = (3.29A)/(tP^2/B)^{1/2} \qquad (21.2)$$

where A is a matrix factor (ZAF product) ~1.0; t is the total peak measurement time (sec); P is the peak intensity of pure element standard in counts per sec (cps) (remember to convert the phosphorus K_α intensity from that of the phosphide to that of a pure element); and B is the background intensity of the pure element standard (cps). Calculate C_{DL} from Equation (21.2) for both WDS and EDS and place the result in Table 21.6.

Table 21.6. Minimum Detectability Limits for Phosphorus

Method	P/B	C_{DL} from Equation (21.1)	C_{DL} from Equation (21.2)
WDS			
EDS			

Table 21.7. Data for Measurement of Phosphorus in a Meteorite

Point	Ni (counts)		Fe (counts)		P (counts)	
	N	N_B	N_B	N_B	N	N_B
1						
2						
3						
4						
5						
6						

Questions
1. Compare C_{DL} from Experiment 21.3 with C_{DL} from Experiment 21.4.
2. Compare the WDS and EDS values for C_{DL}. Why is one method so much better than the other?
3. Can EDS can be used to measure the P concentration in the meteorite?

21.3 Measurement of Trace Element Concentrations

The metallic meteorite phase (bcc kamacite) containing 0.08 wt% phosphorus will be used to illustrate the measurement of phosphorus at low concentrations.

Experiment 21.5: Measurement of P Content in a Meteorite.
  (a) On the low-nickel kamacite phase of the meteorite sample, measure the phosphorus peak intensity N and the background N_B with WDS, and EDS if appropriate. Measure at least four different positions on the sample and record your data in Table 21.7. If time permits measure a WDS wavelength scan as in Experiment 21.2a around the PK_α peak to demonstrate the presence of phosphorus in the meteorite.
  (b) Using a ZAF or $\phi(\rho z)$ technique, calculate the phosphorus concentration in the meteorite.
  (c) Using the value of $3(N_B)^{1/2}$, calculate the C_{DL}, as in Experiment 21.3.
  (d) Express the phosphorus concentration as the value measured in (b) $\pm C_{DL}$

References
[1] H. A. Liebhafsky, H. G. Pfeiffer, and P. D. Zemany, in *X-Ray Microscopy and X-Ray Microanalysis*, eds. A. Engström, V. Cosslett, and H. Pattee, Elsevier/North-Holland, Amsterdam (1960) 321.
[2] T. O. Ziebold, *Anal. Chem.* **39** (1967) 858.

Laboratory 22

Particle and Rough Surface Microanalysis

Purpose

There are two main objectives of this laboratory: (1) to study the differences observed in EDS and WDS x-ray spectra obtained from particles and rough surfaces as compared to bulk targets, and (2) to test the comparative accuracies of (a) conventional quantitative analysis methods, (b) the peak-to-background method, and (c) the particle ZAF method when applied to particles. More details may be found in *SEMXM*, Chapter 7.

Equipment
1. An SEM with EDS and/or WDS x-ray detection system.

Specimens
1. For the experiment on spherical particles or fracture surface analysis (in order of decreasing preference) either:
   (a) A spherical particle (approximately 100 μm diameter) of a Ni-Al or Fe-Al alloy mounted on a SEM stub and a flat polished specimen of the same composition.
   (b) A fractured specimen of a Ni-Al or Fe-Al alloy and a flat polished specimen of the same composition.
   (c) A fractured or roughened specimen of brass or copper alloy and a flat polished specimen of the same composition.
   (d) A carbon-coated flat polished specimen of NBS standard K411 glass and a carbon-coated dispersion of crushed-up particles from a tiny crystal of K411 glass on a smooth (e.g., pyrolitic) graphite planchet. The particles should be crushed up so they range in diameter from < 1 μm to 20 or 30 μm.

2. For the experiment on particle analysis (in order of decreasing preference) either:
   (a) The particle sample and polished specimen of NBS K411 glass listed in 1(d) above.
   (b) A flat polished specimen and particle dispersion of a microprobe standard containing analyzable elements with a range of atomic number of at least 10. Specimens should be prepared as those in 1(d).

Time for this lab session
Two to three hours.

22.1 Microanalysis of Particles

Conventional correction procedures for quantitative analysis with electron beam instruments require that the sample be infinitely thick with respect to electron penetration, be polished flat and smooth, and be mounted normal to the electron beam. When these criteria are met, then it is possible to simply calculate the amount of x-ray emission at any given depth ρz in the sample:

$$I(\rho z) = \phi(\rho z)\exp(-\mu/\rho \csc \psi \rho z)$$

where $\phi(\rho z)$ is the production of x-rays at the depth z for a sample of density ρ, μ/ρ is the mass absorption coefficient of the element's emitted x-rays by the matrix, and ψ is the spectrometer take-off angle. The path length that electrons must travel to leave the sample, when they are produced at depth z, is $z(\csc \psi)$. For a particle or rough surface, the path length that x-rays must travel to leave the sample from their point of origin becomes a very complicated function of particle shape and size. Moreover, because electrons that are energetic enough to ionize inner shell electrons in the sample can also scatter from the sides of the particle or through the bottom of the particle, the expression for $\phi(\rho z)$ also becomes a complicated function of sample size and shape.

Three methods are generally employed in attempting to perform quantitative particle analysis with an electron microprobe or analytical SEM:

Method 1: Ignore Particle Effects. One assumes that the effects of particle size and shape can be ignored. The particles are treated as if they were conventional thick polished specimens and the results are normalized to 100%. Then one hopes for the best.

Method 2: Peak-to-Background Method. One assumes that the characteristic x-ray peak intensity and bremsstrahlung (background) intensity at the same energy were produced with the same relative spatial distribution in the sample and thus the same relative absorption. If this is the case, then the peak-to-background ratio should be independent of the sample size and shape. Then one can calculate K ratios that can be corrected by conventional ZAF procedures to obtain concentrations by measuring peak-to-background ratios in the particle (or rough surface) and in the standard [1]:

$$K = P'_{tps}/P_{std} = (P/B)_{ptc} \times (B'_{tps}/P_{std}) \quad (22.1)$$

where

$$B'_{tps} = \Sigma C_i B_{i,E}$$

where P'_{tps} and B'_{tps} are the calculated peak and background intensities for a hypothetical thick, polished specimen with the same composition as the particle or rough surface, $(P/B)_{ptc}$ is the measured peak-to-background ratio for the particle or rough surface, P_{std} is the measured peak intensity for the standard, and $B_{i,E}$ is the measured (or calculated) background intensity for a pure element thick, polished specimen. In using this method for particles, care must be taken that *all* of the measured background comes from the particle and none from the matrix on which the particle rests.

Method 3: Particle ZAF Correction Method. One measures the particle size and estimates an idealized geometric model that best matches the particle shape (e.g., cube, sphere, cylinder, etc.) and then uses the ZAF corrections developed by Armstrong and Buseck [2] to account for effects of particle geometry and size on emitted x-ray intensities. This procedure does *not* apply to rough surfaces and requires that the electron beam be defocused or rastered to cover the whole particle surface.

This laboratory will provide the opportunity to test all of these ZAF corrections procedures on a rough specimen and on individual microparticles.

22.2 Spherical Particles or Fracture Surfaces

Experiment 22.1: The Microscope-Detector Relationship.

(a) Determine what the x-ray take-off angle is for the EDS and WDS detectors on the instrument with which you are working.

(b) Determine the relative position of the EDS and WDS detector in the image as recorded on the SEM micrograph. Determine the relative position of the Everhart-Thornley detector in the same image.

(c) Determine how to position the sample so that it is at the right height and tilt for the nominal EDS and WDS detector take-off angle.

Experiment 22.2: Rastered Beam and Point Analyses (Method 1).
(a) On the flat polished Ni-Al alloy specimen record and store an EDS spectrum or obtain WDS intensities at 20 kV. Record your data in Table 22.1.
(b) Place the particle or fracture surface of the same Ni-Al alloy in the instrument and obtain the following EDS spectra or WDS intensities at 20 kV.
  1. Overscan the particle with a rastered beam which just brackets the particle or fracture field and obtain a spectrum.
  2. With a point beam, obtain a spectrum on the side of the particle/fracture facing toward the detector.
  3. With a point beam, obtain a spectrum on the side of the particle/fracture facing away from the detector.
(c) Process the EDS spectra or WDS intensities through the conventional ZAF or "standardless" software. How do the results compare for overscanning versus point analyses?

Experiment 22.3: Peak-to-Background Ratios (Method 2). If the EDS or WDS software permits, compare the peak-to-background ratios for AlK_α, NiK_α, and NiL_α, and record the data in Table 22.1. How similar are the ratios? Are they constant, as the peak-to-background method predicts? How does the rastered analysis compare to the point analysis? Use Equation 22.1 to convert the peak-to-background ratios to compositions.

Experiment 22.4: Low Voltage. If time permits, repeat Experiments 22.2 and 22.3 for NiL_α and AlK_α at 10 kV (or 5 kV). Record the data in Table 22.1. How do the results compare with those taken at 20 keV?

22.3 Irregular or Rough Particles

Experiment 22.5: Microanalysis of Irregular Particles.
(a) Mount samples of the polished K-411 glass standard and the crushed particles of K-411 on a carbon planchet. Insert the specimen into the SEM at the proper height and tilt. At 15 keV obtain an EDS spectrum or WDS intensities of the flat polished standard. Process the results through the ZAF or standardless software. The actual composition of the K-411 glass is shown in Table 22.2.
(b) Find one or two particles smaller than 0.5 µm, one or two about 10 µm, and one or two about 20 to 30 µm in diameter. Estimate the size, thickness, and shape of each particle. Record your data in Table 22.3. If the particle has a flat top and flat sides perpendicular to the sample-detector axis, then call it a rectangular prism (Rec.Pr.). If the particle has a flat top and angular or curved sides with respect to the sample-detector axis, call it a tetragonal prism (Tetr.Pr.). If the particle has an angular or curved top, but flat sides perpendicular to the sample-detector axis, call it a triangular prism [Trig.Pr.]. If the particle has angular or curved tops and sides, call it a square pyramid (Sqr.Pyr.). Overscan each particle with a rastered beam hat just bracket the particles and collect EDS spectra or WDS intensities as in Experiment 22.2b(1). Process the EDS spectra with the conventional EDS software to obtain peak intensities above background and compositions.
(c) Determine the K-ratios of each of the elements in the particle to those in the polished standard and then ratio the K-ratios for all of the other elements in the particles to those for silicon ($R_{x/Si}$ factors). Record these results in Table 22.3.

Table 22.1. Data for Spherical Particles or Fracture Surfaces

Take-off angle: _____

20 kV		Flat specimen	Rastered beam	Point analyses Toward detector	Point analyses Away from detector
Intensity:	NiL_α				
	AlK_α				
	NiK_α				
Background:	NiL_α				
	AlK_α				
	NiK_α				
Relative intensity:	AlK_α/NiK_α				
	AlK_α/NiL_α				
Peak/ background:	$P/B\ NiL_\alpha$				
	$P/B\ AlK_\alpha$				
	$P/B\ NiK_\alpha$				
10 kV					
Intensity:	NiL_α				
	AlK_α				
	NiK_α				
Background:	NiL_α				
	AlK_α				
	NiK_α				
Relative intensity:	AlK_α/NiK_α				
	AlK_α/NiL_α				
Peak/ background:	$P/B\ NiL_\alpha$				
	$P/B\ AlK_\alpha$				
	$P/B\ NiK_\alpha$				

Table 22.2. NBS K-411 Glass Composition

Element	Wt%	Oxide	Wt%
Mg	9.17	MgO	15.21
Si	25.52	SiO_2	54.59
Ca	11.02	CaO	15.42
Fe	11.31	FeO	14.55
O	42.83		

Table 22.3. Data for Irregular Particles

Rastered beam analysis
Accelerating potential: _____ kV

Diameter (in µm):	Thick						
Shape (Rec. Pr., etc.):	Flat						

K-ratio:	MgK_α	1						
	SiK_α	1						
	CaK_α	1						
	FeK_α	1						

R-factor:	Mg/Si	1						
	Ca/Si	1						
	Fe/Si	1						

Point beam analyses
Accelerating potential: _____ kV

Diameter:	Thick						
Shape:	Flat						

K-ratio:	MgK_α	1						
	SiK_α	1						
	CaK_α	1						
	FeK_α	1						

R-factor:	Mg/Si	1						
	Ca/Si	1						
	Fe/Si	1						

P/B:	MgK_α							
	SiK_α							
	CaK_α							
	FeK_α							

(d) Use the R_x/S_i correction factors for each element to obtain particle ZAF analyses of the particles measured.

Questions:
1. How does the precision of an EDS measurement compare with that for a WDS measurement?
2. When is the EDS better for particle analysis?
3. When is the WDS superior?

Experiment 22.6: Peak-to-Background Ratios (Method 2). If time permits, analyze some particles with a point beam and evaluate the peak-to-background results as in Experiment 22.3. Record your data in Table 22.3.

References
[1] J. A. Small, K. F. J. Neinrich, D. E. Newbury, and R. L. Myklebust, *Scanning Electron Microscopy/1979* (SEM Inc., AMF O'Hare, IL) **2** (1979) 807.
[2] J. T. Armstrong and P. R. Buseck, *Anal. Chem.* **47** (1975) 2178.

Laboratory 23

X-Ray Images

Purpose
This laboratory demonstrates several important considerations for producing elemental distribution images using WDS and EDS. More detail can be found in *SEMXM*, Chapters 5, 6, and 8; and especially in *ADSEM*, Chapter 5.

Equipment
1. SEM or electron microprobe equipped with EDS or WDS.
2. Hardware and software for digital beam control to produce digital x-ray images.

Specimens
Basalt or Raney nickel specimen polished flat, and left unetched.

Time for this lab session
Two hours.

23.1 Analog X-Ray Dot Maps

Although computer-based x-ray imaging techniques are rapidly gaining favor, the dominant method of recording x-ray images remains the analog technique known as dot mapping (or area scanning). In this method, the output either from a wavelength-dispersive x-ray spectrometer or from an energy-dispersive x-ray spectrometer is used to modulate the brightness of the photographic recording CRT, and the dot map image is recorded on film.

The major difference between analog image recording using x-rays compared with the more familiar electron image is the relative strength of the signals. Electron signals are generally within a factor of 10 of the electron beam current. Characteristic x-ray signals are reduced by 4 or 5 orders of magnitude, relative to the electron beam current, as a consequence of the relatively low probability of electron interactions causing inner shell ionizations, the low x-ray yield for many metals, and the small geometric efficiency of x-ray spectrometers. When the concentration of the constituent of interest is considered, the available x-ray intensity is proportionally further reduced. As a result, for dilute concentrations (10 wt% or less) of an element, the characteristic x-ray signal is so low that many picture points in a single photographic image (100 sec/frame) will not record any x-ray pulse, and most picture points will only record a single x-ray pulse. It is not possible to assign a true gray scale to such an image since the signal is not continuous. The practical solution to this problem is to trigger the CRT to write a "full white" dot at any scan location where one or more x-rays are detected.

Recording useful x-ray dot maps depends on three factors:

1. Proper adjustment of the record CRT. The white "dot" written on the CRT must be sufficiently bright to record the x-ray pulse but not so bright as to bloom into adjacent pixels.
2. An adequate number of x-ray counts must be recorded. For high-quality results, approximately 500,000 counts must be accumulated per frame if fine scale image details and low concentration regions of the image are to be discerned.
3. The magnification must be sufficiently high so that defocusing effects of the wavelength-dispersive spectrometer do not dominate the image. Alternatively, an analog scan correction may be used to compensate for defocusing.

Experiment 23.1: Recording Dot Maps. Select a multiphase specimen such as a polished sample of basalt (or other fine-grained rock, ceramic, or metal alloy) which will produce distinct elemental contrast in an x-ray area scan. Choose a magnification ($\geq 500x$) and a field of view which will place an interface between a high iron and a low iron phase in the image (this may be quickly assessed using the atomic number contrast in the backscattered electron image), preferably crossing the field diagonally. Select beam conditions to produce a useful FeK_α x-ray signal. For a wavelength-dispersive x-ray spectrometer select an accelerating voltage of 20 kV and a beam current of 20-100 nA. For an energy-dispersive spectrometer select 20 kV and a current that produces a dead time of <20% (about 0.1-1.0 nA).

(a) Adjustment of CRT recording dot. Adjust the contrast and brightness controls on the record CRT to obtain a suitable x-ray dot. Photograph a single scan sweep at 100 sec frame time. Can the interface be discerned? Reduce the dot brightness until it is barely visible and record a 100-sec image. Increase the dot brightness until blooming is evident and record a 100-sec image. How do the three images compare?

(b) Effect of the number of x-ray counts. Choose the CRT dot brightness which gave the best result in (a) and record the same field of view for 1000 sec. The best way to implement this scan is to use repeated overscans of 100 sec each and use the film to integrate the result.

(c) Wavelength-dispersive spectrometer defocusing. Repeat the 100-sec, optimum dot brightness map centered on the same field of view but with progressively lower magnifications (400x, 200x, 100x). Is the spectrometer defocusing artifact evident in the low magnification images? If the instrument has the capability to compensate for spectrometer defocusing by adjusting the spectrometer as a function of scan position, apply this correction at 100x and compare the corrected and uncorrected images. Is there any other way to correct for defocusing?

23.2 Digital X-Ray Images

An important consideration for digital x-ray images is the proper number of pixels required to form meaningful images. Due to the relatively large volume from which x-rays are generated within solid specimens, an image formed using x-rays will always have a lower resolution than that using an electron signal such as BSEs or SEs. Furthermore, since the generation of x-rays is 3-4 powers of ten less efficient than for the electron signals, the resulting images will be rather noisy. One can conclude from these facts that fewer pixels should be used to form an image using an x-ray signal.

We can estimate the number of pixels required by the following rationale. First, we calculate the x-ray range for the chosen characteristic x-ray line by use of the Anderson-Hasler equation (incorrect in *SEMXM*, p. 108):

$$R = \frac{0.064}{\rho} \left(E_0^{1.68} - E_c^{1.68} \right) \mu m \qquad (23.1)$$

where R is the depth in μm from which most of the chosen characteristic x-rays are generated, E_0 is the electron beam energy, E_c is the critical ionization energy for the x-ray line of interest, and ρ is the specimen density in g/cm³. Even though the width of the x-ray volume is generally less than its depth, the depth provides a conservative estimate of the lateral resolution.

Using R for the x-ray range, we may calculate the minimum number of pixels needed so that each pixel is smaller than the x-ray range:

$$n > (L/M \cdot R) \qquad (23.2)$$

where n is the number of pixels along a scan line; L is the width of the display or recording screen in micrometers (typically 100,000 μm); M is the image magnification; and R is the x-ray range, in micrometers, as calculated in Equation (23.1).

Finally, it is necessary to consider the optimal number of counts (peak + background) N required in an average pixel. For various visibility and statistical purposes, N should be greater than 8. The dwell time per pixel also affects the counts per pixel. Consequently, we choose n to satisfy both Equation (23.2) and N/pixel > 8. Often the limiting factor is the time available to obtain the image.

The best acquisition strategy is to use the maximum number of pixels available in a particular hardware configuration and then "average" adjacent pixels until the above two conditions are satisfied. Not all systems allow reduction of pixel density by averaging, so the operator must choose a low pixel density (e.g., 128 x 128) or alter the magnification or the dwell time per pixel.

Experiment 23.2: Choice of Digital Image Parameters. (a) *Assuming a density of 5.5 g/cm³ for basalt, a beam voltage of 15 kV, a magnification of 1000x, and an image acquisition time of 15 minutes, determine the optimum number of pixels and the dwell time per pixel for the FeK$_\alpha$ line at the beam current you are using.*

(b) Setup for EDS. Take a 100-sec (live time) EDS spectrum of basalt. Create windows around the major peaks and at least one background region, say near the iron peak (low energy side) with the same number of channels as the iron peak. Use these windows, and the pixel density and dwell time determined in (a), to take a digital x-ray image. Select a beam current that will not produce a system dead time of more than 20% at any point in the scanned area. Check that for the dwell time per pixel selected the counts per pixel exceed 8 in the areas of interest. Store these x-ray images. How does this digital x-ray image differ from a dot map?

(c) Setup for WDS. Set the spectrometer for the x-ray line of interest such as FeK$_\alpha$. Increase the probe current to 20-100 nA. Route the signal from this spectrometer into the MCA computer system which controls the digital stepping of the electron beam. If possible, collect the WDS and EDS digital images simultaneously.

23.3 Background Removal

If the x-ray spectrometer is an EDS, it is particularly important to subtract the background x-rays upon which the characteristic x-rays are added. This follows from the fact that the peak-to-background ratio of the energy-dispersive spectrometer is relatively low compared to the wavelength-dispersive spectrometer and that the intensity of the background is dependent upon the average atomic number of the phase analyzed. If your specimen has phases with sufficient differences in atomic number, you can scan for an element that is not present in any phase and obtain an x-ray image that contains contrast similar to that of a backscattered electron image. For a discussion of the various procedures to remove background see *SEMXM*, Chapter 8.

Experiment 23.3: Background Subtraction. *Choose an area on the basalt specimen where there are large atomic number variations such as a metal and silicate phase. Using the on-the-fly automatic background removal feature provided by the manufacturer of the EDS hardware, image several elements in the basalt. If background removal on-the-fly is not possible with your system take an image for an element and one for an adjacent background region. Subtract the two images (sometimes smooth before subtraction is helpful). Also, collect an image for an element not in basalt such as scandium or chromium and an image for a background region. Use a magnification of <1000x, a pixel density of 128 x 128, and a 15-min frame time, and use a probe current that produces a system dead time of no more than 20% at any point in the scanned area.*

Can you see the element that does not exist in the specimen? What would be the practical implications of the apparent presence of this element in the x-ray image?

23.4 Dead Time Considerations

When the x-ray count rate increases the EDS dead time increases. If this happens in an image and is not corrected, the relative concentrations of an element will not be correct. In the extreme case, regions with high concentrations of an element may be shown as having little or none of that element present.

Experiment 23.4: Test of the Dead Time Circuit. *(a) Low-dead time image. Choose an area on the basalt specimen which will include both a high-iron and a low-iron phase. Acquire an image at a magnification of <1000x (store it for reference), a beam voltage of 15 kV, iron as the analytical element, an acquisition time of 15 min, a probe current which will produce a dead time of <5% on the metal phase, and a pixel density of 128 x 128. Make sure that the low-iron phase contains some iron or else this comparison cannot be made.*

(b) High-dead time image. Under identical conditions acquire another image but use a probe current which will produce a system dead time of 50% on the metal phase. Store the image.

(c) Compare the ratio of the iron counts between high-iron and low-iron phases in the two images. Make sure that the low-iron phase contains some iron or else this comparison cannot be made. Use the "image region" feature of the EDS software to average at least several hundred pixels from these phases. The ratios should be the same. Are they?

Note: If the ratios are not the same then the dead time circuitry may not be correctly adjusted. Consult the hardware manual or the manufacturer for assistance.

23.5 Intensity Measurement

Often, the concentration of an element for which we wish to determine a spatial distribution is simply too low to form a meaningful image. In this case we must remove one dimension by dropping from three dimensions (x-ray intensity versus x and y) to two dimensions (x-ray intensity versus x). The two-dimensional case is called a line scan.

Since it is difficult to correctly estimate relative intensities of pixels by viewing the x-ray image alone (see discussion on this topic in Laboratory 17), line scans are a useful supplement to x-ray images. Unambiguous assessments of intensities in a region of an x-ray image can be made by superimposing the line scan on the image.

Experiment 23.5: X-Ray Line Scans. *For an x-ray image recorded at <20% dead time that shows a strong change in elemental concentration with distance, select a locus of points in the image that runs across this elemental distribution. EDS computer systems configured for* *digital beam control allow the selection of this locus by mouse-selection of endpoints, coordinate specification, etc. If possible, superimpose the x-ray intensity trace (x-ray line scan) along this locus onto the x-ray image. Carefully compare the information content of the two.* Does the line scan give a considerably better idea of relative concentrations along the line than the intensity image?

23.6 X-Ray Image Processing

Since x-ray images are relatively noisy compared to electron images, x-ray images can often benefit from digital image processing methods such as image smoothing and image coloring. There are several pitfalls that should be recognized when interpreting processed images (see the discussion of this topic in Laboratory 17).

Experiment 23.6: Imaging Smoothing. Apply a 3x3 image smoothing kernel available in your EDS computer to one of x-ray images that you have stored. Are there worm-like artifacts in the smoothed image? *Use a smoothing kernel with more elements (e.g., 17x17 or 27x27) to reduce this artifact.*

Experiment 23.7: Primary Coloring. Code the smoothed elemental images from Experiment 23.2b or 23.3 with primary colors (refer to the EDS manufacturer's instructions). For example, let red = silicon, green = potassium, and blue = iron. Add these images together in pairs and all three together. How does primary coloring aid in phase identification?

Part IV

ANALYTICAL ELECTRON MICROSCOPY

Laboratory 24

Scanning Transmission Imaging in the AEM

Purpose
The aim of this laboratory session is to introduce the scanning transmission electron microscope (STEM), used in this and other analytical electron microscopy (AEM) laboratory sessions, and to demonstrate the various imaging modes available. The experimental control that the operator has over the information in the image will be emphasized. More details may be obtained in *PAEM*, Chapter 3 and D. B. Williams, *Practical Analytical Electron Microscopy in Materials Science*, Philips Electron Optics, Mahwah, New Jersey, 1984.

Equipment
1. A STEM of either the TEM/STEM type or the dedicated (D) STEM type.
2. Device for measuring specimen current such as a Faraday cup specimen holder and picoammeter or calibrated exposure meter.

Specimens
1. Electropolished disk of aluminum or any crystalline specimen exhibiting diffraction contrast.
2. Microtomed thin section of a two-phase polymer such as polybutadiene in a thermoplastic matrix stained with OsO_4, or any specimen exhibiting mass-thickness contrast.
3. Latex particles on a carbon film shadowed with an evaporated metal film, such as gold or Au-Pd.

Time for this lab session
Two hours.

24.1 Characteristics of the STEM Electron Probe

The characteristics of the STEM probe are important, not only for imaging, but also for microanalysis and microdiffraction. In a TEM/STEM it is possible to project an image of the STEM probe onto the TEM screen, either as a stationary probe or as a scanned raster. The steps to carry out this procedure are given in the manufacturer's manual. In a DSTEM it is only possible to measure the probe diameter indirectly by scanning the beam across a sharp edge (such as a gold island on a carbon film or an MgO cube). The probe size can be determined by measuring the change in the annular dark field (ADF) signal intensity. (A similar SEM experiment is described in Laboratory 2.)

Experiment 24.1: Effect of C_1 Lens Strength. Vary the strength of the first condenser lens and note the changes in probe size and shape on the TEM screen. Measure changes in probe current with a Faraday cage/picoammeter or indirectly with the TEM exposure meter. On a DSTEM, observe changes in the signal profile across a sharp edge, by changing C_1. Should the same probe size be used for both imaging and microanalysis?

Experiment 24.2: Effect of C_2 Aperture Size. Insert a large and a small second condenser aperture and note the changes in probe size and probe current. Misalign the aperture and note the change in probe shape. On a DSTEM change the size of the virtual objective aperture (VOA).

Experiment 24.3: Effect of Condenser and Objective Stigmators. Misadjust the stigmator controls one at a time and note the effects on probe shape and probe current. On a DSTEM, observe the changes in the ADF image rather than the probe profile.

Experiment 24.4: Effect of Objective Lens Focus. Defocus the objective lens and note the effects on probe size and probe current.

24.2 STEM Image Formation

STEM images are formed by scanning a small beam across the specimen and detecting the electron signal as a function of beam position on either a bright field (BF) or annular dark field (ADF) detector beyond the specimen. Some instruments collect these signals when the microscope is operated in the normal image mode and some in diffraction mode. Consult the manufacturer's manual for the specific steps required to switch from TEM to STEM. (In a DSTEM, the following experiment is redundant, since TEM images cannot be obtained.)

Experiment 24.5: Image Relationships. Using the specimen of polycrystalline aluminum, obtain an image in TEM mode with some recognizable feature near the edge of the foil. Insure that this same feature reappears in the STEM image. What is the relationship between features in the TEM image and the STEM image? How does the beam convergence angle differ between TEM and STEM?

24.3 Measurement of Convergence and Collection Angles

By stopping the STEM probe (spot mode) and observing the TEM screen in diffraction mode, a convergent beam electron diffraction (CBED) pattern should be visible. From this pattern, the probe convergence angle $2\alpha_s$ and the STEM detector collection angle $2\beta_s$ can be calculated. These parameters are important in STEM imaging, in addition to their significance in convergent beam electron diffraction and microanalysis. In a DSTEM, $2\alpha_s$ can be measured from a CBED pattern on the TV image of the diffraction screen. To measure $2\beta_s$, use a rocking beam pattern such that the EELS entrance aperture limits the size of the diffraction disks.

Experiment 24.6: Calculation of $2\alpha_s$ and $2\beta_s$. Tilt the aluminum specimen to a low index zone axis such as [001] where the diffraction spots can be easily indexed. Select a C_2 aperture small enough so that the disks (spots) do not overlap. Photograph or estimate the size of the following: a = the diameter of the diffraction disk, b = the distance between the 000 and the hkl diffraction disks, and c = the estimated projected diameter of the bright field STEM detector. Calculate the convergence angle and collection angle. Hint: Set up the proportions $2\alpha_s/2\theta_B = a/b$ and $2\beta_s/2\theta_B = c/b$ (where θ_B is the Bragg angle).

24.4 Contrast Effects in STEM Images

Contrast in a typical STEM image is either due to diffraction effects or mass-thickness effects. Most phase contrast effects are on too fine a scale to be routinely observed and usually require a field-emission gun to obtain enough signal. Diffraction contrast effects will be observed in a thin foil metal specimen and mass-thickness effects in a polymer sample.

Experiment 24.7: Diffraction Contrast. In bright field TEM mode or on a DSTEM (BF image) tilt the metal foil specimen to obtain an image which exhibits strong diffraction contrast such as bend contours. On the TEM/STEM overfocus the second condenser lens to approximate parallel illumination. (On a DSTEM, only carry out experiments b-d, since a is not possible.)

(a) Compare the contrast of the bend contours in TEM and STEM bright field modes. Use a small probe size (5-10 nm) and relatively small convergence and collection angles for the STEM image. Why is the STEM contrast different? Observe the same image with STEM annular dark field mode. Does annular DF give an image equivalent to TEM diffraction contrast DF mode?

(b) Increase the probe size by altering the C_1 lens setting. Why does the STEM BF image resolution degrade?

(c) Increase $2\alpha_s$ and explain the changes in the BF STEM image.

(d) Increase $2\beta_s$ and explain the changes in the BF STEM image.

Experiment 24.8: Mass-Thickness Contrast. Using the two-phase polymer specimen, set up a bright field TEM image of a region showing mass-thickness contrast.

(a) Obtain STEM BF and annular DF images of the same region. Why should BF and ADF now show complementary contrast which was not the case in Experiment 24.7?

(b) Increase the probe size using the C_1 lens. What is the effect on the ADF image?

(c) Increase $2\alpha_s$ and explain the changes in the ADF STEM image.

(d) Increase $2\beta_s$ and explain the changes in the ADF STEM image.

Experiment 24.9: Phase Contrast Imaging. In a DSTEM, or a TEM/STEM with a FEG source, it is possible to obtain high-resolution phase contrast images. Use a graphitized carbon sample and an objective aperture large enough to include both 000 and 0002 from the basal planes. The basal planes should be imaged with a 0.34-nm spacing. Why cannot a thermionic source instrument show high-resolution images?

24.5 Scanned Images Using Other Electron Signals

As long as a small probe has sufficient current in it, a scanned image may be formed with any signal that can be collected with some detector and amplified sufficiently to modulate the brightness of the CRT.

Experiment 24.10: Secondary Electron Imaging. Using the specimen of shadowed latex spheres, obtain a secondary electron image. Why does this SE image have resolution better than a conventional SEM image taken with a thermionic tungsten gun?

Experiment 24.11: Backscattered Electron Imaging. Using the same specimen, take a backscattered electron image. Why does this image have better resolution, but poorer contrast, than a BSE image taken on a conventional SEM?

Laboratory 25

X-Ray Microanalysis in the AEM

Purpose
   The aim of this laboratory is to introduce the principles and practice of quantitative x-ray microanalysis in the AEM using an energy-dispersive spectrometer (EDS). Since it is important to recognize the limitations of the technique as well as the relative ease of quantification, some effects due to spurious x-rays in the EDS spectrum will be identified in the first experiment. More details may be found in *PAEM*, Chapters 4 and 5.

Equipment
   1. STEM equipped with an EDS system.
   2. Software for quantitative thin-film analysis by the Cliff-Lorimer ratio method including an absorption correction.
   3. Low-background specimen holder with beryllium or graphite specimen cup.

Specimens
   1. Cr thin film evaporated onto a carbon film partially covering a heavy metal grid (e.g., 400-mesh gold) or heavy metal aperture (e.g., molybdenum).
   2. Fe-Ni thin foil (or any alloy containing elements of similar z) fabricated by electropolishing or ion-beam milling.
   3. Ni-Al thin foil (or any alloy containing elements of widely differing z) fabricated by electropolishing or ion-beam milling.
   4. Biotite mica crystals (or any multielement specimen) ground and dispersed onto a carbon film.
   5. Copper or nickel thin foil (or any simple metal/alloy foil with large grain size) fabricated by electropolishing or ion-beam milling.

Time for this lab session
   Two to three hours.

25.1 Sources of Spurious X-Rays

   Ideally the x-rays reaching the detector should be generated only from the volume of the thin foil illuminated by the electron probe. Unfortunately, unwanted x-rays can also be generated. For example, spurious x-rays from locations on the specimen remote from the electron beam may be generated by high-energy continuum x-rays or uncollimated electrons from the illumination system of the microscope (the "hole count"), or as the result of electron probe-specimen interactions. These sources of spurious x-rays must be understood and minimized before quantitative analysis is attempted.

Experiment 25.1: The "Hole Count." Using the chromium film on the molybdenum aperture or gold grid specimen in the low background holder, observe the presence (or absence) of stray radiation and/or electrons from the illumination system by comparing the spectrum obtained in the center of an open grid (the hole) square to that obtained on the chromium film. Ratio the full width AuL_α or MoK_α peak taken in the hole to the full width CrK_α peak taken on the film and multiply by 100 to obtain the "hole count." <u>Approximate conditions</u>: Use a spot size (C_1 setting) that produces about 0.5 nA of probe current; probe size = 2-50 nm depending on the electron source; "hard x-ray" C_2 aperture = 50-70 µm; objective aperture out; counting time = 100 live sec.

(a) Compare the hole count obtained with the thick platinum C_2 "hard x-ray" aperture to that obtained with a conventional thin aperture (if available). Record data on table below. The thick aperture should be used for the rest of this laboratory.

(b) Change to the smallest spot size but one and repeat the hole count. What causes the observed result?

Data sheet for hole count

Instrument:_____ kV:_____ STEM or TEM mode:_____
Emission current (µA): _____ Probe current (nA): _____ Probe size (nm): _____
CrK_α window: _____ keV to _____ keV; AuL_α window: _____ keV to _____ keV

	Spot size	Thick C_2 aperture?	Live time (sec)	Film CrK_α (counts)	Hole AuL_α (counts)	% Hole count (Au/Cr)x100
Expt. 25.1a		Yes				
Expt. 25.1a		No				
Expt. 25.1b		Yes				
Expt. 25.1b		Yes				

Experiment 25.2: Postspecimen Scatter. Obtain a spectrum from the chromium film with the specimen holder at 0° tilt angle. Tilt the specimen about 30° toward the detector and observe any change in the count rate on the AuL_α or (MoK_α) peak. What are the causes of the difference in the count rate? Return to 0° tilt.

AuL_α count rate at 0° tilt =
AuL_α count rate at 30° tilt =

25.2 X-Ray Data Collection

Quantitative analysis requires a knowledge of the specimen detector geometry as well as accurate data for the number of characteristic x-ray photons collected from the specimen above background. Simply stated, quality analysis requires quality data.

Experiment 25.3: EDS Detector Position with Respect to the Image (for a side entry goniometer microscope only). Insert the Fe-Ni thin foil specimen. Determine the orientation of the STEM image of the specimen by gently pushing the end of the specimen holder rod and noting the direction of image motion. The image will move along the axis of the specimen rod. From the location of the EDS detector on the column determine the direction of the detector on the STEM (or TEM) image. On which side of the specimen hole should the analysis be taken?

Note that there is no image rotation when the magnification is changed in STEM mode. The STEM image may be inverted with respect to the TEM image on some AEMs. When analyzing planar interfaces in the AEM, the interface plane should be oriented to point at the EDS detector. Why is this necessary?

Experiment 25.4: Collection of X-Rays. Collect an x-ray spectrum using the following conditions: maximum kV; 10-nm probe size (thermionic source); thick C_2 aperture; objective aperture out; 50 sec counting time; specimen thickness which gives about 20% dead time. Determine the count rate in the FeK_α (no need to subtract background at this stage) and note the dead time when:
   (a)   A lower kV is used.
   (b)   The probe size is doubled (TEM/STEM C_1 current decreased; for DSTEM increase size of virtual objective aperture).
   (c)   The probe size is halved (TEM/STEM C_1 current increased; for DSTEM, decrease size of virtual objective aperture).
   (d)   The C_2 aperture size is increased (TEM/STEM only).
   (e)   The C_2 aperture size is decreased (TEM/STEM only).

	FeK_α count rate	Percent dead time	
Maximum kV			
_____ kV			Return to max.
2 x probe size _____			
1/2 probe size _____			
C_2 aperture _____			
C_2 aperture _____			

Experiment 25.5: Background Subtraction. Collect a spectrum from the Fe-Ni thin foil until the NiK_α peak has about 11,000 counts in it, including the background. Obtain the Fe and NiK_α intensities by subtracting the background using whatever method the multichannel analyzer (MCA) permits. Compare the intensities obtained by the computer with estimates based on subtracting the average background intensities in "windows" on either side of the Fe and NiK_α peaks.

5.3 Quantification of Data

The k-factors for the Cliff-Lorimer ratio method may be obtained either by direct measurement on standards of known composition or, in cases where standards are unavailable, by calculation from first principles. Although the latter method is generally less accurate, it is often used in the quantitative software that manufacturers supply with the EDS system.

Experiment 25.6: Measurement of the k-Factor. *Given that the Fe-Ni specimen is 49 wt% Ni, use the intensities obtained in Experiment 25.5 to determine the k_{FeNi} from the equation:*

$$\frac{C_{Fe}}{C_{Ni}} = k_{FeNi} \frac{I_{Fe}}{I_{Ni}}$$

Experiment 25.7: Errors in k-Factor Determination. *Assuming the composition of the standard is accurate to ±2% relative, estimate the error in the k-factor. Assume a "3σ" error in I_{Fe} and I_{Ni} where the estimate of the standard deviation $s_n = \sqrt{N}$ and N = number of counts. Thus,*

$$\text{Percent error in } \frac{C_{Fe}}{C_{Ni}} = 2\%$$

$$\text{Percent error in } \frac{I_{Fe}}{I_{Ni}} = \frac{3\sqrt{N_{Fe}}}{N_{Fe}} \times 100 + \frac{3\sqrt{N_{Ni}}}{N_{Ni}} \times 100$$

$$\text{Percent error in } k_{FeNi} = 2\% + \frac{3\sqrt{N_{Fe}}}{N_{Fe}} \times 100 + \frac{3\sqrt{N_{Ni}}}{N_{Ni}} \times 100$$

However, when the number of separate measurements is less than 30, a correction to the normal distribution, called the "student's t" distribution is used to estimate the error.
Repeat the determination of the k-factor using four separate and independent measurements and determine the error in k_{FeNi} at the 99% confidence limit using the "student t" distribution:

$$\text{Percent error in } k_{FeNi} = \frac{t_{99}^{n-1}}{\sqrt{N}} \frac{s_n}{k_{FeNi}} \times 100$$

where $n = 4$, $t_{99}^{n-1} = 5.841$, and s_n is the standard deviation calculated for four values of k_{FeNi} of average value $\overline{k_{FeNi}}$.

Simple binary samples like Fe-Ni can be quantified without resorting to the computer, because of the ease of background subtraction and the lack of absorption. In fact a rough estimate to within ±10%-20% relative can be made just by ratioing the peak heights or peak intensities. However, only integrated peak intensities should be used for quantitative analysis and multielement spectra generally require a computer.

Experiment 25.8: Multielement Quantification. *Test the accuracy of the quantification routines available in the MCA computer software using the specimen of biotite (nominal composition 17.3 wt% Si; 15.1 wt% Fe; 10.5 wt% Al; 6.3 wt% K; 5.4 wt% Mg; 0.9 wt% Ti, balance oxygen). In order to minimize absorption choose a thin flake, but try to obtain at least 10^3 counts in all major peaks to be quantified. If you have an ultra-thin-window EDS, then detect the oxygen signal also.*

Experiment 25.9: Absorption Correction. *Using the sample of stoichiometric NiAl (33 wt% Al) obtain two x-ray spectra, one from a thin region and one from a thick region, ea* containing 10,000 counts in the AlK_α peak. Using a k_{NiAl} value of 1.2, or the calculated value stored in the MCA/computer software, determine the apparent wt% aluminum from each spectrum. Now use the absorption correction routine in the minicomputer to correct the data by choosing various thickness values until the correct answer is given (to within ±10%).

25.4 Coherent Bremsstrahlung Effects

In addition to the small peaks that are artifacts due to the EDS detector (escape peaks, sum peaks, and the silicon internal fluorescence peak), the specimen itself may be responsible for some small unidentifiable peaks arising from coherent bremsstrahlung effects.

Experiment 25.10: Detection of Coherent Bremsstrahlung. Tilt the copper or nickel foil sample to a low index zone axis, place a large probe (20 nm) on the specimen and count for 300 sec, making sure dead time is <25%. Examine the spectrum, and account for all the characteristic peaks and EDS artifacts. Any small gaussian peaks that are unidentified may be due to coherent bremsstrahlung effects. Collect a spectrum under similar conditions but reduce the kV or change the specimen orientation substantially. The characteristic peaks and artifacts should not change position, but the coherent bremsstrahlung peaks will move. In what situations would coherent bremsstrahlung peaks cause a problem for quantitative microanalysis?

25.5 X-Ray Emission Imaging (Optional)

Digital control of the electron beam allows reasonable elemental images or maps to be obtained from characteristic x-ray emission signals. The old x-ray "dot maps" usually did not have very good contrast between features in the image and the overall x-ray background, whereas digital x-ray images may be displayed in a background-subtracted form with excellent contrast. However, the signal from a thin specimen is usually about 10^5 times smaller than the signal emitted from an electron microprobe or an SEM. Thus, long dwell times are necessary to produce digital x-ray images, especially using a thermionic electron source (several hours may be required for a single frame).

Experiment 25.11: X-Ray Elemental Image. Insert the Cr/Au grid hole count specimen. Set the EDS system to collect 128x128 digital x-ray images of chromium and gold with a 50-msec dwell time per pixel at 5000x or 10,000x magnification. Go to an area where the chromium film covers only part of a grid square and the gold grid is partially in the image. Start the digital x-ray image collection.

Laboratory 26

Electron Energy Loss Spectrometry

Purpose
The aim of this laboratory session is to demonstrate the use of a magnetic prism electron spectrometer to perform electron energy loss spectrometry (EELS). The main characteristics of the energy loss spectrum will be discussed as well as the effect of instrumental and specimen parameters on the spectrum. Quantitative elemental analysis will be demonstrated and if time permits an example of EELS imaging will be shown. More details can be found in *PAEM*, Chapters 7 and 8.

Equipment
A dedicated STEM or a TEM/STEM equipped with a magnetic prism spectrometer such as the Gatan 607 serial EELS (SEELS) or the Gatan 666 parallel EELS (PEELS).

Specimens
1. Holey carbon film (amorphous) with very thin graphite flakes over holes on a copper grid.
2. Aluminum thin foil fabricated by electropolishing or ion-beam milling.
3. Very thin boron nitride particles on a holey carbon film on a copper grid.

Time for this lab session
Two to three hours (SEELS); one and a half to two hours (PEELS).

26.1 Characteristics of the Energy Loss Spectrum

SEELS and PEELS differ only in the method and time of spectral acquisiton. SEELS gathers the spectrum one channel at a time, and may take several tens or hundreds of seconds to acquire the spectrum. PEELS gathers the whole spectrum simultaneously and usually in a matter of a few seconds or a few tenths of seconds. The final SEELS and PEELS spectra are indistinguishable when displayed. All spectra shown in this lab were acquired with a SEELS.
A thin film of amorphous carbon (<20 nm) can be used to show the three main spectral features of the loss spectrum: (1) the zero loss peak E_o; (2) the carbon plasmon peak; and (3) the carbon K edge. Features (2) and (3) are superimposed on a rapidly decreasing background intensity.

Experiment 26.1a: Serial Collection Conditions. Collect the energy loss spectrum with the following conditions: TEM imaging mode at a moderate magnification (e.g., 10-30,000X), large C_2 aperture (150 μm), no objective aperture, No. 3 spectrometer entrance aperture (3 mm), and spectrometer slit adjusted for moderate resolution conditions (small flat top on zero-loss peak on monitor). Scan the spectrum over an energy range from -100 to +500 eV energy loss. Set up two regions of interest on the MCA such that the energy region from ΔE = -100 to +100 eV has a short dwell time (about 5 msec/channel) and the region from +100 eV to +500 eV loss has a long dwell time (about 100-200 msec/channel). Leave one channel between the two regions of interest. Use dwell times such that the whole spectrum can be collected in about 60-100 sec. To simplify spectrum collection, turn the auto gain change off for these experiments. Identify the zero loss peak, the plasmon peak, and the carbon K edge.

Experiment 26.1b: Parallel Collection Conditions. Use the same conditions as in Experiment 26.1A above (note however that there is no entrance slit to the spectrometer). Select a relatively coarse display resolution (~0.5-1 eV/channel). Preset two acquisition modes, Mode 1 for the intense zero and low loss region of the spectrum and Mode 2 for the high loss region. Insure the voltage scan module is switched to MCS. Typical conditions are:
  Mode 1, integration time 0.025 sec, 1 acquisition, zero voltage offset on the drift tube, attenuator on, normal acquisition, no correction.
  Mode 2, integration time 1 sec, 1 acquisition, 250 V offset (to view the C K-edge at 284 eV), attenuator off, normal acquisition, no correction. Acquire a spectrum in each mode, using the C2 lens to spread or condense the beam on the sample if the zero loss peak is too intense (i.e., the symmetrical peak is "clipped" at the top) or the high loss spectrum is too weak. Store each acquisition. Observe the intense zero loss and the first plasmon peak at ~25eV in the low loss spectrum and the C K-edge (characteristic sawtooth intensity distribution) in the high loss spectrum.

26.2 Calibration of the Spectrometer

Experiment 26.2a: SEELS Energy Calibration. Calibrate the spectrometer energy scale using the E_o peak (0 eV loss) and carbon K-edge (284 eV) as two points of known energy. Use the inflection point of the carbon edge as the "true" position of the edge since the energy resolution with a large spectrometer entrance aperture will be several eV.

Experiment 26.2b: PEELS Energy Calibration. Use the drift tube voltage offset to displace the zero loss peak by a known amount (e.g. -100-200 V) and calibrate the energy scale on the display in this manner. If necessary store the 0-V and applied offset voltage spectra in different memories then overlap the displays. Note: Observe the zero loss peak in "view" mode (i.e., repeated displays) and use the focus controls to insure that the zero loss peak is symmetrical and of maximum height before each calibration.

26.3 Measurement of the Energy Resolution of the Spectrometer

Experiment 26.3: Determination of the Energy Resolution from the Zero-Loss Peak. Measure the full-width-at-half-maximum (FWHM) in eV, the intensity of the peak, and background intensity either side of the zero loss peak. Enter these values in the table below. To insure optimum display resolution, recalibrate the display to 0.1eV channel in either SEELS or PEELS. In PEELS, adjust the beam intensity to avoid "clipping" of the zero loss peak, and in both cases, acquire sufficient count to give good statistics (~500,000 in maximum intensity channel).

26.4 Spectrometer Variables

Experiment 26.4: Effect of Entrance Aperture Size on Resolution. Collect a spectrum with both the E_o peak and the carbon K-edge (3 mm entrance aperture). Enter estimated values for FWHM, intensity, background and jump ratio (edge height/bg) in the table below. Insert a smaller spectrometer entrance aperture (e.g. 1 mm). Note the changes in resolution and intensity of the E_o peak and the carbon K-edge. The smaller aperture would normally be used for high-resolution operation. Should the microscope objective aperture have an effect? *In PEELS, observe the spectra in "view" mode while changing the entrance aperture size to see the effect dynamically.*

Experiment 26.5: (SEELS Only) Effect of Spectrum Scan Direction. Reverse the spectrum scan direction (i.e., collect the spectrum from high loss to low loss) if the software permits this and repeat these measurements. Compare the two spectra. Why is the background different when the spectrum is scanned from high-loss to low-loss? Enter FWHM, intensity, background, and jump ratio (edge height/bg) data in the table below.

	Condition	E_o				Carbon edge		
		FWHM	Intensity	Bg.	P/B	Edge Ht.	Bg.	Jump Ratio
E26.3	3-mm aperture							
E26.4	1-mm aperture							
E26.5	Reverse scan							
E26.6	Reduced slit width							

Experiment 26.6: (SEELS Only) Exit Slit Width. Reduce the slit width to attain high-resolution conditions. You may wish to return to the larger spectrometer entrance aperture to observe this effect within a reasonable time. At this point the energy resolution on the zero loss peak should be nearly optimal for this type of system.

Note: The $\pi*$ transition peak on the rise of the K edge should be resolved after Experiment 26.6, and during all PEELS acquisitions.

26.5 Near-Edge Fine Structure

When an inner-shell electron is ionized by the electron beam, it may not completely escape the atom but reside in an outer energy level still associated with that atom. These outer energy levels can be perturbed by the atomic environment. Thus, the low-energy side of the edge can possess a fine structure of resolved and unresolved features related to chemical bonding effects.

Experiment 26.7: Near-Edge Fine Structure of Carbon. Move to an area of the holey carbon film containing a thin graphite flakes hanging over a hole. Collect two spectra:

one from an area of graphitized carbon and one from an area of amorphous carbon film only. What are the differences in the near edge fine structure? Could these differences be used to routinely distinguish graphite from amorphous carbon in a "real" specimen?

26.6 Specimen Thickness Effects

Two limitations of the EELS technique are that ionization loss edges can be observed clearly only in very thin specimens (usually << 50 nm at 100 kV) and the intensity of the edges may be very low. Features in the low-loss region of the spectrum, particularly the plasmon losses of certain metals, can be observed in thicker foils and are usually much more intense. In fact, the ratio of the first plason loss to the E_o peak provides a convenient way of assessing specimen thickness.

Experiment 26.8: Shape of the Carbon Edge. *Collect a spectrum at the carbon edge from a thick region of carbon. Compare the shape of the carbon K-edge with the previous spectra from a thin carbon film. What causes the second hump on the edge?*
In PEELS, observe the spectra in "view" mode and simply translate the sample from a thin to thick region and observe the change in the carbon K-edge shape.

Experiment 26.9: Plasmon Losses in Aluminum. *Change to the wedge-shaped disc specimen of pure aluminum. Using entrance aperture 2 (2 mm), collect a spectrum showing plasmon loss peaks (at 15, 30 eV, etc.) and the aluminum $L_{2,3}$ ionization edge (at 73 eV). Move to a thicker portion of the specimen and collect another spectrum. What effect does specimen thickness have on (a) the intensity and number of plasmon loss peaks and (b) the relative intensity of the L2,3 edge compared to background?*
In PEELS, as in Experiment 26.8, observe the spectrum in "view" mode and translate the sample from a thin to a thick region to observe the change in the low-loss region.

Experiment 26.10: Measurement of Specimen Thickness. *Estimate the specimen thickness t in the two areas for which you just collected spectra by using the fact that the mean-free-path L_p for plasmon excitation in aluminum is 120 nm at 100 kV, 140 nm at 120 kV, 180 nm at 200 kV and 220 nm at 300 kV. For reasonably thin specimens the following relationship may be used:*

$$t = L_p \left(\frac{I_p}{I_o}\right)$$

where I_p and I_o are the intensities of the first plasmon loss and E_o peaks, respectively.

Experiment 26.11: Thickness Required for Quantitative Microanalysis. *Insert the specimen of fine BN particles supported on a carbon film. Collect a spectrum from a thin flake of BN hanging over a hole in the carbon film. Collect a spectrum from a thicker particle of BN. What changes occur in the spectrum? Find a particle of thickness such that the low-loss region is approximately one-tenth the intensity of the zero loss peak. This is the maximum thickness for microanalysis using ionization loss edges.*

26.7 Quantification of the Ionization Loss Spectrum

This section demonstrates the procedures by which quantitative EELS data may be obtained and analyzed.

Experiment 26.12: The Atomic Ratio of Boron to Nitrogen in BN.
(a) Find a thin flake of BN hanging over a hole in the carbon film.
(b) Set up the spectrometer to collect a relatively high spectrum intensity in the ionization loss region. Select a large spectrometer entrance aperture (3 or 5 mm). Open the slit so that the zero loss peak has a small flat top when viewed on the video monitor.
(c) Collect two spectra including the B and N edges (scan the energy loss range -100 to +500 eV) with collection angles at the specimen of (1) $2\beta_s \sim 100$ mrad (no objective aperture) and (2) $2\beta_s \leq 10$ mrad. Since only the B/N atomic ratio will be sought in this experiment, data in the low loss range ($\Delta E = <50$ eV) will not be needed.
(d) Subtract the background from each spectrum using the quantitative EELS software of the MCA/computer system.
(e) Estimate the B/N ratio by obtaining the intensities (I_K) in the B and N edges and using the relationship

$$\frac{N_B}{N_N} = \frac{\sigma_{N_K}(\Delta,\beta)}{\sigma_{B_K}(\Delta,\beta)} \frac{I_{B_K}(\Delta,\beta)}{I_{N_K}(\Delta,\beta)}$$

Where $\sigma_{ik}(\Delta,\beta)$ is the K-shell ionization cross-section (usually in dimensions of cm²/atom) for element i and N_i is the number of atoms (per cm²) in the analyzed volume. Note that $\sigma_{ik}(\Delta,\beta)$ is given as a partial ionization cross section, i.e., a function of the energy window (Δ) and the collection angle (β). Use $\Delta=20$ eV first and then 80 eV. Partial ionization cross-sections calculated on a "hydrogenic" model can be obtained from a computer program called SIGMAK [1,2]. Enter your B/N ratios in the table below:

	$\Delta = 20$ eV	$\Delta = 80$ eV
$2\beta = 100$ mrad	B/N =	B/N =
$2\beta = 5.8$ mrad	B/N =	B/N =

How well does the B/N ratio agree for the two collection angles and two energy windows?

26.8 Energy Loss Imaging (Optional, SEELS Only)

Experiment 26.13: False EELS Imaging. The EELS system may be used to collect an image using boron K, nitrogen K, carbon K, and background electrons collected within windows over the regions of interest. Does this "EELS image" represent true variations in composition or just thickness variations?

References
[1] R. F. Egerton, *Ultramicroscopy* **3** (1978) 243.
[2] P. Rez and R. Leapman, *Analytical Electron Microscopy*, San Francisco Press (1981) 181.

Laboratory 27

Convergent Beam Electron Diffraction

Purpose

The objective of this laboratory session is to introduce the principal method of obtaining electron diffraction information from regions smaller than the limit of conventional selected area diffraction (SAD), about 0.5 µm in diameter. The technique is termed convergent beam electron diffraction (CBED). We will see that CBED patterns can be used to generate much more information than can be obtained from SAD patterns.

Equipment

1. A STEM (or TEM) that can form CBED patterns.
2. Double-tilt specimen holder.

Specimen

Thin foil specimen of austenitic stainless steel (annealed before thinning to produce large grains), or thinned disc of silicon.

Time for this lab session

Two to three hours.

27.1 CBED Pattern Formation

The most useful CBED information is obtained from zone axis patterns (ZAPs) where the electron beam is directed down a relatively low-index crystallographic direction (for example, <001>, <011>, <111>, <114>, etc.). The specimen should be tilted until an appropriate orientation is obtained. Whenever a convergent STEM electron beam (large $2\alpha_s$) can be focused onto the specimen, a CBED pattern will be formed in the back focal plane of the objective lens. A CBED pattern with nonoverlapping discs and fringe contrast variations within the discs is called the Kossel-Möllenstedt or K-M pattern. Our task is to view this and other types of CBED patterns at appropriate camera lengths to see the rich information they contain.

Experiment 27.1: Setup in STEM Mode. Set the sample of austenitic stainless steel in the eucentric position, and obtain a STEM BF image of a reasonably thick (0.1-0.2 µm) region using a 10-20 nm probe size, a camera length of about 800 mm, and a C_2 aperture size of <50 µm. Focus the image carefully. Remove the STEM detector, and if necessary, lower the TEM screen. Stop the probe to observe the CBED pattern on the TEM screen. An array of CBED discs should be visible. If possible, tilt the specimen to a low index zone axis. Use the binoculars to focus the detail in the central spot. Defocus the pattern with the objective lens and observe a shadow image of the specimen. Go through focus and observe the image inversion. Return to the convergent beam pattern by adjusting the focus such that the shadow image information expands to infinite magnification.

In a dedicated STEM with no TEM screen, the pattern is viewed through a TV camera focused onto a small screen that intercepts the pattern beyond the objective lens. Otherwise all aspects of this CBED laboratory are the same as for a TEM/STEM or DSTEM.

Experiment 27.2: Effect of Electron Optical Variables. With a zone axis pattern on the viewing screen perform the following adjustments and note their effect:
 (a) Observe the effect on the zone axis CBED pattern of small adjustments to the specimen traverses and the STEM probe traverses (x-y beam deflection potentiometers).
 (b) Move the C_2 variable aperture off-axis and observe the effect on the pattern symmetry. [In a DSTEM, move the real objective aperture (if inserted) or the virtual objective aperture (VOA).]
 (c) Change the probe convergence angle $2\alpha_s$ by changing the C_2 aperture (or VOA) size. Which aperture produces a K-M pattern in which the discs just touch?
 (d) Change the camera length to larger and smaller values. Which camera length is best to view the zero-order Laue zone (ZOLZ)? the higher-order Laue zones (HOLZs)?
 (e) In a TEM/STEM, activate the C_2 lens (which is usually off in STEM mode on a TEM/STEM) and change the value of $2\alpha_s$ by changing the C_2 lens strength. Compare the maximum convergence angles in STEM mode (C_2 off) and TEM mode (C_2 on).

27.2 Kossel-Möllenstedt Fringes for Thickness Determination

Dynamical diffraction information is observed within the ZOLZ discs if the specimen is thick enough. In thicker specimens tilted to a two-beam condition, the spacing of K-M fringes within the ZOLZ discs may be used to measure thickness.

Experiment 27.3: Thin and Thick Foils. Move the specimen to a very thin region and then to a very thick region and observe the change in information in the diffraction discs. It may be necessary to operate at the maximum camera length to demonstrate these effects. Choose an intermediate thickness and remain there for the rest of the demonstration.

Experiment 27.4: K-M Fringes for Thickness Measurements. Tilt the foil to the [001] or [011] zone axis. If the chosen region of the specimen is defect-free, of suitable thickness (~100 nm), and tilted slightly off the zone axis to a two-beam diffraction condition (say 200 or 220), it is possible to observe parallel bright and dark intensity fringes in the strongly excited hkl diffraction disc. The value of $2\beta_s$ must be adjusted to be $< 2\theta_B$ so the diffraction discs do not overlap. The fringes are termed "Kossel-Möllenstedt fringes," and they arise from the same dynamical scattering effects that cause bend contours and thickness fringes in TEM images. The fringes increase in number with increasing thickness. From the fringe spacing both the extinction distance ξ_g (hkl) and the foil thickness t can be determined.

27.3 Higher-Order Laue Zone Lines (Defect HOLZ Lines)

If zone axis conditions are set up again (a relatively low symmetry axis such as <114> is a good choice) and a high camera length condition chosen, similar K-M type dynamical intensity fringes can be observed in the 000 (BF) disc. Furthermore, sharp dark lines may be observed crossing the 000 disc. These dark lines are termed "higher order Laue zone" (HOLZ defect lines or simply HOLZ lines. They are due to elastic scatter from hkl planes not in the zero order layer of reciprocal space (i.e. planes not parallel to the beam). These lines contain three-dimensional symmetry information about the specimen and are very sensitive indicators of the exact crystal orientation and lattice parameters.

Experiment 27.5: Observation of HOLZ Lines. At high camera length, tilt the specimen and observe the HOLZ line behavior. Observe the effect of specimen thickness on the HOLZ line intensity. Change the accelerating voltage and see what happens to the HOLZ lines.

27.4 Higher-Order Laue Zone Diffraction Maxima (HOLZ Rings)

The diffraction spots in the first Laue zone form a ring where the Ewald sphere cuts the next layer up in the reciprocal lattice. The diameter of this ring can often provide d-value data for crystal planes perpendicular to the beam. From the ZOLZ, and all SAD patterns, d-values are only obtained for planes parallel to the beam. Each bright (excess) HOLZ maximum can be associated with one of the dark (defect) HOLZ lines in the 000 disc. However, excess HOLZ maxima are of low intensity and generally require overexposure of the ZOLZ region in order to see them. Cooling the sample to liquid N_2 temperatures also improves the intensity in the HOLZ rings.

Experiment 27.6: Observation of HOLZ Rings. Under zone axis conditions, return to a low camera length setting. If a suitable low-symmetry, high-index (e.g., <114>) zone axis has been selected, and the C2 aperture is smal enough that it may be possible to observe a ring of spots around the basic zero order (ZOLZ) pattern. This ring of spots is a HOLZ ring containing diffracted discs from HOLZ planes in which $hU + kV + lW > 0$ where <UVW> is the zone axis. By measuring the diameter of the ring of spots the lattice spacing in the beam direction may be obtained [1].

27.5 Kossel Patterns

All the previous information has been obtained with a choice of C_2 such that $2\alpha_s < 2\theta_B$ so that disc overlap did not occur (so-called K-M conditions). With a very large convergence angle, the discs overlap and contain information from upper (higher-order) Laue zones. This pattern is often termed a "Kossel pattern," because it is identical in symmetry to the x-ray Kossel pattern from the same specimen.

Experiment 27.7: Observation of a Kossel Pattern. Use a large C_2 aperture (>100 μm diameter); then all the diffraction discs overlap and the resultant pattern contains just HOLZ lines which exhibit the full three-dimensional crystal symmetry. Are there rotational symmetry axes or mirror planes in the Kossel pattern?

27.6 How to Record Convergent Beam Diffraction Patterns

Since most AEMs still record diffraction information on photographic film, with the inherent problems of limited dynamic range, several exposures of each of the types of CBED patterns must be taken.

Experiment 27.8: Take a CBED Pattern. To obtain all the information available in CBED patterns it is necessary to:

(a) Obtain the appropriate zone axis pattern and make the pattern as symmetrical as possible.

(b) Record a CBED pattern at low camera length and at high camera length to obtain both the HOLZ ring and the defect HOLZ lines in the 000 disc. What exposure times should be used?

(c) Record patterns with both low $2\alpha_s$ and high $2\alpha_s$ to get K-M and Kossel conditions, respectively.

(d) Tilt to two-beam conditions to permit determination of the specimen thickness. Record the pattern.

27.7 Determination of Point Group Symmetry

Any crystal may be classified as belonging to one of the 32 crystal point groups. For a single crystal analyzed in the AEM, the three-dimensional diffraction information derived from the CBED pattern may be placed in categories called "diffraction groups" [2]. Diffraction groups determined for different orientations may be cross-referenced to yield a unique point group for the crystal. A simple example of such an analysis is given below.

Experiment 27.9: Point Group of Austenitic Stainless Steel. Tilt a large grain of the crystal to an orientation with fourfold symmetry, that is down the [001] zone axis. Take CBED patterns of the bright field 000 disc and the Kossel pattern. From the symmetry of the two patterns determine the diffraction group for the [001] zone axis from Table 27.1 (from Buxton et al.). Repeat for the [111] and [011] zone axes. Take the six candidate diffraction groups to Table 27.2 (from Buxton et al.) to determine the one point group that correlates to a candidate diffraction group for each of the orientations analyzed.

References

[1] M. Raghavan, J. C. Scanlon, and J. Steeds, *Metallurgical Transactions* **15A** (1984) 1299.

[2] B. F. Buxton, J. A. Eades, J. Steeds, and G. Rackman, *Philosophical Transactions of the Royal Society of London* **281** (1976) 181.

Part V

GUIDE TO SPECIMEN PREPARATION

Laboratory 28

Bulk Specimens for SEM and X-Ray Microanalysis

Purpose
　　The purpose of this laboratory is to prepare samples of metallic, ceramic, polymeric, and biological specimens for examination and analysis in the SEM. The organization is such that under each type of material sample preparations are discussed for surface topography (e.g., fracture surface), microstructural analysis (e.g., phase morphology), and x-ray microanalysis. Special procedures for semiconductor devices, polymers, and biological samples are also considered. The objective is to provide a brief outline and enough general references to enable the reader to produce all of the specimens used in this workbook. The outlined methods should not be considered comprehensive, however, and the reader is strongly urged to consult the references listed. For further discussion, see *SEMXM*, Chapters 9-12.

28.1 Metals, Ceramics, and Minerals

Equipment
1. Slicing and wafering equipment.
2. Cut-off saws, wire saws, diamond wheel wafering saws, etc.
3. Facilities for mounting small specimens in polymeric compounds.
4. Metallographic/petrographic polishing equipment (lapping wheels, polishing wheels, polishing compounds, electropolishing equipment, etching supplies).
5. Conductive paint such as silver paint or colloidal graphite.
6. Coating equipment such as evaporator and sputter coater (see Laboratory 30).

Specimens
　　Examples of specimen types will be drawn from Laboratories 1-23.

Time for this lab session
　　Approximately one hour per specimen, unless otherwise noted, after satisfactory technique has been developed. Technique development, however, may take weeks or even months.

28.1.1 Surface Topography

Typical Specimen: Aluminum fracture specimen used in Laboratories 2 and 4.

　　These samples are among the easiest to prepare for examination in the SEM. The key consideration is to insure that the sample surface be clean and undamaged (see *SEMXM*, pp. 447-449). The following steps will serve as a guide:

Step 1: Cut large specimens to fit specimen holder.

Step 2: Degrease the specimen in a solvent such as clean acetone. An ultrasonic cleaner is useful. A final wash with methanol will remove any remaining surface film. Ensure that the solvent does not compromise the integrity of the surface. *Warning: Flammable solvents in an ultrasonic cleaner may be hazardous. Read all safety and disposal information pertaining to the solvent.*

Step 3: Mount specimen on specimen stub either mechanically or with glue, conductive paint or sticky tape. Run a track of conductive paint from the specimen to the stub to insure good electrical contact.

Step 4: Dry specimen in a clean, low-temperature (75°C) oven. The sample should never be "pumped dry" in the SEM chamber or airlock.

Step 5: Coat specimens that are elecrical insulators with a thin conductive layer. See Laboratory 30 for procedures. These procedures are summarized schematically in Figure 28.1

28.1.2 Specimens for Microstructural Morphology and Metallography

Typical Specimen: Multiphase tool steel, basalt, or Raney nickel specimen used in Laboratories 3, 8, 14, and 17.

The SEM may be used as an extension of the light optical microscope on typical metallographic specimens. Often the same specimens prepared for light metallography may be examined directly in the SEM.

Step 1: Slice specimen into pieces small enough to be placed into an appropriate 1-1.5" metallurgical mount. If the cut surface is the one to be ultimately polished and examined, the cut should be made with a slow speed diamond saw or slurry wire saw.

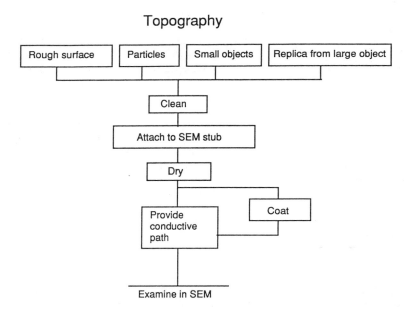

Figure 28.1. Flow chart showing the preparation of bulk specimens to examine topographical features. Inspired by Goodhew [21].

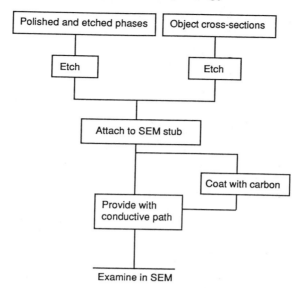

Figure 28.2. Flow chart showing the preparation of bulk specimens for microstructural morphology studies. Inspired by Goodhew [21].

Step 2: Mount using standard metallographic practice [1-3]. Either an epoxy, cold mount, Bakelite or in some cases fusible metal alloy should be employed. Although a specimen itself may be conducting (e.g., an alloy steel), it may be useful to mount the specimen in a conductive epoxy to prevent charging during examination.

Step 3: Polish using standard metallographic practice [1-6] and appropriate polishing compounds. A typical grinding and polishing sequence might be 320, 400, 600 SiC papers followed by 3 μm, 1 μm, 0.25 μm alumina or diamond polishing compounds. This work may be done by hand or by using rotating wheels. Be sure to carefully clean the entire mount in soap and water before moving to the next smaller polishing compound. Electropolishing may be needed to remove surface damage and smearing of one phase over the next.

Step 4: Etch the surface to bring out the phase structure using standard chemical, electrochemical, or ion etching procedures [1, 3, 6]. The detail in these samples can be as small as a few tens of nanometers in the secondary electron mode (depending on the material). However, often the degree of etching required for SEM observation is less than for light metallography. Heavy etching may generate artifacts which could be confused with the true microstructure. Even without etching, the phase structure may be apparent in backscatter images as a result of atomic number contrast. Resolution in BSE imaging mode (100-300 nm) is generally inferior to that obtainable with secondary electrons on etched samples, but still better than that obtainable by most light optical metallography (500 nm). Etched samples should never be used for x-ray microanalysis (see Section 28.1.3). Rinse and dry thoroughly.

Warning: The solutions used for chemical and electrochemical polishing and etching may be dangerous. They may be highly corrosive, flammable, or explosive. Use only in a hood, with appropriate safety glasses and with proper training. Contact your local safety personnel for detailed instructions for handling these materials.

Step 5: Attach the specimen to the SEM stub either mechanically or by an adhesive cement, conductive paint, or tape. Run a track of conductive paint to the stub and from the stub to the stage to insure good electrical contact. Dry thoroughly before placing in the SEM.

Step 6: Coat the surface of insulating specimens with a thin conductive film to prevent charging. See Laboratory 30. These procedures are summarized schematically in Figure 28.2.

28.1.3 X-Ray Microanalysis

Typical Specimen: Specimens for microanalysis used in Laboratories 5-7, 18-21, and 23.

For qualitative analysis the procedure is similar to Section 28.1.1. However, to eliminate systems peaks of aluminum and copper, it is useful to use carbon planchets or stubs for specimen mounting. For quantitative analysis the procedure is the same as described in Section 28.1.2, except that these samples must be "perfectly" flat, which precludes etching (see *SEMXM*, pp. 449-453). If etching is needed to allow one to find regions of interest in the sample, the specimen should be etched, marked with microhardness indentations, photographed, and repolished. The region of interest may then be found between the microhardness marks by using atomic number contrast with the BSE detector. The conductive coatings should be of a light element to reduce absorption effects. Carbon and aluminum are good choices. Care is required to prepare the unknown sample and the microanalysis standards in an identical manner. The best way to do this is to mount the unknown and the standard in the same mount if possible. See Figure 28.3.

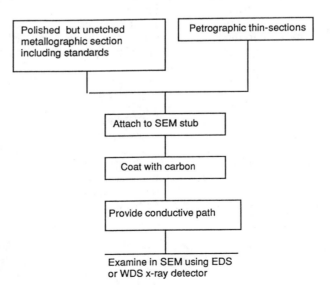

Figure 28.3. Flow chart showing the preparation of metallographic and petrographic samples for x-ray microanalysis. Inspired by Goodhew [21].

28.1.4 Petrographic Thin Sections

In ceramics and geological materials, some phases are transparent at optical wavelengths. Thin sections, which are typically 25 to 50 µm thick, allow one to examine these materials with transmitted light in the SEM or microprobe with a light optical microscope and simultaneously take x-ray data. The samples are prepared by slicing a wafer to the appropriate thickness, polishing one side, and attaching that side to a glass slide with a glue which is stable in a vacuum, and then polishing the opposite side [8-11]. These specimens must be coated, usually with carbon so as not to interfere with microanalysis (see Laboratory 30).

28.1.5 Semiconductors and Devices

Most semiconductor specimens can be prepared using the general methods of Section 28.1.2. However, special surface preparation is often needed to see certain effects in voltage contrast, electron channeling, and charge collection microscopy [7]. Electrical connections to devices are described in Laboratory 15.

Additional Equipment: Specialized coating facilities for Schottky barriers (UHV deposition of gold, titanium).

Electron Channeling (Laboratory 13). A "perfectly" flat, strain free surface is required for electron channeling. To achieve flatness, it is often necessary to polish the sample metallographically through 0.06-µm alumina (or other abrasive) as described in Section 28.1.2. The mechanical polishing will leave a strained surface layer (Bilby layer) which will destroy the channeling effect. This layer must be removed either chemically or by electropolishing. The chemical etching solution required depends on the exact nature of the material. Dilute solutions of HF can often be used for silicon.

Inactive Devices (Laboratory 15). There are two principal reasons to image inactive devices: to examine p-n junctions with EBIC and to examine the packaging details of the circuit (bond pads, interconnect metallization lines, passivation layer integrity, etc.). Clean loose dust and grease from the sample. If the passivation layer is to be examined, a conductive coating may be required (see Laboratory 30), although it may be possible to examine the specimen at low voltage (about 1 kV) without a coating. For analysis of p-n junctions, metallizations lines, etc., it will be necessary to remove the passivation layer. The removal technique depends on the nature of the passivation layer. HF is often used for glass and organic solvents for polymeric based passivation layers. Plasma etching is also occasionally used. The sample must be properly attached to the stage and grounded. In some cases it is possible to examine metallization lines through the passivation glass by using the high-energy backscattered electron signal. An operating voltage in excess of 25 kV is required. Resolution of such images is poor, but will allow identification of voids in metals.

Active Circuits (Laboratory 15). Active circuits are often observed with voltage contrast. The specimen must be electrically isolated from the specimen stage and activated by leads reaching the specimen from the electrical feedthrough. The activation of the circuit may be static (conventional voltage contrast) or dynamic (stroboscopy) by connection to an external signal generation.

Schottky Barriers for EBIC Imaging of Semiconductors (Laboratory 15). If EBIC is to be used to examine a semiconductor material that is not in a device with a depletion region, a metal layer called a Schottky barrier may be used to create a depletion region near the top surface. The deposition of high-quality Schottky barriers requires great care. The surface must be properly cleaned and polished, similar to that required for electron channeling, and the barrier metal must be deposited in a very clean evaporator, free from any trace of hydrocarbon contamination (use liquid N_2 trapped diffusion pump or turbo pump). The barrier must be continuous, but thin (20-40 nm). For intrinsic or n-type material, gold seems to be the best barrier material. Titanium seems to work best for *p*-type material. A significant amount of experimentation may be required to produce a suitable Schottky barrier for a given situation. A detailed description of the procedure used to deposit good Schottky barriers is described in *ASEMXM*, page 78, Table 2.3.

28.2 Polymers

Polymers, plastics, and other nonhydrated or partially hydrated organic samples require special handling for examination in the SEM to:
1. Expose the interior of the sample, if necessary;
2. Remove water without "deflating" the polymer;
3. Reduce electron beam damage of the specimen;
4. Increase contrast in second phases, if present.

Equipment
1. Microtome, razor blades and scalpels.
2. Fine forceps.
3. Glassware and chemicals found in a general electron microscope preparation laboratory.

Typical Specimens: (1) Single-phase polymer to be fractured or polished as used in Laboratories 10, 11, and 16; (2) Two-phase polymer to be fractured or polished as used in Laboratories 8 and 24.

Many polymers contain water. Simple air drying can be used for rigid polymers. However, soft and pliable polymers will tend to distort as the water evaporates since water has a relatively high surface tension. In these cases, the water may be removed by dehydrating through a series of ethanol or acetone solutions (care should be taken not to dissolve the polymer) or alternatively by critical point drying or freeze drying (see Section 28.3). Cryofracture and replication may be used to observe the polymer interior and avoid the drying step altogether. General procedures for preparing polymers may be found in reference [12].

28.2.1 *Specimens for Surface Morphology*

Both dehydration in vacuum and electron beam damage can severely alter the morphology of a polymer surface. At low SEM magnifications both problems can be avoided by coating the specimen with a thick (>20-nm) self-supporting layer of gold or Au-Pd (see Laboratory 30). While the basic surface morphology is preserved in the gold casing, all fine details of the surface are lost. To retain fine surface details the following steps will be helpful:

Step 1: Expose the interior surface for examination. Methods for this include the following:
Fracturing. Brittle polymers will fracture along a surface of least resistance. The samples can be trimmed by repeated, more precise fracturing. Soft materials must be cooled well below their glass transition temperature to allow them to be fractured. This is best achieved by immersing samples in liquid nitrogen and fracturing by impact. Tough polymers may be split (fractured) and peeled back along their long axis.
Polished Bulk Samples. Hard polymers and plastics can often be polished [13-14]. Samples, if very small, may first need embedding in an epoxy resin. It is important to check the solubility of the material in the resin. The hardened material is cut with a diamond saw to produce a flat surface. This surface is then polished by using standard metallographic procedures with graded bits of silicon carbide, aluminum oxide, chromium oxide, or diamond grits in water-based slurries (see Section 28.1.2).

Polymers and Biological Samples

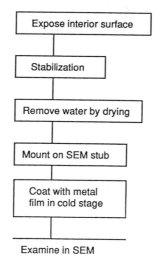

Figure 28.4. Flow chart showing the preparation of polymers and biological samples for the SEM. Inspired by Goodhew [21].

Sectioning. The aim of this procedure is to produce sections of the sample which are thin enough to be examined by a transmitted beam of electrons. Polymers and plastics have a very wide range of cutting characteristics. It may be necessary to embed small samples and fibers in suitable resins. Sections may be cut dry or wet using metal, glass, or diamond knives. Soft plastics, emulsions, elastomers, and polymers which absorb water, may be cut by using cryomicrotomy methods. The specimen in the chuck from which the sections have been cut has been planed to a smooth finish and may be observed in the SEM, while the sections themselves may be observed in the TEM (see Laboratory 29).

Etching. There are physical and/or chemical procedures which selectively remove one or more of the components in a polymer mixture. Whole molecules of material may be removed by dissolution. The physical procedures involving plasma, ion, and electron beam etching are generally less satisfactory than chemical procedures since they cause many uncontrollable artifacts. The effectiveness of solvent etching will depend largely on the polymer or plastic being studied and there is no one general method which can be recommended.

Step 2: Dry the polymer to remove water. Be careful about polymer solubility in organic fluids.

Step 3: Mount the specimen on an SEM stub with silver epoxy, conductive paint, or double-sided adhesive tape painted with a drop of conductive paint. Be careful that solvents in these preparations do not rise up onto the specimen by capillary action and degrade the surface to be imaged.

Step 4: Coat the specimen with a thin metal film to provide a conductive path to electrical ground (see Laboratory 30). Since many polymers are heat sensitive, they should not be exposed to excessive heat during the coating process. This is a particular problem with carbon evaporation. An Au-Pd sputter coater with a cold stage is often useful. See Figure 28.4.

28.2.2 Specimens for Analysis of Second Phases

Since the average atomic numbers of various polymeric materials (containing largely carbon, nitrogen, oxygen, and hydrogen) are very similar, second phases that contain different bonding arrangements of these same atoms cannot be distinguished by atomic number contrast. While heavy metal staining of the second phase is possible, second phases are most often observed in the SEM as morphological features in fracture surfaces, and the methods of Sections 28.2.1 should be used.

28.2.3 X-Ray Microanalysis

Specimen preparation here should follow the scheme outlined in Section 28.1.3. However, analysis of a heavy metal stain in the polymer that tags only one phase may be a way of locating the distribution of a second phase that selectively absorbed the stain.

28.3 Biological Samples

Very few bulk biological specimens may be placed directly into the high vacuum of an electron beam instrument and immediately provide significant information. Depending on the inherent stability of the specimen, varying degrees of sample preparation will be necessary to ensure that the maximum amount of information may be derived. As with other bulk SEM specimens, biological specimens must be free of foreign particles, stable in vacuum, stable in the electron beam, electrically conductive, and must be unaltered in chemistry and morphology. It is difficult to meet all these criteria at the same time and this guide does not intend to be all inclusive. Please consult the references. The discussion which follows provides an introduction to the main features of these preparative procedures. The rationales behind the use of particular procedures are discussed in the textbooks: *SEMXM* in Chapters 11 and 12 and *ASEMXM*, Chapter 8.

Equipment
1. Freeze dryer, critical point dryer.
2. General glassware and chemicals associated with an electron microscope laboratory.

Time for this lab session
Time is highly variable depending on the specimen. For example, from a few minutes (mounting an insect) to several hours (preparing tissue).

28.3.1 Skeletal and Hard Tissues

Typical Specimen: An insect with a hard exoskeleton, a piece of wood, or a piece of paper.

A popular SEM image is the head and body of an insect. This may be quite simply prepared and gives reasonably good details of the *surface features* of the insect.

Step 1: Mount the nonliving specimen on a specimen stub. For insects allow the legs to touch the stub surface that has been precoated with a thin layer of glue or conductive paint. For a sliver of wood or paper join the specimen to the stub with a layer of conductive paint.

Step 2: Coat the specimen with a relatively thick layer of gold or Au-Pd for low magnification observation (see Laboratory 30). Other hard tissues may be prepared in a manner similar to polymers (Section 28.2) or ceramics (Section 28.1).

28.3.2 Soft Tissue Preparation

Considerably more effort is involved in preserving a soft tissue specimen that may require fixation and removal of water [15-18]. Typical preparation of this type consists of the following:

Step 1: Selection and Cleaning. Most samples may be cut, sliced, sawed, or fractured, and in addition to reducing samples to a suitable size, these procedures also provide one of the best ways of exposing a clean surface which has the characteristics of the bulk sample. The natural or artificially exposed surface may require further cleaning.

Gentle cleaning is needed with biological and hydrated organic material where the main contaminants are usually derived from the sample itself, i.e., pieces of the specimen, mucus, body fluids, and exudates. Gentle washing in clean growth medium or a balanced organic buffer solution such as PIPES or HEPES will usually remove most surface contaminants. Care has to be taken that these treatments do not cause chemical losses from the sample. More persistent material may be removed by gentle washing in weak detergent solutions or solutions of an appropriate enzyme. Great care must be taken not to damage the surface and underlying tissues.

Step 2: Structural Stabilization. This forms the central part of the procedures used in preparing most biological and hydrated material for microscopy and analysis. For structural studies, the aim is to preserve the macromolecular architecture of the specimen and methods based on chemical fixation usually provide the best results. For analytical studies it is necessary to retain the complete chemical identity of the sample and the best results are obtained using low-temperature fixation which avoids the use of disruptive chemicals. There are many different recipes and those methods which work well for a particular specimen examined in the TEM will usually work equally well for the SEM. Refer to Chapters 11 and 12 in *SEMXM* for more details.

Step 3: Drying. Nearly all biological specimens will need drying before they may be examined in the electron microscope. The principal liquid which has to be removed is water, although some materials samples may have other organic liquids. Unless the specimen is very tough and rigid, e.g., wood, bone, some seeds, air drying should not be used. Water has a high surface tension and as the last traces of the liquid are removed the surface tension forces which develop will seriously distort soft and pliable surfaces. Samples may be dried by solvent drying or by critical point drying. Refer to Chapters 11 and 12 in SEMXM for more details, or consider environmental SEM (see Laboratory 16).

Critical point drying works on the principle that the dehydrating solvent which remains in the sample is replaced with a compound which is liquid at high pressures and room temperatures and turns to a gas as the temperature is raised slightly. If this replacement process takes place in a pressurized container, the pressure will increase as the temperature is raised and the compound will pass through its critical point. At the critical point, the phase boundary between the gas and liquid no longer exists and surface tension is zero. Under these conditions the drying occurs without sample distortion. The most convenient compound to use is carbon dioxide. The critical point drying process is carried out in specialized apparatus.

Warning: It is important that one is familiar with the operating procedures for the particular critical point dryer to be used. It is a potentially dangerous procedure as it involves the use of gases at pressures up to 1200 psi. Be sure to read the safety instructions supplied by the manufacturer. Contact your safety officer.

Step 4: Attaching the Specimen to the SEM Stub. Most specimens, down to a size of 1-2 mm, can be effectively attached to the sample stub by using conductive silver paint for morphological studies or carbon paint if elemental analysis is to be carried out. Organic glues such as epoxy adhesives or "super-glue" may also be used. Specimen stubs are made either of aluminum or copper alloys and provide good supports for morphological studies. If critical analytical studies are to be carried out, either the specimen support should be made of carbon or the samples should be attached to carbon discs which may in turn be fixed to metal stubs.

Step 5: Coating the Specimen to Prevent Charging. See Laboratory 30 for coating procedures appropriate for the low-magnification imaging (<10,000x) and high-magnification imaging (>10,000x).

28.4 Particles and Fibers

Equipment
1. Conductive paints (i.e., silver, aluminum, carbon).
2. Particle dispersant equipment (evaporative fluid, e.g., freon).
3. Coating equipment (evaporator, sputterer).

Specimens
The procedures described here will give satisfactory results for most specimens under most conditions. Each specimen is unique and will ultimately require specialized procedures, the details of which cannot be fully described here. Some of the specimens for Laboratory 22 may be prepared with these methods.

Time for this lab session
Usually less than one hour.

28.4.1 Particulates as a Special Case

The imaging and analysis of particles and fibers is a special case of sample preparation. Since the electron interaction volume is often larger than the particle itself, routine microanalysis of particles and fibers is not usually possible and special techniques are required (see Laboratory 22). Specialized specimen preparation techniques are also needed.

The techniques for the preparation of particle and fiber samples for scanning electron microscopy and x-ray microanalysis are to a great extent determined by the following characteristics of these materials:

(1) Because of the shape of particles and fibers, only a relatively small region of contact may exist between a particle or fiber and the substrate on which it rests. The particles or fibers may therefore not be stable mechanically unless special precautions are taken. Without a suitable method of attachment or trapping, particles may be lost during preparation, handling, and examination in the microscope.

(2) Particles and fibers are often composed of materials which are nonconducting, e.g., minerals, glass, textiles, wood fibers, or microorganisms. Exceptions exist, such as metal powders, but these are not the usual case. Thus, special attention must be paid to the problem of charging. When combined with condition (1), the need to minimize or eliminate charging becomes critical if actual loss of particles or fibers due to mutual repulsion of similarly charged objects is to be avoided.

(3) Particles and fibers often have one or more dimensions which approach the dimensions of the electron beam interaction volume in bulk material of the same composition, e.g., on the order of 1 µm. As a result, beam electrons are likely to penetrate through a particle

or fiber or scatter through the sides. The sharply tilted or curved sides of a particle or fiber may also enhance backscattering. After electrons have left the particle or fiber, they may interact with the sample substrate or other surroundings and generate spurious signals which may adversely affect imaging and/or x-ray microanalysis.

Taken together, these conditions suggest that suitable sample preparation techniques for particles and fibers must provide for trapping of the objects or at least provide a strong mechanism of adhesion in a medium which is conducting and which adds minimal detectable electron and/or x-ray signals (see *SEMXM*, pp. 453-458). Three general methods will be described in the following sections.

28.4.2 Large Particles

Particle collection techniques are described by DeNee [19]. A simple but effective technique for entrapping free standing particles is to place a drop of carbon paint on a carbon substrate and spread the drop to form a layer. The solvent of the carbon paint is then allowed to evaporate to near dryness. While the paint is still slightly tacky, the particles or fibers are simply dropped on the surface. The momentum from falling will embed the particles into the carbon paint. Alternatively, particles and fibers may be placed on the surface using fine forceps or an eyelash probe viewed through a binocular microscope. It is important that the carbon paint should not be so wet that the solvent and colloidal carbon can wick up onto the particle surfaces. With practice, the microscopist will develop a sense of timing to produce relatively clean surfaces. This method works best with large particles (dimensions > 10 mm). For microscopy, the carbon paint will provide a low backscattering/low secondary electron environment. For x-ray analysis, it is important to determine, as a blank, the x-ray spectrum generated by the carbon paint as it is dried on the surface of the carbon stub. In addition to the colloidal carbon, there may be additional elements at a trace level in the carbon paint. While the x-ray signals from these trace levels can usually be safely ignored, it is important to realize that the x-ray spectrum obtained from a small particle may be dominated by that from the substrate. While the characteristic x-ray peaks from the carbon paint/carbon planchet may be negligible if the carbon is of high purity, it is important to note that the bremsstrahlung continuum in the particle x-ray spectrum may arise from the paint/substrate, especially for particles with dimensions in the micrometer or submicrometer range. This effect will be important if a quantitative analysis procedure such as the peak-to-background method is used (Laboratory 22).

28.4.3 Small Particles

For more careful work, particularly with very small (< 5 μm diameter) particles, there is an alternative but still fairly simple approach for mounting the particles on carbon planchets. For particles smaller than about 40 μm in diameter, the adhesion due to surface charge in addition to the adhesion provided by the carbon coat is adequate to keep most particles in place. A bulk sample of fine particles can be dispersed on a carbon planchet by (a) placing the particles in a very dilute suspension in a fast-evaporating solvent such as Freon TF, (b) ultrasonicating the solution to keep the particles in suspension, and (c) pipeting a small aliquot of the particle suspension onto the carbon planchet and allowing the solution to evaporate [20]. The faster the solution evaporates, the less particle aggregation will occur. You can assist evaporation by placing the planchet and particle suspension under a heat lamp, although this may damage organic and biological samples. When the planchet is completely dry, it should be carbon coated (see Laboratory 30). The planchet should be made of high-purity graphite and polished to as smooth a surface as possible to act as a fine particle substrate. Pyrolitic graphite planchets are particularly good as substrates for particles.

28.4.4 Mounting on TEM Grids

An elegant solution to the problem of mounting small particles is to use a method for preparing particles for transmission electron microscopy (TEM). Small particles can be dispersed from a solvent such as water, alcohol, or freon, onto thin carbon films (~20 nm thick) carried on 3 mm diameter electron microscope grids. The deposited particles can be coated with a further thin carbon layer to trap them in a carbon film sandwich, which provides a good measure of mechanical stability and charge dissipation. In order to eliminate beam electrons which may penetrate the thin carbon film and the particle itself and thereby continue to generate spurious signals, the grid may be placed over a blind hole slightly smaller than the grid diameter. This blind hole is drilled into a thick (1-cm) carbon block. The resulting grid mount/carbon block is then placed in the SEM. Alternatively, this TEM sample preparation is ideal for examination in a scanning transmission electron microscopy (STEM) experiment carried out in an SEM (see Laboratory 9). In this case, the specimen grid is placed in a special STEM stage which permits an open area underneath the specimen plane for mounting of the STEM detector.

References

[1] G. F. vanderVoort, *Metallography: Principles and Practice*, McGraw-Hill, New York (1984).
[2] L. E. Samuels, *Metallographic Polishing by Mechanical Methods*, Third Edition, American Society for Metals, Metals Park, Ohio (1982).
[3] J. L. McCall and W. M. Mueller, *Metallographic Specimen Preparation: Optical and Electron Microscopy*, Plenum Press, New York (1974).
[4] W. Tegart, *Electrolytic and Chemical Polishing of Metals in Research and Industry*, Second Edition, Pergammon, New York (1959).
[5] *ASM Metals Handbook, Metallography and Microstructures*, **5**, ASM International, Metals Park, Ohio (1985).
[6] ASTM, *Methods of Metallographic Specimen Preparation*, ASTM, STP 284, American Society for Testing Materials, Philadelphia, Pennsylvania. (1960).
[7] D. B. Holt and D. C. Joy, eds., *SEM Microcharacterization of Semiconductors*, Academic Press, New York (1989).
[8] Petrographic Sample Preparation, AB Met. Dig. **12/13**(1), Buehler Ltd., Evanston, Illinois (1973).
[9] D. G. W. Smith, ed., *Microbeam Techniques*, A Short Course Handbook, **1**, Minearological Association of Canada (1976).
[10] D. E. Caldwell and P. W. Weiblem, *Economic Geology*, **60** (1965) 1320.
[11] C. M. Taylor and A. S. Radtke, *Economic Geology*, **60** (1965) 1306.
[12] L. C. Sawyer and D. T. Grubb, *Polymer Microscopy*, Chapman and Hall, New York (1987).
[13] A. S. Holik et al., Grinding and polishing techniques for thin sectioning of polymeric materials for transmission light microscopy, *Microstructural Science*, **7** (1979) 357-367.
[14] U. Linke and W. U. Kopp, Preparation of polished specimens and thin sections of plastics, *Praktische Metallographie*, **17** (1980) 479-488.
[15] D. G. Robinson et al., *Methods of Preparation for EM*, Springer-Verlag, Berlin (1987).
[16] M. A. Hayat, *Fixation for Electron Microscopy*, Academic Press, New York (1982).
[17] M. A. Hayat, *Basic Techniques for Transmission Microscopy*, Academic Press, New York (1986).
[18] A. M. Glauert, ed., *Practical Methods in Electron Microscopy*, **3** and **5**, Elsevier-North Holland (1975-1977).
[19] P. B. DeNee, *Scanning Electron Microscopy/1978*, O. Johari ed., SEM, Inc., AMF O'Hare, Illinois (1978) 479.
[20] D. A. Walker, *Scanning Electron Microscopy*, O. Johari ed., SEM, Inc., AMF O'Hare, Illinois (1978) 185.

[21] P. J. Goodhew in "*Specimen Preparation for Transmission Electron Microscopy of Materials*, Materials Research Society Symposium Series Vol. 115, J. G. Bravman, R. M. Anderson, and M. L. McDonald eds. (1988) 51.

Laboratory 29

Thin Specimens for TEM and AEM

Purpose

The purpose of this laboratory is to prepare samples of metals, ceramics, and geological and microelectronic specimens for examination and analysis in the transmission electron microscope (TEM) and analytical electron microscope (AEM). Material forms include bulk, thin films, and particles. To achieve suitable electron transparency, and meet various requirements for electron diffraction and microanalysis, the thin region of the specimen must be <100 nm thick (or <20 nm thick in the case of EELS). The specimen may be self-supporting with thin areas around a hole in the center or it may be thin fragments or particles supported on a carbon-coated grid. Note that, in general, thin specimen preparation is a difficult *art* and that volumes have been written on the subject. This guide should only be used as a starting point. A serious user of the AEM *must* refer to the references given.

29.1 Metals, Ceramics, and Minerals

For penetration of a 100-kV electron beam, a typical metal or ceramic specimen must be <100 nm thick. However, the specimen usually is required to represent the unaltered bulk material in terms of structure, chemistry, and content of defects. If one also wishes this specimen to have large amounts of electron-transparent thin area, be flat and unbent, and be rugged enough to be easily handled, the task of making such a specimen from an arbitrary material is very difficult. Some specimens such as thin films may be examined directly with very little specimen preparation. For brittle materials, crushing or repeated cleaving may produce small electron-transparent regions. Obtaining thin specimens from bulk materials generally requires an initial mechanical thinning to the size and shape required for final thinning by chemical polishing, electropolishing, or ion milling. Ultramicrotomy is the standard thin sectioning technique in biology, but it is also useful in AEM investigations of metals and ceramics where chemical alterations of the sample may occur with other final thinning methods.

Equipment
1. Slicing and wafering equipment (cut-off saws, wire saws, etc.).
2. Specimen mounting facilities (mechanical fixtures, epoxies, and other polymers).
3. Metallographic/petrographic polishing equipment (lapping wheels, polishing wheels, polishing compounds).
4. Dimpler (especially for nonconducting specimens).
5. Ultramicrotome (for polymers, biological specimens, and very small pieces of inorganic materials).
6. Thinning equipment (ion beam thinners, etc.).
7. Electrochemical baths (jet polishers).
8. Crushing/cleaving equipment for brittle solids.
9. Coater for nonconductive specimens.

Specimens

Examples of specimen types will be drawn from Laboratories 24-27. The procedures described here will give satisfactory results for most specimens under most conditions. Each specimen is unique and will ultimately require specialized procedures, the details of which can not be fully described here.

Time for this lab session

Approximately one to several hours after a suitable thinning technique has been developed. Technique development may take a considerable length of time, weeks or even months.

29.1.1 Bulk Materials

Typical specimen: Electropolished aluminum foil used in Laboratories 24 and 26. Electropolished stainless steel used in Laboratory 27.

Each different material has special thinning requirements [1-4]. Figure 29.1 shows a schematic flow chart for the preparation of thin specimens from various starting forms. The reader is encouraged to seek additional information in the references. A brief outline of typical steps is as follows:

Step 1: Starting material. If possible start with a rod 3 mm in diameter or a sheet 200-300 µm thick.

Step 2: Make a disc. Slice the rod material into wafers 200-300 µm thick with a slow-speed diamond saw. Mechanical damage produced in this step can be minimized by using an acid saw to cut discs from a rod. Alternatively, use an ultrasonic disc cutter to produce 3-mm discs from 200-300 µm thick sheet material.

Step 3: Initial thinning. Grind the 3-mm disc to approximately 100 µm on papers of 400 and 600 grit followed by 3 µm, 1 µm, and 0.25 µm alumina or diamond polishing compound. Note that the damaged region beneath the surface extends to a depth about three times the diameter of the grid used. Therefore, the next grit must remove the damage from the previous grit. Turn specimen over. It is important to produce discs with smooth surfaces and parallel sides to facilitate later stages of thinning.

Step 4: Dimple the center of the disc 20-30 µm on each side with a commercial dimpling apparatus. This will insure that the perforation will be in the center of the disc. This step may not be necessary for electropolishing.

Step 5: Final thinning. This may be done by electropolishing (metals only) or by ion-beam milling (most materials).
  Electropolishing: A twin-jet electropolisher splashes acid electrolyte onto the center of the specimen while the specimen is held at a positive potential (anode) with respect to cathodes placed in front of each side of the specimen. Polishing conditions are established for each metal specimen by selecting the appropriate combination of electrolyte, temperature, potential (voltage), and flow rate (consult references and operation manual).
  Ion-beam milling: A 2-6-kV ion beam impinges on the specimen at a low angle (<12°) while the specimen is rotated. For metals and many ceramics, argon ions are satisfactory, while iodine ions are often used for semiconductors. Since the ion beam can generate crystalline defects in the specimen, it is advisable to start with the thinnest possible polished disc to reduce ion milling time to a minimum. Many specimens should be kept at low temperatures to achieve the best results (consult references and operation manual).

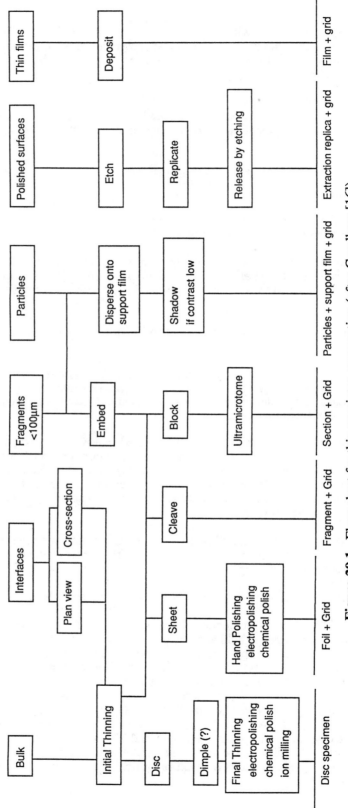

Figure 29.1. Flow chart for thin specimen preparation (after Goodhew [16]).

Step 6: Coat nonconducting specimen with carbon to avoid charging (see Laboratory 30).

*Warning: The solutions used for chemical and electrochemical polishing may be dangerous. They may be highly **corrosive, flammable,** or **explosive**. Use only with proper training and in appropriate facilities. Contact your local safety personnel for detailed instructions in handling these materials.*

29.1.2 Thin Films and Electronic Devices

Thin films deposited by sputtering, evaporation, CVD, etc., may be examined in plan view or in cross section. Electronic devices are essentially composites manufactured of thin films. To examine in plan view, the film is thinned from both sides first mechanically then with the ion mill. Depending on the film and the substrate materials, it may be necessary to remove the film from the substrate prior to final thinning. A variety of mechanical and chemical means may be used to remove the film from the substrate. Preparing a cross-sectional specimen is more difficult [5-6]. One method is to glue two samples together face-to-face with epoxy and then mount the sample on epoxy. The mounted sample is then sliced to form a 200-300 µm thick cross-section wafer which is ground to a thickness of 100 µm with fine abrasive paper. After dimpling, the final thinning is done with an ion thinner.

29.1.3 Cleavage

Certain ceramics will cleave with a razor blade to reveal thin areas along specific crystallographic directions. The specimen is first prethinned to <59 µm using fine 600 grit grinding paper. Score one side along the cleavage direction and place the scored side down on several layers of filter paper. Cleave by pressing down behind the scoring with a sharp point. The cleaved piece may be glued to an electron microscope grid to view the thin area at the very edge.

29.1.4 Particles

Small particles can be examined by placing them directly on a carbon-coated TEM grid [7]. Use a beryllium or carbon grid for analytical work, not copper. Crushed samples of brittle materials may be examined in this way if enough thin area is obtained at the edges. Occasionally, they can be sprinkled directly on the grid, but it is often better to ultrasonically disperse them in a fluid, such as freon, then place a droplet on the grid. When the liquid evaporates, the particles should be widely dispersed and not agglomerated. For larger particles, it may be desirable to cross-section them. One procedure is to mix the particles into epoxy and centrifuge the mixture to get a high particle density. After hardening, the epoxy can be sliced, dimpled, and ion thinned. Small particles embedded in a matrix may be extracted and analyzed using extraction replica techniques.

29.1.5 Ultramicrotomy

Soft metals (magnesium, aluminum) may be cut directly with a diamond-knife microtome [8]. Harder metals and ceramics may also be cut with a microtome if very small (<100 µm) pieces are embedded in epoxy first (see Section 29.2.3).

Warning: Diamond knives are expensive and easily damaged. Do not use a diamond knife without proper instruction.

29.1.6 Specimens for Microanalysis

Typical specimen: Fe-Ni alloy used in Laboratory 25.

Artifacts from specimen preparation often alter the elemental compositions of the thin specimen [8]. For instance, elements may be differentially redeposited on the surface during electropolishing. This problem may be corrected by a few minutes of ion beam cleaning in a clean vacuum. The problem may be avoided by using cleavage or ultramicrotomy to obtain thin specimens instead of either electropolishing or ion beam thinning.

29.2 Polymers

Equipment
1. Ultramicrotome with diamond or glass knife, and possibly a cryo-attachment.
2. Dehydration and embedding facilities (for bulk samples).
3. Staining media (e.g., OsO_4).
4. Coating facilities (evaporator/sputter coater).
5. Freeze dryer or critical point dryer.

Specimens
The procedures described here will give satisfactory results for most specimens under most conditions. Each specimen is unique and will ultimately require specialized procedures, the details of which cannot be fully described here.

Time for this lab session
Extremely variable, from minutes (for particles in suspension) to several days for successful thinning of bulk polymers.

29.2.1 Dispersions of Particulates

Solutions of polymers or suspensions of polymer particles may be examined by dispersing the liquid on a carbon-coated grid [9]. While spraying and atomizing may be used, the simplest method is to place a drop of liquid on the carbon and allow the solvent to evaporate. If the particles are agglomerated, it may be useful to ultrasonically disintegrate the agglomerate before dispensing onto the carbon film. Evaporated metal (e.g., Au-Pd) deposited at an angle of 30-40° will provide additional contrast by "shadowing" the particles.

29.2.2 Thin Film Coatings

Thin films of polymer may be cast on a glass slide by dipping the slide into the solution or by casting the film on the surface of an immiscible liquid [9]. Electron transparent films may also be drawn from molten polymer on a glass slide.

29.2.3 Ultramicrotomy

Typical specimen: Thin section of stained two-phase polymer used in Laboratory 24.

The bulk structure of a polymer may be examined in thin sections cut to thicknesses of 30-100 nm with glass or diamond knives in an ultramicrotome [10-11]. The technique is similar that used in biology (see Section 29.3). However, polymers have a wide range of cutting characteristics. They may be soft and brittle like waxes, soft and tough like rubbers, hard and tough like nylon, or hard and brittle like Lucite. It may be necessary to embed small samples and fibers in epoxy resin of similar hardness. Soft plastics, emulsions, elastomers, and polymers which absorb water may be cut using cryomicrotomy methods.

Microtomy is a specialized procedure and cannot be fully described here, but the specimen usually is processed through the following steps:

Step 1: Starting material. Generally, small blocks of polymers are cut from the bulk. A typical site would be a cube 1-3 mm on a side.

Step 2: Staining. Most polymers are composed of low atomic number elements with little variation in mass-density to provide contrast. Electron dense stains which chemically bond to certain characteristic regions of the polymer provide enhanced contrast. Osmium tetroxide is often used to react with the carbon-carbon double bonds of rubber phases providing both enhanced electron contrast and hardening (cross linking) of the rubber. Many other stains are used to stain specific functional groups. Staining may also be performed on the final cut sections of polymer.

Step 3: Drying. While many polymers may be dried in air without an alteration in their structure, some polymers tend to collapse if water is removed rapidly in air. In the latter cases the water may be replaced by epoxy by vacuum impregnation, or removed by critical point drying or freeze drying.

Step 4: Embedding. A small piece of polymer is placed in the bottom of a plastic embedding capsule. Epoxy resins may be altered in formulation to match the hardness of the polymer. Curing times and temperatures vary with individual products. The hardened epoxy block may be removed from the capsule.

Step 5: Block trimming. The plane of the specimen is revealed at the end of the block by trimming with a razor blade. The epoxy is then trimmed away to leave the specimen at the tip of a small flat-topped pyramid.

Step 6: Ultramicrotomy. The specimen block is mounted in the chuck of the specimen holder which moves up, down, and forward toward the stationary knife. A glass knife may be used to face the end of the block at the specimen. Either a glass or diamond knife is used to cut thin sections which float onto a water bath behind the knife. The sections are then picked up on grids. For soft polymers microtomy at liquid nitrogen temperatures is often helpful.

Step 7: Coating. To avoid charging effects in the electron beam the entire specimen is usually coated with carbon or a metal (see Laboratory 30).

Warning: The staining media used such as OsO_4 and others can be highly dangerous. Careless use may cause blindness or may be fatal. Liquid nitrogen may kill body tissue. The inexperienced user should not try to prepare thin sections without proper training. Contact your local safety personnel for detailed instructions in handling these materials.

29.3 Biological Tissue

Equipment
1. Freeze dryer or critical point dryer.
2. Dehydration and embedding facilities.
3. Staining media (e.g., OsO_4).
4. Ultramicrotome with diamond or glass knife.

Specimens

The procedures outlined in this section are general methods that will give satisfactory results most of the time. Each specimen will ultimately require its own specialized preparati procedure.

Time for this lab session

Typically several days from start to finish.

29.3.1 *Ultramicrotomy*

Thin section preparation for biological tissue is similar to that described for polymers except some specimen handling, stabilization, and drying procedures may differ [11-13]. Typical steps include the following:

Step 1: Sample selection and cleaning. Most tissue may be cut or sliced into samples abou mm^3 to 5 mm^3 which are representative of the structures to be observed. The main contaminants are usually derived from the sample itself, i.e., mucus, body fluids, etc. Gentl washing in clean growth medium or in a balanced organic buffer such as PIPES or HEPES will usually remove most surface contaminants.

Step 2: Structural stabilization. This forms the central part of the procedures used in preparing most biological tissues. This involves prefixation in, for example, 2% glutaraldehyde in an isotonic buffer for 2-12 hours, followed by rinsing several times in buff solution alone. Fixation is continued using 1% OsO_4 in the same buffer solution for up to three hours, followed by rinsing in buffer and several rinses in distilled water. There are a wide range of fixation procedures and most must be adapted to a given specimen.

Step 3: Dehydration. The fixed samples are gradually dehydrated over a two-hour period a room temperature in a graded series of ethanol or acetone, i.e., 30, 50, 75, 90, 95, 100% and finally 100% solvent dried over molecular sieve. Samples must remain immersed. Change liquids by decanting. Store sample in 100% dehydrating solvent. There are a wide range of fixation procedures and most need adopting to a given specimen.

Step 4: Embedding. Transfer sample to a 1:2 mixture of a prepared epoxy resin and dehydrating solvent. Each resin formulation has its own instructions which should be followed. Agitate resin-solvent mixture containing the sample for one hour. Transfer sample through a sequence of 1:1 resin-solvent, 2:1, two changes of pure resin, each for one 1 hour. Transfer resin-infiltrated sample to a fresh pure resin mixture containing the appropriate accelerator and leave overnight at room temperature with gentle agitation. Transfer samples to a fresh resin-accelerator mixture place in capsules and polymerize for 12-48 hours at 60°C. Remove block from capsule.

Step 5: Block trimming. The plane of the specimen is revealed at the end of the block by trimming with a razor blade. The epoxy is then trimmed away to leave the specimen at the tip of a small flat-topped pyramid.

Step 6: Ultramicrotomy. The specimen block is mounted in the chuck of the specimen holde which moves up, down, and forward toward the stationary knife. A glass knife may be used to face the end of the block at the specimen. Either a glass or diamond knife is used to cut thir sections which float onto a water bath behind the knife. The sections are then picked up on grids.

Step 7: Coating. The entire specimen may be coated with carbon to avoid changing effects i the beam (see Laboratory 30).

9.3.2 Specimens for Microanalysis

Immobilization of ions in their natural locations is imperative for the elemental analysis of biological tissues [14-15]. One method of achieving this is to rapidly freeze fresh tissue and keep the specimen at liquid nitrogen temperatures while transferring it into the microscope. This is a specialized procedure that yields unexpected results in a number of situations.

References

[1] J. C. Bravman, R. M. Anderson and M. L. McDonald, *Specimen Preparation for Transmission Electron Microscopy of Materials*, Materials Research Society Symposium Proceedings, **115**, Materials Research Society, Pittsburgh, Pennsylvania (1988).

[2] P. J. Goodhew, *Specimen Preparation for Transmission Electron Microscopy of Materials*, Microscopy Handbook, **3**, Royal Microscopical Society, Oxford University Press, London, UK (1984).

[3] P. J. Goodhew, *Thin Foil Preparation for Electron Microscopy*, Practical Methods in Electron Microscopy, Elsevier, Amsterdam, Netherlands (1985).

[4] K. C. Thompson-Russell and J. W. Edington, *Electron Microscope Specimen Preparation Techniques in Materials Science*, MacMillan Philips Technical Library, Eindhoven, Netherlands (1977).

[5] R. B. Marcus and T. T. Sheng, *Transmission Electron Microscopy of Silicon VLSI Circuits and Structures*, John Wiley, New York (1983).

[6] J. C. Bravman and R. Sinclair, The preparation of cross-section specimens for transmission electron microscopy, *J. Electron Micros. Tech.*, **1** (1984) 53.

[7] I. M. Watt, *The Principles and Practice of Electron Microscopy*, Cambridge University Press, Cambridge, UK (1985).

[8] T. F. Malis, AEM Specimens: Staying One Step Ahead, *Microbeam Analysis - 1989*, P. E. Russell, ed., San Francisco Press (1989) 487.

[9] L. C. Sawyer and D. T. Grubb, *Polymer Microtomy*, Chapman and Hall, New York (1987).

[10] N. Reid, Ultramicrotomy, in *Practical Methods in Electron Microscopy*, 3, A. M. Glavert, ed., Elsevier-North Holland, Amsterdam (1974).

[11] M. A. Hayat, *Basic Techniques for Transmission Microscopy*, Academic Press, New York (1986).

[12] M. A. Hayat, ed., *Principles and Techniques of Electron Microscopy (Biological Applications)*, Van Nostrand Reinhold, New York, **1-9** (1970-1978).

[13] A. M. Glauert, ed., *Practical Methods in Electron Microscopy*, Elsevier-North Holland, **3** and **5** (1975-1977).

[14] T. E. Hutchinson and A. P. Somlyo, *Microprobe Analysis of Biological Systems*, Academic Press, New York (1981).

[15] A. J. Morgan, *X-Ray Microanalysis in Electron Microscopy for Biologists*, Microscopy Handbook, Royal Microscopical Society, Oxford University Press, London, UK, **5** (1985).

[16] P. J. Goodhew in *Specimen Preparation for Transmission Electron Microscopy of Materials*, Materials Research Symposium Series, J. C. Bravman, R. M. Anderson, and M. L. McDonald eds., **115** (1988) 51.

Laboratory 30

Coating Methods

Purpose
This lab should serve as a guide for specimen coating. These procedures are designed increase the surface conductivity of nonconductive SEM samples to prevent surface charging which can seriously deteriorate the image. Sometimes metallic coatings are used to conduct heat away from the sample during imaging. Heavy metal coating increases the secondary electron emission from the sample. Metallic coatings also improve images from metallic specimens by covering the thin oxide layer that may be present and providing a surface that h a high secondary electron yield. Since the resolution of secondary electron imaging in the SE has improved to the point where the grain size of conventional metal coatings can easily be resolved, better coating techniques have been developed. Sections of this laboratory address the coating requirements of both low- and high-magnification imaging and for x-ray microanalysis. For more details see *SEMXM*, Chapter 10, and *ASEMXM*, Chapter 7.

Equipment
1. Magnetron sputter coater.
2. Evaporative coater.
3. Ion beam sputter coater.

Warning: Read instruction manual for each piece of equipment. The comments in this guide are not sufficient to operate these devices. Contact your safety personnel.

Specimen
Any nonconductive sample. Plastic foam, chocolate, or a piece of paper are good test samples.

Time for this lab session
Depending on the procedure used coating may take from about 10 min to about an hour

30.1 Metal Coating of SEM Specimens

The coating required to prevent charging on insulators must be continuous, electrically conducting, thin, stable, and have a high secondary emission coefficient. The coating ideally should be smooth and featureless. It is difficult to obtain all these qualities simultaneously in coating. Originally gold, then Au-Pd, was thermally evaporated in a vacuum onto a rotating specimen to produce a coating with islandlike features about 10 nm in diameter. This was no problem when the SEM had a resolution of 20 nm, but as the resolution of secondary electron imaging improved the need for smoother coatings became apparent. Currently, sputter coatin of Au-Pd in a poor vacuum to a 10 nm thick coating is the most common method of producin SEM specimens for low-resolution work. For the high-resolution SE imaging described in Laboratory 11, metal coatings on the order of 1-2 nm thick are required to utilize the high-resolution capabilities of the modern LaB$_6$ or field emission source SEM. Many specialized methods of depositing these thin films such as ion beam sputtering are currently under development. For x-ray microanalysis carbon is still the coating of choice because of its low absorption of x-rays emerging from the sample. Carbon can generally only be evaporated by resistance heating of carbon rods in a good vacuum. Thus, a laboratory bell jar evaporator is adequate for the deposition of most carbon films.

30.2 Coatings for Low-Magnification Imaging

For low-magnification imaging (<10,000x) any conductive coating with features less than about 20 nm may be used. 80Au-20Pd has a finer grain than pure gold. The diode sputter coating device is the most popular way of applying these coatings. In this device, argon ions are accelerated to a cathode containing the target metal to be sputtered. The argon ions knock off (sputter) the target atoms which then migrate through the rather high pressure (0.1 Torr) of argon gas to the workpiece containing the specimen as shown in Figure 30.1.

The relatively high pressure usually assures that the sputtered metal atoms will be scattered to the specimen surfaces that do not have a direct line of sight to the target. The procedure is outlined in the following steps (be sure to read manufacturer's instructions before using):

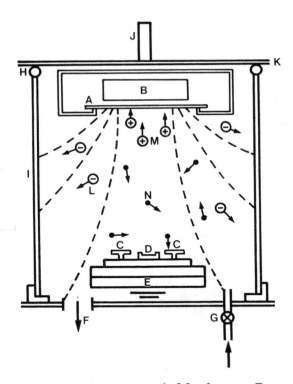

Figure 30.1. Cooled magnetron sputter coater. A, Metal target; B, permanent magnet; C, specimen; D, thin film monitor; E, Peltier cooled specimen table; F, vacuum pump out port; G, needle valve; H, vacuum seal; I, glass cylinder; J, lead to high-voltage supply; K, removable top plate for sample loading; L, negatively charged plasma particles swept away from sample along lines of force from the permanent magnet; M, positively charged argon ions attacking metal target; N, multiply scattered target atoms eroded from target.

Step 1: Mount specimen (C) on the sample stage (E). Adjust thin-film thickness monitor (D) for the target material to be used (e.g., Au-Pd) and zero the meter. Check that the metal target (A) is free of contaminants.

Step 2: Evacuate the work chamber to about 0.1 Torr with a clean rotary pump. Some sputter devices allow the sample stage to be cooled (E) to about 280 K (7°C) during operation.

Step 3: Flush chamber with argon by opening the argon needle valve (G) to partially fill the chamber. Use clean, dry argon gas at a pressure recommended by the manufacturer. Do not attempt to sputter coat with air. Close needle valve and allow pressure to recover to 10^{-1} Torr. Repeat several times. Set argon pressure to about 100 Torr.

Step 4: Turn on high voltage to about 1.5-2.0 kV. A deep purple plasma should form in the space between the target and the specimen.

Step 5: Set argon pressure to give a plasma current of no more than 20 mA.

Step 6: Apply coating of 15-20 nm thickness according to thin film thickness monitor. Typically, 2-3 min are required to deposit 15-20 nm of Au-Pd at 1.8 kV and 20 mA.

Step 7: When coating is complete turn off high voltage and vacuum pump and allow the cooling stage to return to room temperature. Open the argon needle valve to return the chamber to atmospheric pressure.

Step 8: Remove specimens, turn off gas supply, and return sputter unit to standby condition.

Run a track of silver paint or carbon paint from the specimen to the stub. If charging of the specimen still persists repeat this procedure to deposit additional Au-Pd.

30.3 Coatings for Medium-Magnification Imaging

For imaging in the medium-high magnification range of 10,000-50,000x thin metal coatings of finer grain size are used to improve surface contrast and to provide sufficient electrical conductivity. If problems with specimen charging still occur, the SEM accelerating voltage should be reduced rather than the coating thickness increased because the metal coating may be resolved and contribute to the image.

Since gold or Au-Pd film structure is usually resolved above 10,000x, Au-Pd coating must be replaced by platinum, which has a finer grain size. However, platinum deposition requires accurate thickness measurement, improved vacuum quality, and special deposition technology.

Practical platinum coatings can be produced by magnetron sputtering in a tabletop device provided with a turbomolecular pump. The high-vacuum capability is used to cleanse the system of water vapor and hydrocarbon contaminations. The reduced grain size of platinum coatings can be established if an additional LN_2 cold trap is installed above the turbo pump and if an oil-free roughing pump is used instead of the conventional oil vane pump. Also, these thinner films require additional means for even distribution. A specimen rotation stage is essential and is positioned underneath the target so that the rotating specimen is oriented at about 0° tilt angle to one side of the target surface.

If the specimen is positioned under the center of the cathode, the quartz sensor of the thickness monitor may be positioned at either side. The actual amount of metal being deposited on the quartz and on the specimen has to be experimentally determined.

Step 1: Insert the specimen and choose platinum as the target metal. Pump down the system and cool the LN_2 cold trap.

Step 2: After a vacuum approximately 0.08 Torr (10 Pa) is established for approximately 15 min for outgassing the chamber and the specimen, backfill with ultrapure argon and initialize the quartz thickness monitor.

Step 3: Establish a stable ion current and observe the deposition rate. A reduced rate indicates the presence of contaminants. Deposition times are considerably longer than in all other deposition techniques and are on the order of 15-20 min for 2 nm thin films.

Step 4: After the film deposition is finished, warm up the cold trap with warm air, switch off the pumps, vent the chamber, and remove the specimen. The vacuum chamber should be pumped down to 1-10 Torr before the pump stand is switched out.

30.4 Coatings for High-Magnification Imaging

High-magnification imaging (>50,000x) uses the SE-I signal and requires very special metal deposition techniques for films of only 1-2 nm thickness [1].

Very thin metal films can be deposited by ion beam sputtering. A high-energy ion beam is produced in small ion guns and directed onto a metal target. The sputtered metal atoms adhere to the specimen surface and produce very fine-grained continuous films. High-energy ion beam sputtering can produce 1-2 nm thin continuous films at room temperature. In evaporation, liquid-nitrogen specimen temperatures are required to produce such films. Conventional thick film coatings with platinum will not generate a SE-I signal contrast and will not allow full utilization of the ultrahigh resolution power of modern field-emission SEMs. For high-resolution SEM images of rough surfaces, low-Z metals (chromium, titanium) must be used for background signal reduction. Such ultrathin films do not prevent charging of inhomogeneous low-conductivity samples. Additional procedures for increase of conductivity may be required on plastics or biological materials (e.g., osmium vapor treatment or osmium impregnation).

A schematic of an ion-beam sputtering device is shown in Figure 30.2. Usually ion beam sputter coaters are pumped by turbomolecular pumps and oil-free roughing pumps for reduction of hydrocarbon contamination. Additionally, a liquid-nitrogen-cooled cold trap is necessary for the sputtering of low sputter yield metals (titanium, chromium, tantalum). Special attention must be directed to specimen movement and thickness measurement during the metal deposition. The specimen must tilt through 0-90° and rotate. Under this condition only ~33% (deposition factor) of the metal deposited into the specimen plane will accumulate at the specimen surface. The reduced metal accumulation rate necessitates long deposition times and requires thickness measurement with a quartz thickness sensor.

Typical steps in operating an ion sputtering device for production of 1-2 nm thick chromium films are as follows:

Step 1: Start the system and let it pump down and outgas. The argon inlet valves should be closed and the high-voltage supply for the ion guns should be switched off.

Step 2: Never touch parts of the vacuum cleaner chamber with unprotected hands. Wear disposable plastic gloves (clean room gloves). Mechanically mount the specimen on a clean specimen holder. Do not use carbon paste without extensive vacuum drying before coating. Dry mounting with double stick may be applied. If possible, mount beside the specimen a piece of clean and dry filter paper for film quality control.

Step 3: Vent the ion beam sputter coater, mount the specimen holder into the specimen table, check that the tilt-rotate stage operates and close the shutter if provided. Pump down the system below 10^{-5} Torr.

Step 4: If provided, fill the cold trap with liquid nitrogen and wait until vacuum is better than

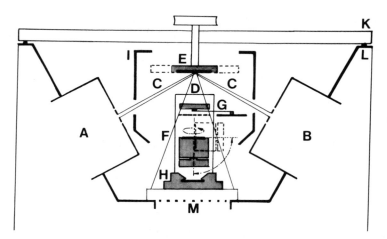

Figure 30.2. Ion beam sputter coater. Detail of vacuum trough provided with two ion gun A, B, Ion gun; C, focused ion beams; D, metal atom beam; E, interchangeable metal targets; specimen table shown in 0° and 90° (dashed lines) eucentric tilt positions; G = S, shutter (dashed lines indicate closed position); H, quartz crystal sensor of film thickness monitor; I, liquid-nitrogen-cooled cold trap; K, transparent plexiglass lid; L, vacuum trough; M, port for turbomolecular pump.

4×10^{-6} Torr. Set the quartz thickness monitor to the appropriate values for metal density and film thickness as described in the operation manual.

Step 5: Switch on the power supply of the ion guns and set the operation voltage (e.g., 4 kV). Open the argon gas supply until the desired ion current is obtained (e.g., 30 mA for each gun). Observe fluorescence within the guns. Bring the potassium chloride crystal into the target position and check the focus of the ion beam in the target plane. Readjustment is normally not required.

Step 6: Open shutter. Wait until reading of quartz monitor is constant and at expected value. Zero the quartz sensor reading and start the specimen movement.

Step 7: Deposit the desired amount of metal. Then, close the shutter, shut down the ion gun power supply, stop the specimen movement, and warm up the cold trap if used. Finally, close the argon supply and vent the vacuum chamber with dry nitrogen.

Step 8: Remove the specimen and close the vacuum chamber. Return the system to standby. Check the filter paper for metal film quality. Typical film color for most metals is a neutral gray tone. A yellowish color indicates deposition of hydrocarbon contaminations or metal oxide. In such cases, the vacuum plant must be thoroughly cleaned before the next coating is performed.

30.5 Coatings for X-Ray Microanalysis

It is not possible to sputter coat carbon using laboratory sputter coaters. This material must be applied by high-vacuum evaporative techniques. Evaporative coaters may also be used to coat the sample with metal coatings if no sputter coater is available. There are a number of different evaporative coaters and you should familiarize yourself with the operating instructions for the piece of equipment you will be using.

The basis of the method is to resistively heat the sharpened section of a carbon rod to white heat. This heat causes carbon atoms to escape the rod and travel to the specimen. The specimen must be tilted and rotated since the carbon atoms reach the specimen by line-of-sight only. A schematic of a vacuum evaporator is shown in Figure 30.3. An outline of the general procedure is in the following steps:

Step 1: Start the system. Check that the evaporative coater is in the correct standby condition. The gas inlet valve (D) should be closed and all vacuum seals (B) should be clean and in position. The evaporation chamber (A) should be clean and the electrodes (J) should be fitted with carbon rods and connected to the appropriate power supply. Start the pumping procedure to ensure that the pumps and backing lines are at the correct operating vacuum.

Step 2: Only use spectroscopically pure carbon rods, which should be shaped to the appropriate configuration for evaporation (I). The two most useful configurations are the following:
  (a) A finely pointed spring-loaded carbon rod pressing lightly against the flat face of a stationary carbon rod.
  (b) A thin 1.5 mm x 6 mm spigot cut in the end of a spring-loaded carbon rod pressing

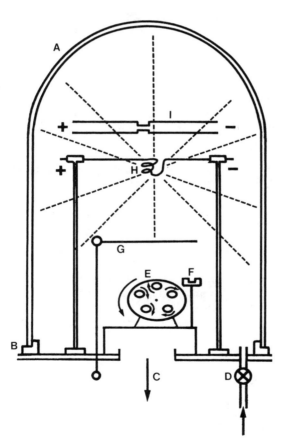

Figure 30.3. High-vacuum evaporation unit. A, Glass bell jar; B, vacuum seals; C, port-to-vacuum pumps; D, dry nitrogen gas inlet valve; E, samples on a planetary-tilting holder; F, thin film monitor; G, shutter; H, refractory metal assembly for evaporating metal; I, carbon rod assembly for evaporating carbon.

30 COATING METHODS

lightly against the flat face of a stationary carbon rod.
If the evaporative coater is being used to coat the sample with a thin metal film, a tungsten wire basket containing metal chips (H) or a V-shaped tungsten wire holding a loop of metal should be used in place of the carbon configuration.

Step 3: Load specimens on the specimen table (E), which should rotate and tilt in a planetary motion. Check that the motions operate properly.

Step 4: The evaporation chamber is closed and evacuated to a pressure of 10^{-5} Torr. During the pumpdown procedure it is useful to check that the carbon electrodes are in electrical contact. The liquid-nitrogen cold trap above the diffusion pump should be filled to prevent backstreaming once the working pressure has been reached.

Step 5: Once the required pressure is reached, the shutter (G) should be moved into position between the samples and the carbon source. The power supply to the two carbon electrodes (I) is increased to the point at which they are heated to a cherry red color. This is done in order to outgas the carbon rods. The outgassing is signified by a small transient rise in the pressure of the evaporation chamber.

Step 6: The samples should be rotated at about 100 rpm and tilted between ±45°. The shutter (G) between the samples and carbon source is now swung back.

Step 7: The power supply to the carbon electrodes should be increased quickly to the point at which carbon evaporation occurs (white heat). This process should only be observed through dark glass. Evaporation is complete when a few sparks are emitted from the electrodes (within 5-7 sec). The power to the electrodes should be returned to zero and the rotating sample table turned off.
    It is only necessary to apply a very thin carbon coating layer of between 3 and 5 nm. This may be measured either by a thin-film thickness monitor (F) or by observing the color of the carbon film deposited on a small piece of smooth white paper placed next to the specimens. A chocolate brown color is slightly thinner than the required thickness while a grey metallic sheen is somewhat thicker.

Step 8: The appropriate baffle valves to the evaporation chamber should be closed and the chamber returned to atmospheric pressure, preferably by using dry nitrogen gas (D).

Step 9: Remove the samples and return the system to standby.

Reference
[1] K.-R. Peters in *Advanced Techniques in Biological Electron Microscopy III*, J. K. Koehler, ed., Springer-Verlag, New York (1986) 101-166.

Part I
SCANNING ELECTRON MICROSCOPY AND X-RAY MICROANALYSIS

SOLUTIONS

Laboratory 2

Electron Beam Parameters

2.1 Gun Saturation and Alignment

Without proper filament saturation and gun alignment, the maximum probe current will not be obtained, as shown in Figure A2.1.

Experiment 2.1: Filament Saturation and Experiment 2.2: Gun Misalignment. Either image intensity (secondary electron signal) or beam current may be used for setting the filament saturation. The emission will be somewhat less sensitive although it also may be used to set the saturation point.

In a cold field-emission gun, the maximum current is usually limited by the gun vacuum. The more current emitted, the more gas atoms are desorbed from the metal gun chamber that may settle on the field emitter tip, causing instability and, in extreme cases, tip failure.

2.2 Beam Current

It is important to distinguish between gun emission current (usually on the order of tens of microamperes) and the beam current that generates images (usually on the order of nanoamperes).

Experiment 2.3: Beam Current and Emission Current. The beam current at the specimen is much less than the emission current because much of the current from the gun is intercepted by the anode, the spray apertures, and the walls of the liner tube. The emission

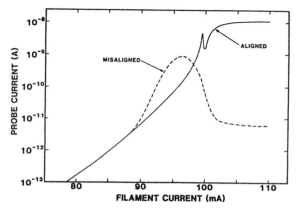

Figure A2.1. Typical tungsten filament saturation behavior at 20 kV, 80-μA emission current, and 10-μA beam current.

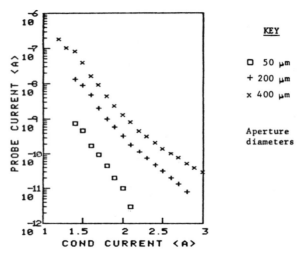

Figure A2.2. Beam current (probe current) versus condenser lens setting for several apertures. Aperture diameters: ☐ 50 µm; + 200 µm; x 400 µm.

current did not change during this experiment because the gun parameters were constant. The emission current may be varied by changing (1) the height of the filament (rarely done), (2) the gun bias voltage, or (3) the filament heating current. For a particular accelerating voltage there should be only one gun setup that provides optimum performance since the bias voltage should be set to obtain maximum brightness and the filament heating should be set to saturation for maximum stability (see *SEMXM*, Figure 2.3).

Experiment 2.4: Effect of Lens Settings on Beam Current. While the emission current is set and fixed by the saturation of the gun, the beam current is under the control of the operator and should be changed according to the type of image or information desired. The probe current decreases with both stronger condenser lens current (smaller spot size) and smaller objective lens apertures. The data for the plot in Figure A2.2 were generated on a typical tungsten thermionic source SEM with a 20-kV accelerating voltage and a gun emission current of 120 µA.

For higher condenser lens currents (smaller beam sizes), the focusing action of the lens increases the divergence (convergence) angle of the beam, causing electrons to be sprayed onto the apertures and the walls of the liner tube. Variation of the objective lens current while keeping the condenser lens fixed should not alter the beam current significantly. The objective lens only varies the vertical position of the final crossover.

2.3 Beam Size

The data below, taken across the edge of a fractured silicon wafer, show that beam size measurements down to a few tens of nanometers are possible. However, this measurement is best performed in transmission to avoid high secondary electron emission at the silicon fracture edge.

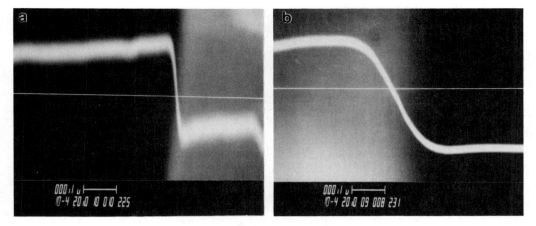

Figure A2.3. Beam size measured at silicon fracture edge. (a) Waveform of a 100-pA beam measuring 23 nm in diameter. (b) Waveform of a 10-nA beam measuring 130 nm in diameter. Marker = 0.1 μm.

Experiment 2.5: Beam Diameter. Note that while the same magnification is shown for the two examples in Figure A2.3, a magnification should be chosen that allows convenient measurement of a waveform change at least several millimeters wide.

When the beam current is plotted against beam size, as shown in Figure A2.4, a dramatic reduction in beam current is observed for the smaller beam sizes.

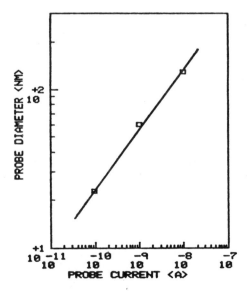

Figure A2.4. Beam diameter (probe diameter) versus beam current (probe current) for an SEM with a tungsten filament.

2 ELECTRON BEAM PARAMETERS

2.4 Beam Convergence

This parameter can be measured in all microscopes by the method described, but it is dependent on the quality of the specimen used for beam size measurements. However, reasonable values can be obtained if the beam sizes are not too small and the change in specimen height z is large enough (e.g., 1 mm).

Experiment 2.6: Beam Convergence. For a 200-µm aperture, the initial beam size d_i was measured to be 30 nm. After increasing z by 1 mm, the final beam size d_f was 19,000 nm. Therefore,

$$\alpha \approx \frac{19{,}000 \text{ nm} - 30 \text{ nm}}{2 \times (1 \text{ mm})} = 9.5 \times 10^{-3} \text{ radian}$$

Note that a reasonable estimate of the convergence may be obtained by neglecting the small, difficult-to-measure d_i. The convergence of the beam with the 400-µm aperture may be estimated as 2×9.5 mrad $= 19$ mrad.

2.5 Brightness

Experiment 2.7: Brightness. For a 200-µm aperture, typical values for beam current, beam size, and beam convergence for a thermionic tungsten gun are 0.1 nA (10^{-10} A), 30 nm, and 9.5 mrad, respectively. Thus, from the brightness equation we see that

$$\beta = \frac{10^{-10} \text{ A}}{\pi^2 (300 \times 10^{-8} \text{ cm})^2 (9.5 \times 10^{-3})^2}$$

$$\beta = 1.2 \times 10^4 \text{ A/(cm}^2 \text{ sr)}$$

The maximum brightness at the gun will not be measured at the specimen because of beam current lost through lens aberrations. Typical brightnesses for various electron sources are roughly as follows:

Tungsten hairpin gun: $10^4 - 10^5$ A/cm² sr
LaB_6 gun: $10^6 - 10^7$
Field-emission gun: $10^8 - 10^9$

For thermionic guns at lower voltage, the resolution degrades because the low brightness forces a much larger spot size in order to obtain enough current for an acceptable image. Also, chromatic aberration is a more serious image defect at lower voltage.

2.6 Depth-of-Field

To obtain images with high depth-of-field (everything in focus) a small beam convergence angle α is required. This small beam angle can be obtained by using either a small objective aperture or a long working distance, or both. The images in Figure A2.5 were obtained with the relatively short working distance of 7 mm and two different aperture sizes.

Experiment 2.8: Effect of Aperture Size. At a working distance of 5 mm, the depth-of-field may be calculated for a 100-µm aperture at 50x as follows:

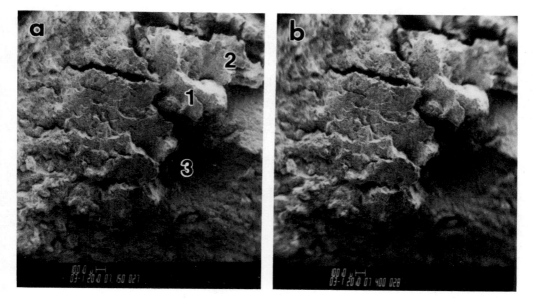

Figure A2.5. Depth-of-field at 7-nm working distance. (a) 150-µm aperture, (b) 400-µm aperture.

$$\alpha_{5\,mm, 100\,\mu m} = (\alpha_{10\,mm, 200\,\mu m}) \left(\frac{100}{200} \times \frac{10}{5}\right)$$

$$= (9.5 \times 10^{-3}\,\text{rad})(200)$$

$$D = \frac{0.2\,\text{mm}}{(9.5 \times 10^{-3}\,\text{rad})(50x)} = 0.42\,\text{mm} = 420\,\mu m$$

The calculated depths-of-field for the combinations of aperture and magnification requested are as follows:

	Depth-of-field at 5-mm working distance	
Aperture	50x	5000x
100 µm	420 µm	4 µm
400 µm	105 µm	1 µm

Experiment 2.9: Effect of Working Distance. Increasing the working distance decreases the convergence angle and therefore increases the depth-of-field as shown in Figure A2.6. Long working distances are also used to obtain the lowest magnifications possible, e.g., 5-20x. Scaling the convergence angle to the longer working distance with the 400-µm aperture, the depths-of-field at 50x and 5000x may be calculated to be 600 µm and 6 µm, respectively.

The above data show that the largest depth-of-field is obtained when a small objective aperture is used together with a long working distance. The vertical distance considered as the

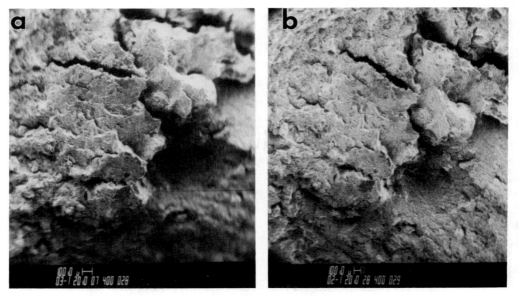

Figure A2.6. Depth-of-field with a 400-μm aperture. (a) 7-mm working distance; (b) 28-mm working distance.

depth-of-field can be determined by refocusing the out-of-focus region with the z-drive and noting the change in height or by measurements from a stereo pair (see Laboratory 4). The height distances between the regions in the image taken with the 150-μm aperture (Figure A2.5) are as follows:

Region 1 to region 2: 0.7 mm = 700 μm
Region 1 to region 3: 1.2 mm = 1200 μm

Experiment 2.10: Image Rotation. At long working distances, particularly at low magnifications, the image may be observed to rotate several degrees about the electron optical axis of the microscope. This is a demonstration of the relative rotation of an image traversing magnetic lens as the strength of the lens changes. This rotation occurs because electrons follow a helical path through a magnetic field.

2.7 Small Beam Size versus High Beam Current

It is clear that electron probes lose current rapidly as the spot size is reduced. Since the instrumental resolution of most secondary electron images depends upon the beam size (small probes give better resolution), the highest-resolution images may even be quite noisy because the smallest probe contains very little current to generate the electron signal. Long photographic exposures (60-100 sec) must be used to compensate. Laboratory 3 deals with this problem in greater detail.

Experiment 2.11: Small Beam Size. Note that the detail in the center of Figure A2.7 is quite sharp.

The contrast of electron images increases with higher probe currents. Thus, high probe

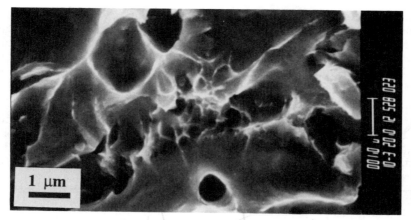

Figure A2.7. A high-resolution image at 10,000x taken with a highly excited condenser lens.

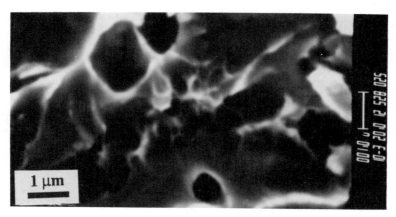

Figure A2.8. Same area as Figure A2.7 (10,000x but with 100 times higher beam current).

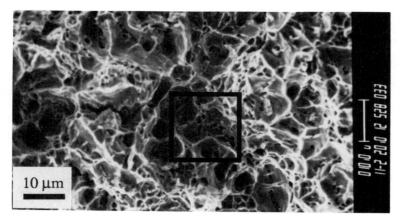

Figure A2.9. Same high-beam current conditions but at a lower magnification (1100x). Box in center shows area of Figure A2.8. Note that the image is again sharp.

currents (and large probe sizes!) will often produce better images when the specimen has low inherent contrast. At lower magnifications the better resolution expected from smaller probes cannot be observed. Thus, it is wise to use larger spot sizes and higher currents at low magnifications. X-ray microanalysis also usually demands higher probe currents than those obtained under high-resolution imaging conditions.

Experiment 2.12: High Beam Current. Note that the larger beam size required under high beam current blurs the image in Figure A2.8. However, this blurring will not be observable at lower magnifications as shown in Figure A2.9. Of course, with an FEG very small spot sizes can be obtained with relatively high probe currents. Thus, good images may be obtained at high magnification and at low voltages.

Laboratory 3
Image Contrast and Quality

3.1 Atomic Number Contrast

Atomic number contrast may be observed from the polished sample using several different detectors.

Experiment 3.1: Z Contrast Using a Dedicated BSE Detector. The dedicated BSE detector directly above the specimen is very efficient and gives quality, high-contrast images (see Figure A3.1a). Variations in composition (average Z) are sensed by the numbers of BSEs which are emitted and hit the detector. Therefore, atomic number contrast can be considered pure "number" contrast. The image resolution estimated by scanning the beam across a phase boundary is generally on the order of 0.2 μm.

Experiment 3.2: Z Contrast Using the E-T Detector. The E-T detector biased positively is designed to gather secondary electrons (SEs) (see Figure A3.1b). Any BSE whose trajectories happen to strike the ET will also be incorporated into the SE image. However, the E-T detector collection efficiency for BSE is very low. With low signal and low contrast, the BSE image of a flat polished sample is noisier, unless the beam current is increased, since the measured SEs should not produce much contrast in this sample. Most Z contrast in the SE image is due to BSEs that hit the lens pole piece and generate SEs. The E-T detector as a BSE detector is discussed in more detail in Laboratory 8.

Experiment 3.3: Z Contrast Using the SC Detector. Absorbed specimen current images have contrast opposite (inverted) compared to emissive BSE images. Topography visible in Figure A3.1b disappears in the SC image (Figure A3.1c) although some edge and corner effects remain visible.

3.2 Topographic Contrast

If the area of interest is positioned so that part of it is shielded from the E-T detector, it will be easy to distinguish the number and trajectory components of the topographic contrast.

Experiment 3.4: Topographic Contrast from SEs. The positively biased E-T detector collects most of the emitted SEs, and gives a good image (see Figure A3.2a) of the surface topography. Poor signal intensity only comes from regions where SEs were not collected because a hole was too deep for the SEs to escape. From such an image, it is easy to deduce the sense of topography in the sample. The signal also contains a strong BSE component from regions of the fracture surface facing towards the E-T detector.

Figure A3.1. Images of a polished section of multiphase basalt using different detectors. (a) Solid state BSE detector directly over specimen. (b) E-T detector positively biased. (c) Specimen current signal. Note depressions in the sample surface, especially in (b), caused by sample porosity. Tungsten filament, 20-kV accelerating voltage, 80-µA emission current, 5-nA beam current.

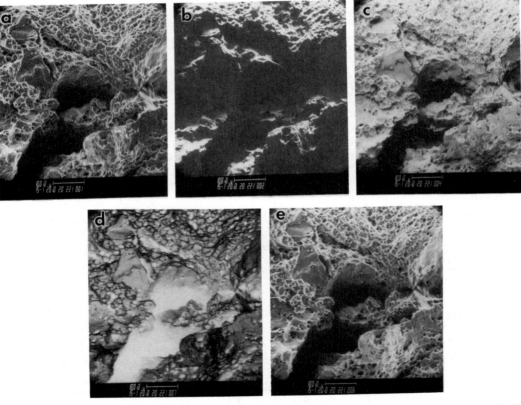

Figure A3.2. Images of an aluminum fracture surface using different detectors. (a) E-T detector positively biased (50-pA beam current). (b) E-T detector negatively biased (2-nA beam current). (c) Solid state BSE detector (0.5-nA beam current). (d) Direct specimen current image (2-nA beam current). (e) Inverted specimen current image (2-nA beam current). Note the inceases in beam current necessary for some detectors.

Experiment 3.5: Topographic Contrast from BSEs. When the E-T detector is biased negatively or turned off, no SEs enter the detector. The only electrons to be registered are those BSEs which travel directly into the detector (pure trajectory contrast). The negatively biased E-T detector has a tremendous sensitivity to the trajectories of the BSEs. As such, only a few regions of the fracture surface will be favorably oriented for direct backscatter of electrons toward the detector. These surfaces appear as strong highlights. Because the E-T detector is a small detector, it only detects a small fraction of BSEs and thus is very inefficient. Therefore, the beam current has to be increased to provide reasonable intensity in the image. Note that although the contrast is stronger in Figure A3.2b than Figure A3.2a, Figure A3.2a contains far more useful information since SEs are also observed from regions of the fracture surface which are not facing the E-T detector. However, from Figure A3.2b, it is clear that only the bright regions are in direct line of sight of the detector.

The BSE detector which sits above the specimen gathers electrons scattered from all areas of the sample surface, so the topographic effects tend to be "washed out" (see Figure A3.2c). Therefore, the strong trajectory effects visible in Figure A3.2b are lost. Because the dedicated BSE detector is more efficient (it collects a larger fraction of BSEs) than the negatively biased E-T detector, an image can be obtained with less beam current.

Experiment 3.6: Topographic Contrast from the Specimen Current Signal. The direct specimen current image is a pure number contrast image with no trajectory component. The contrast in the image is inverted with respect to the image from the solid state BSE detector or the SE detector since $i_{sc} = i_b - i_{SE} - i_{BSE}$ (compare Figures A3.2a,b with A3.2d). When the specimen current image contrast is inverted electronically (Figure A3.2e), the image looks very similar to the SE image and one can immediately get a sense of the topography of the sample.

Experiment 3.7: Stereo Images. The true topography can only be observed by using a stereo pair (Figure A3.3) of the same region. If the tilt axis of the microscope stage lies parallel

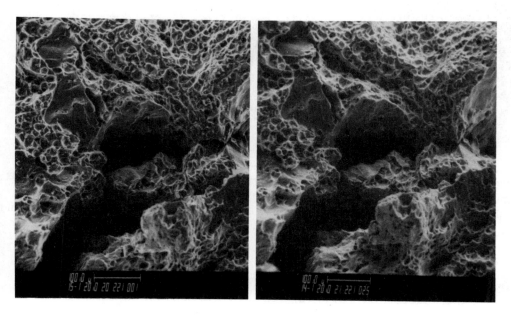

Figure A3.3. Stereo pair using E-T detector, positively biased. Right-hand photo tilted 4°.

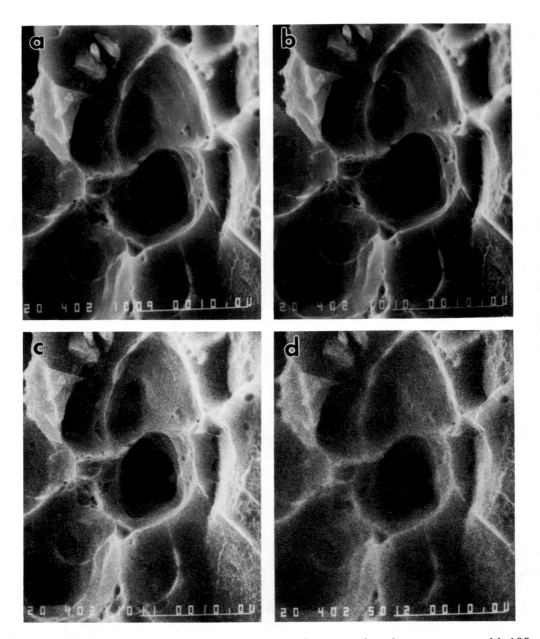

Figure A3.4. Images of an aluminum fracture surface at various beam currents with 100-sec frame time. (a) 1-nA beam current. (b) 100-pA beam current. (c) 10-pA beam current. (d) 5-pA beam current. Note loss of visibility as beam current decreases. Tungsten filament at 20 kV

to the bottom edge of the CRT screen, the micrographs must be rotated through 90° to orient them correctly. If the magnification is not too high, the stereo pair should show that many of the brightest facets on the fracture surface are closer to the observer. If the photos are reversed left-for-right, the stereo effect is lost. Stereo images are discussed in detail in Laboratory 4.

3.3 Image Quality

Experiment 3.8: Effect of Beam Current on Image Quality. As the beam size is decreased and the probe current drops, two conflicting things happen. With a smaller beam size, the theoretical resolution of the image improves. However, as the probe current decreases, the image quality degrades for a constant frame time (see Figure A3.4). In practice with a thermionic source, the loss of signal dominates any improvement in resolution. For the case illustrated, a beam current of 100 pA gives acceptable detail in the flat regions of the fracture surface. Thus, it is expected that the threshold current has been exceeded. This is confirmed by calculation:

$$i_{th} > \frac{4 \times 10^{-12}}{\varepsilon C^2 t_f}$$

For $\varepsilon = 0.25$, $C = 0.1$ and $t_f = 100$ sec $i_{th} > 1.6 \times 10^{-11}$ A = 16 pA. Thus at least 16 pA is required to observe contrast differences on the order of 10%.

Experiment 3.9: Effect of Scan Rate on Image Quality. A similar effect is noted as the frame time is decreased for a fixed probe current of 100 pA (see Figure A3.5). At 100 sec frame time the image quality is high, but at 10 sec and 1 sec the fine-scale detail in the image is progressively lost. Thus, the threshold frame time must be greater than 10 sec. This is confirmed by solving for t_f in the threshold equation:

$$t_f > \frac{4 \times 10^{-12}}{e C^2 i_{beam}}$$

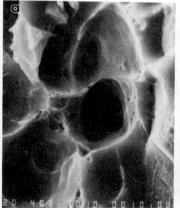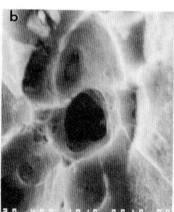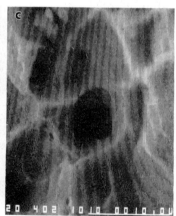

Figure A3.5. Images of same area as Figure A3.4 at various recording frame times, all with a beam current of 100 pA. (a) 100-sec frame time. (b) 10-sec frame time. (c) 1-sec frame time. Note loss of image detail as the frame time decreases.

Thus for $\varepsilon = 0.25$, $C = 0.1$ and $i_{beam} = 10^{-10}$ A, $t_f > 16$ sec.

3.4 Signal Processing

Experiment 3.10: Analog Image Processing. (a,b) Black level adjustment. The simplest analog signal processing is black level suppression and differential amplification. This processing technique is used when parts of the image are too bright or too dark. The black level (detector dc output level) can be used to reduce a bright image or enhance a dark image to give acceptable overall intensity. If the resulting contrast is too low, then differential amplification (the detector gain) can be used to expand the contrast range to cover the visible gray scale. The process also amplifies any noise in the image. Figures A3.6b,c show images with too little and too much black level. The black level should be adjusted to give an image similar to Figure A3.6a with the detector gain adjusted to give an appropriate gray scale.

(c) Gamma processing (nonlinear amplification) is used to enhance contrast in parts of the image which are much darker or much brighter than the average. As shown in Figure A3.6d, signal intensity from down the hole that was not visible in the unprocessed image

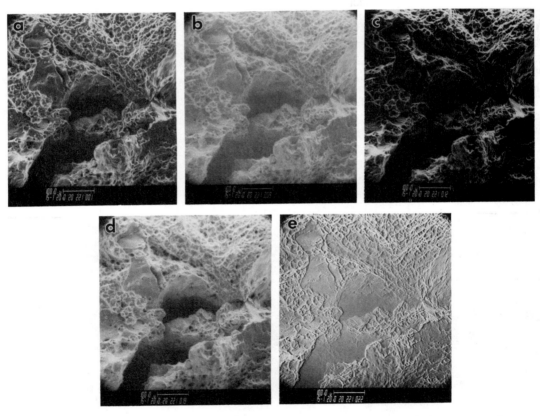

Figure A3.6. Images of an aluminum fracture surface using various image processing techniques. (a) Image without processing. (b) Insufficient black level suppression. (c) Excessive black level suppression. (d) Gamma processing applied. (e) Signal differentiation (100% derivative).

Figure A3.6a) can be detected using the gamma control, without causing the rest of the image to become too bright.

(d) Signal processing differentiation, not to be confused with differential amplification, involves transforming the image contrast to enhance edges in the image where the contrast is greatest (i.e., where the signal intensity changes most rapidly). The net result is an image Figure A3.6e) in which all sense of topography is lost, but the edges of features are enhanced.

3.5 Digital Images

Digital image collection has many advantages over analog image collection, particularly in the area of signal processing. At high pixel densities (256x256 and above), the appearance of a digital image on the viewing screen is very similar to that of an image scanned in an analog manner.

Experiment 3.11: Digital Image Collection. For normal viewing the frame time for a 256x256 image obtained with a dwell time per pixel of 10 µsec is about 0.7 sec. This frame rate is comparable to a fast, non-TV scan rate in conventional analog mode. But for thermionic electron guns, a 10-µsec dwell time may not provide a noise-free image except at low magnifications and with large electron beams. For photography, of course, longer dwell times (e.g., 1 msec) may be used, making the frame time about the same as for analog photos (about 1 min). With higher-brightness electron sources such as an FEG, the same beam size carries more current than for a thermionic source, so shorter dwell times may be used. The magnification for a 10-nm-diameter beam to just fill an image pixel at 256x256 may be calculated as follows:

$$\text{Magnification} = \frac{\text{Screen width}}{(\text{Pixels per line})(\text{Beam size})}$$

$$= \frac{10 \text{ cm}}{(256)(10 \text{ nm})} = 39{,}063\text{x}$$

At magnifications higher than 39,000x, the specimen area will be oversampled (or overscanned), increasing the information obtained from each point on the specimen. At magnifications lower than 39,000x, the specimen area will be undersampled (or underscanned). In this case it is useful to increase the beam size so that each pixel is filled. Individual pixels, which can just be seen in photographs of 256x256 images, can be made smaller by using a higher pixel density such as 512x512 or 1024x1024, but these images take much longer to scan.

Experiment 3.12: Digital Image Processing. Examples of digital image processing are given in Laboratory 17.

Laboratory 4
Stereo Microscopy

4.2 Stereo Photography

Experiment 4.1: The Stereo Effect. Two examples of stereo pairs are presented in Figures A4.1 and A4.2. Figure A4.1 depicts the topography of crystals of native silver. The left-hand image is recorded at a nominal stage tilt of 0° while the right-hand image is recorded at a stage tilt of 8° for a difference in tilt angle of 8°. Note that if the tilt angle is less than 5° the stereo efect is reduced. Figure A4.2 shows the surface relief on a crystal of the mineral galena. The left-hand image is recorded at a nominal stage tilt of 0° while the right-hand image is recorded at a stage tilt of 8° for a difference in tilt angle of 8°. (Note that the SEM used to prepare these images was configured such that the tilt axis was *vertical* on the recorded image with the E-T detector located on the upper right-hand side. The sense of illumination in these images thus appears to come from the upper right.) At higher magnifications the surface relief is likely to be less. Thus, to obtain a stereo effect on a surface with only a small amount of topography, the tilt angle must be increased.

4.3 Quantitative Stereo Microscopy

Experiment 4.2: Height Measurements. As a detailed example, we shall determine the height difference between two steps A and B on the galena crystal in Figure A4.2. The

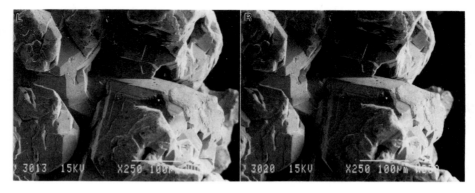

Figure A4.1. Example of stereo pair with the left image recorded at 0° tilt and right image recorded at 8° tilt toward the E-T detector. Magnification reduced to 160x for reproduction.

Figure A4.2. Annotated stereo pair for calculation of height differences. Magnification reduced to 160x for reproduction.

reference point is indicated by 1 in the two images and 1 is located on one plane of interest (A). The difference in tilt angle between the two images is θ = 8°. The left-image (Figure A4.2a) x coordinate of point 2 is:

$$x_L = -32 \text{ mm}$$

The right-image (Figure A4.2b) x coordinate of point 2 is:

$$x_R = -38 \text{ mm}$$

The parallax is then:

$$P = (x_L - x_R) = (-32 - [-38]) = +6 \text{ mm}$$

The z coordinate is therefore:

$$z = P/[2M \sin(\Delta\theta/2)]$$

$$= +6 \text{ mm}/[2 \times 160 \times \sin(8/2)]$$

$$= +0.270 \text{ mm} = +270 \text{ μm}$$

Since the parallax is positive, crystal face B is therefore 136 μm *above* crystal face A. The sense of the topography calculated mathematically is confirmed by qualitative inspection of the surface with the stereo viewer.

In Figure A4.2 the tilt axis was rotated clockwise, as viewed from the detector, to produce the right hand image. This clockwise rotation is defined to give positive Δθ.

Location	x_L	x_R	P	z_i
1	0	0	0	0
2	-32 mm	-38 mm	0 mm	+270 μm
3	-26	-26	0	0
4	-37	-42	+5	+220
5	+16	+15.5	+0.5	+22
6	-43	-50	+7	+310
7	+13.5	+14.5	-1	-44

Laboratory 5
Energy-Dispersive X-Ray Spectrometry

5.1 Spectrometer Setup

Experiment 5.1: Spectrum Acquisition. A typical spectrum from copper is shown in Figure A5.1. The low-energy line is the compressed *L*-series and the two higher-energy lines are the CuK_α and K_β.

Experiment 5.2: Energy Calibration Check. The CuL_α and CuK_α lines in Figure A5.1 are within one channel (10 eV or 20 eV) of their correct positions. This level of calibration is adequate for most purposes.

5.2 Beam Current and Dead Time

These experiments show how the count rate is limited in EDS systems and give some indication as to the appropriate count rate to use in various situations.

Experiment 5.3: Comparing Live Time and Real Time. Note in the table below that the real time to achieve a particular live time, or number of counts, becomes larger as the beam current increases.

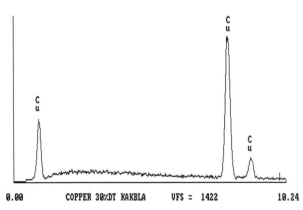

Figure A5.1. A typical spectrum from copper at a beam energy of 20 keV.

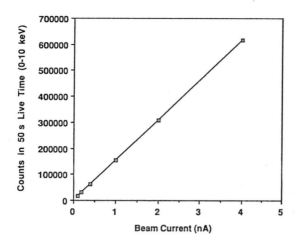

Figure A5.2. Collected x-ray counts 50 sec live time versus beam current at 20 keV.

Dead time	Live time	Real time	Beam current
10%	30 sec	35 sec	16 pA
40%	30 sec	50 sec	254 pA
80%	30 sec	140 sec	1 nA

Experiment 5.4: Testing the Dead Time Correction Circuit. Figure A5.2 shows linear relationship between collected counts and beam current at least up to 10,000 counts/sec

Experiment 5.5: Maximum Output Count Rate. The counts collected per second of real time go through a maximum as shown in Figure A5.3.

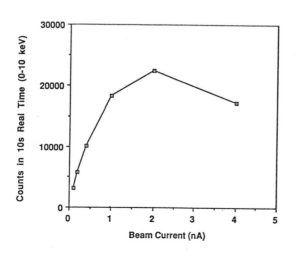

Figure A5.3. Collected counts in 10 sec real time versus beam current at 20 keV. Note that the maximum throughput of x-ray counts occurs for a beam current of about 2 nA.

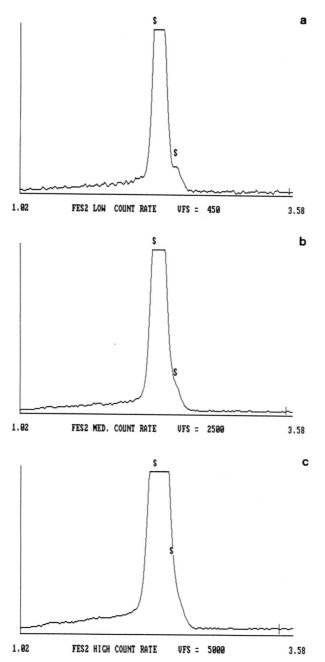

Figure A5.4. Resolution versus amplifier time constant for the sulfur K_α and K_β lines. (a) Long time constant provides some resolution of the K_β line. (b) Medium time constant. (c) Short time constant (used for high count rates) provides no resolution of K_β line.

Experiment 5.6: Detector Resolution. The degradation of resolution with increasing count rate capability is clearly shown in Figures A5.4a,b,c. The spectrum with the best resolution of the sulfur K_α and K_β (Figure A5.4a) was taken with a long amplifier time constant which limited the total number of counts processed to about one-tenth that of Figure A5.4c, in which there is no resolution of the K-series peaks. The setting to use depends upon the particular analyst's situation. The FWHM of the SK_α is less than the resolution value stamped on the detector dewar because the standard resolution for a detector specification is always measured at 5.9 keV.

5.3 Spectral Artifacts

Experiment 5.7: Stray Radiation. If any x-rays are excited when the probe is in the hole (e.g., platinum or aluminum), there must be some stray x-rays or electrons exciting regions far from the beam. Figure A5.5 shows an example.

Experiment 5.8: Escape Peaks, Sum Peaks, and System Peaks. Figure A5.6 shows the most common spectral artifacts. Escape peaks will always have an energy equal to the energy of the parent line minus the energy of the silicon x-ray (1.74 keV). Moreover, the magnitude of the escape peak relative to the parent peak is fixed for a particular parent line. Sum peaks are due to the simultaneous arrival of two TiK_α photons in the detector crystal creating electron-hole pairs in numbers corresponding to a single photon with the sum energy. By reducing the count rate and counting for a longer time, this artifact can be reduced to a low

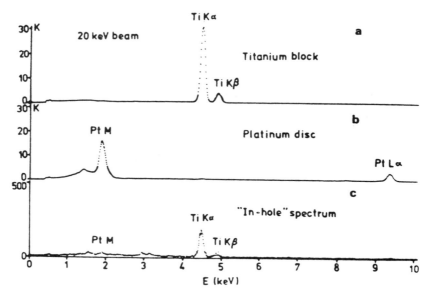

Figure A5.5. Spectra obtained from the components of a Faraday cup. (a) Titanium block directly excited by electrons. (b) Platinum disc. (c) "In-hole" spectrum.

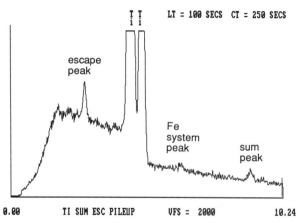

Figure A5.6. Spectrum from pure titanium showing a silicon escape peak for TiK_α, a small iron system peak generated by backscattered electrons hitting the polepiece, and a TiK_α sum peak.

level. The Fe system peak may be reduced by covering the offending parts of the specimen chamber with a low-Z element such as carbon (consult the microscope manufacturer).

Experiment 5.9: False Peaks: Figures A5.7a,b show the danger of not taking enough counts and trying to smooth the spectrum mathematically. Random background counts can look like small peaks after a smoothing operation. In general, smoothing is an unnecessary and potentially misleading operation. If one feels compelled to have a spectrum with a smooth background, don't smooth--take more counts!

Experiment 5.10: Pulse Pileup. Magnesium pulses may be so close to the noise level that the discriminator cannot reject them. This allows a pulse pile-up continuum to build up just below the MgK_α sum peak.

Experiment 5.11: Pile-up Correction Failure. The counts lost from the peak which appear in the pile-up continuum represent a failure in the pile-up correction which arises from a fundamental limitation in the circuitry. Around the energy of MgK (1.25 keV), the pulse heights delivered by the fast channel are falling into the electronic noise, thus defeating the efforts of the circuits to make an accurate pile-up correction. For peaks with energies below 1.25 keV, there is no pile-up rejection, so these peaks should be acquired at low count rates. This phenomenon has serious implications for quantitative analysis of light elements, and is one more reason that quantitative light element analysis should be carried out with wavelength-dispersive spectrometry.

ENERGY-DISPERSIVE X-RAY SPECTROMETRY

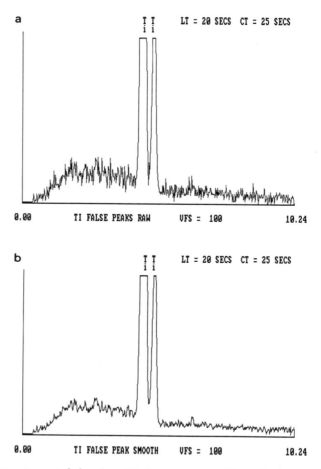

Figure A5.7. Spectrum containing insufficient counts collected over too short a time. (a) Raw unsmoothed spectrum. (b) Mathematical smoothing of spectrum (a). Small apparent peaks arise that are not present in Figure A5.6. (Note also that the sum peak of Figure A5.6 completely absent while there is evidence for the escape peak and the iron system peak.)

Experiment 5.12: Incomplete Charge Collection. The shape of the CuK_α peak is nearly an ideal gaussian, whereas the K_α peak typically shows a slightly higher tail on the low energy side. If your MCA can generate a gaussian peak for the K_α, assess the deviation from the ideal gaussian.

Laboratory 6

Energy-Dispersive X-Ray Microanalysis

6.1 Families of X-Ray Spectra

With increasing atomic number each x-ray family tends to spread out revealing individual lines that were unresolved for lower-Z elements.

Experiment 6.1: K, L, and M Spectra. For the lighter elements ($Z < 26$) only a K series (unresolved for $Z < 16$) is detected with a Be window detector. The aluminum spectrum in Figure A6.1a shows a single peak containing the unresolved K_α and K_β lines). Copper will exhibit an L peak in which the three main L lines are not resolved (see Figure A6.1b). For heavier elements, the L series is more widely spaced in energy such that the separate L_α, L_β, L_γ peaks are resolved (see the gold spectrum in Figure A6.1c). Note the weak peak on the low-energy side of the L_α, the Ll. For the heaviest elements a single strong M line (with weak M lines on either side) can be observed in addition to the L-series (see Figure A6.1d). It is important to be aware of the weak lines that accompany the L and M series so that they will not be identified as another element.

Experiments 6.2: K Family X-Rays. These spectra show how the K spectra appear for elements of increasing atomic number. The lines spread out and are resolved for the heavier elements. For Ti, Fe, and Cu, note the presence of well-resolved K_α and K_β peaks with an approximate peak height ratio of 10:1. Note the difference in the appearance of the K family for Al, Si, and S. The K_β peak is not resolved for Al, and it only forms a slight distortion on the high-energy side of the K_α peak for Si. For sulfur, the K_β forms a significant distortion on the high energy side of the K_α peak but is not completely resolved.

Experiment 6.3: L Family X-Rays. For Cu, only one peak is observed. For Zr, the single peak is slightly distorted on the high-energy side. For Ag the existence of several L peaks can be observed. For Ba, all of the significant L-family lines are resolved. For Au, the lines are completely resolved, and the complexity of the L family for high-atomic-number elements is well-illustrated.

Experiment 6.4: M Family X-Rays. For Ta, only one slightly distorted peak is observed. For Au and Pb, the unresolved M_α and M_β peaks produce a distinctly asymmetric peak. Finally, for uranium, the separation of M_α and M_β is sufficient to observe them as separate peaks.

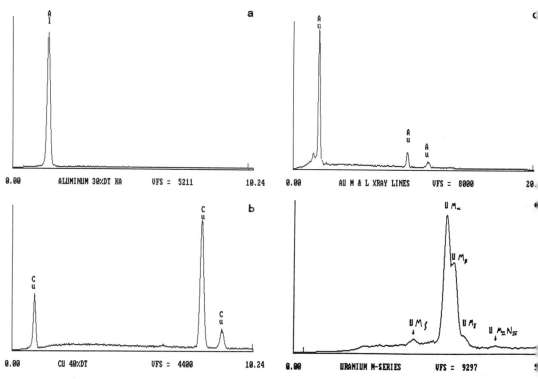

Figure A6.1. Examples of K, L, and M spectral families for selected elements. (a) Aluminum; (b) copper; (c) gold; (d) uranium.

6.2 Qualitative Analysis

Experiment 6.5: Qualitative Analysis of a Simple Spectrum. Figure A6.2 shows an EDS x-ray spectrum containing four peaks. Using the qualitative analysis guidelines, we start with the highest energy peaks. In this case, the K family of iron is recognized by the K_α peak (6.40 keV) and the K_β peak (7.05 keV) at about 10% of the K_α intensity. Next we look for the L series for Fe (0.70 keV). While there is a peak at about 0.55 keV, no peak is apparent at 0.70 keV. There is a slight bump in the background at 4.65 keV which is the Si escape peak for Fe K_α (6.40 keV - 1.74 keV = 4.66 keV). Next we analyze the largest peak in the spectrum at 2.30 keV. This peak could either be the sulfur K_α line, the molybdenum L_α line, or the lead M_α line. This peak is very strong and is not associated with other weaker peaks; therefore, it must be the unresolved sulfur $K_{\alpha,\beta}$ pair. If small amounts of Mo or Pb were present in this specimen they could not be identified without examining higher energy lines in the 10-20 keV range. The small peak at 0.55 keV can be recognized as the Si escape peak for sulfur K_α (2.30 keV - 1.74 keV = 0.56 keV). Thus the specimen contains only iron and sulfur (FeS_2).

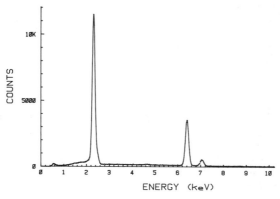

Figure A6.2. Simple spectrum for qualitative analysis. Specimen is iron pyrite (FeS$_2$).

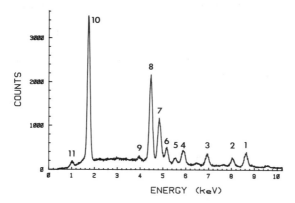

Figure A6.3. Complex spectrum for qualitative analysis. Specimen is a glass containing many dissolved metal atoms. Peaks are numbered for identification in Table A6.1.

Table A6.1. Analysis of a Complex Spectrum

Peak No.	Measured Energy	Identification (Energy)	Remarks
1	8.62 keV	ZnK_α (8.63 keV)	K_β found at 9.58 keV
2	8.04 keV	CuK_α (8.04 keV)	K_β found at 8.90 keV
3	6.92 keV	CoK_α (6.93 keV)	K_β found at 7.62 keV
4	5.90 keV	MnK_α (5.90 keV)	K_β found at 6.50 keV
5	5.54 keV	BaL_γ (5.53 keV)	small peak difficult to identify by itself
6	5.18 keV	Ba$L_{\beta 2}$ (5.16 keV)	
7	4.84 keV	Ba$L_{\beta 1}$ (4.83 keV)	
8	4.46 keV	BaL_α (4.47 keV)	
9	3.96 keV	BaLl (3.95 keV)	
10	1.74 keV	SiK (1.74 keV)	K_β not resolved
11	1.04 keV	NaK (1.04 keV)	K_β not resolved

Experiment 6.6: Qualitative Analysis of a Complex Spectrum. Figure A6.3 shows a spectrum containing at least 10 recognizable peaks. Again working from high energy to low energy for the large peaks, we note the energies of these peaks in Table A6.1.

There are two approaches to this spectrum. Starting at the highest energy peak, which is several times the background intensity (peak 1), we can identify this as ZnK_α both by energy and by the fact there is a small peak at the appropriate energy for ZnK_β. Peaks 2, 3, and 4 may be identified in a similar manner. Peak 5 is more than 130 eV away from the nearest K_α line, whereas peaks 1-4 could be identified to within about 30 eV. Since peak 5 is not near a K_β peak and is at an energy too high to be an M line, it may be part of an L-series (skip this peak for now). Peak 6 is also more than 100 eV away from any K-line (skip this peak also). Peak 7 and 8 can be identified as the BaL_β and L_α peaks, respectively. At this point it is obvious that peaks 5, 6, and 9 are the BaL_α, $BaL_{\beta 2}$, and $BaLl$, respectively. Peaks 10 and 11 can be identified as the SiK and NaK peaks for which the K_β lines are not resolved.

Experiment 6.7: Automatic Qualitative Analysis. It should be clear from a comparison of your manual peak identification with the elements identified by the automatic peak search that caution is always necessary with such automatic routines. In cases where the automatic search yields ambiguous results the manual methods must still be used. When the peaks are too small or the spectrum too noisy, the automatic search may fail to recognize a peak. In the worst case, the data file could even be damaged, leading to incorrect identifications. The value of experience, intuition, and common sense in the analysis of spectra cannot be overemphasized.

6.3 Quantitative Analysis

Experiment 6.8: Establishing Proper Working Conditions. The new copper spectrum should have the same shape as the library standard Cu spectrum although the relative intensity may be slightly different. The quantitative software will scale the new spectrum vertically to match the stored spectrum. Any large deviations in calibration or peak shape may cause difficulty in peak stripping and background subtraction.

Experiment 6.9: ZAF versus Standardless. Table A6.2 below shows a comparison of the ZAF and standardless output.

Table A6.2. Comparison of ZAF and Standardless Methods

Method	Line	K ratio	[Z]	[A]	[F]	At.%	Wt.%
ZAF	NiK_α	0.6700	1.025	1.001	1.000	51.05	68.70
	AlK	0.1277	0.970	2.447	1.000	48.95	30.28
						100.00	98.98
Standardless	NiK_α	0.6760				50.85	69.24
	AlK	0.1240				49.15	30.76
						100.00	100.00

As expected, the largest ZAF correction factor is the absorption correction for aluminum. The two elements emit x-rays of widely different energy so the fluorescence correction factor is negligible. There is nothing significant about the fact that the concentrations of the elements add up to 100%. In fact, if they do, it may indicate that the measurements are not independent. The precision can only be assessed by knowing the total number of counts accumulated in the peaks. Thus, the precision implied by the four significant figures may not be justified. This particular standardless routine uses the stored library spectra for pure Ni and Al to produce a K ratio, and then uses the ZAF program to calculate compositions. This works well as long as the library spectra and the unknown spectrum were taken under identical conditions.

Experiment 6.10: Effect of Take-off Angle. A small change in take-off angle due to a different specimen tilt can result in significant error if the absorption correction is large as it is in Ni-Al. Table A6.3 shows the magnitude of this effect.

Table A6.3. Effect of Take-Off Angle Changes on Composition

Take-off Angle	Element	K ratio	[Z]	[A]	[F]	At.%	Wt.%
30°	NiK_α	0.640	1.027	1.001	1.000	46.96	65.83
	AlK	0.122	0.970	2.893	1.000	53.04	34.17
40°	NiK_α	0.677	1.024	1.001	1.000	51.06	69.42
	AlK	0.123	0.968	2.571	1.000	48.94	30.58
50°	NiK_α	0.701	1.022	1.001	1.000	53.76	71.67
	AlK	0.124	0.967	2.369	1.000	46.24	28.33

Note that the change in take-off angle from 30° to 50° causes a 17% change (error) in Al wt% composition. Since the microprobe used did not allow specimen tilting, the above results were generated by changing the take-off angle in the ZAF program.

Experiment 6.11: Effect of Beam Energy. Both ZAF and standardless programs will give inaccurate results if the voltage used in the calculation is not the same as the voltage used in taking the data. In fact, it is possible for the actual accelerating voltage to be in error by several hundred volts without the operator being aware of it. But the effects of voltage differences are most serious in standardless analysis, where library spectra taken at a different voltage will produce results that are seriously in error. For example, a standardless analysis calculated using 15 kV when the data was taken at 20 kV yields 82.86 wt% Ni instead of 69.24 wt% Ni (Table A6.2). This is an error of nearly 20%! Spatial resolution and element excitation considerations often dictate an accelerating voltage other than 20 kV, rendering the library spectra and the resultant standardless analysis useless.

References

[1] C. E. Fiori and D. E. Newbury, *Scanning Electron Microscopy/1978*, vol. I, SEM Inc., AMF O'Hare, IL, p. 401.

[2] J. A. Bearden, "X-Ray Wavelengths and X-Ray Atomic Energy Levels," NSRDS-NBS 14, National Bureau of Standards, Washington (1967). Also published in recent editions of the *CRC Handbook of Chemistry and Physics*, The Chemical Rubber Company, Cleveland, Ohio.

Laboratory 7

Wavelength-Dispersive X-Ray Spectrometry and Microanalysis

7.1 WDS Operating Conditions

Experiment 7.1: Setting Beam Current and Focusing Beam. From either Figure A2.5 of Laboratory 2 or Figure 2.16a of *SEMXM*, a 30-nA beam may be estimated to be about 200 nm in diameter. The condenser lens controls the beam current available at the specimen, while the focus of the electron beam on the specimen is controlled by the objective lens.

Experiment 7.2: X-Ray Spectrometer Setup. The diffraction angle 2θ for FeK_α can be found from $n\lambda = 2d \sin\theta$. Thus, to detect first-order ($n = 1$) FeK_α, $2\theta = 57.45°$ and to detect second-order ($n = 2$) FeK_α, $2\theta = 148.00°$. For quantitative analysis, the same x-ray take-off angle (the angle between the sample surface and the x-ray beam to the analyzing crystal) must be used at each analysis point to ensure an accurate absorption correction.

Experiment 7.3: X-Ray Intensity versus Sample Current. X-ray intensity varies directly with specimen current. If the specimen current varies over time, then standard x-ray intensities will not be reproducible. For x-ray count rates exceeding about 25,000 counts/sec, the x-ray detector exhibits a dead time. After a photon enters the detector, there is a "dead" interval during which the system cannot respond to another pulse. From a plot such as Figure A7.1 the dead time may be calculated (see *SEMXM*, p. 427). Note that the maximum count rate achievable with the WDS is many times greater than that typically achievable for the EDS. In fact, the beam current typically would be reduced to 1 nA or less for the EDS to be used at all.

7.2 Characteristics of the WDS

Experiment 7.4: Energy Resolution of the WDS. Energy resolution of the WDS is ~10 eV. Energy resolution for the EDS is ~180 eV. Thus, for the WDS the $K_{\alpha 1}$, $K_{\alpha 2}$ doublet of iron ($\Delta E = 13$ eV) can be resolved as shown in Figure A7.2.

WDS resolution depends on Bragg's Law. As the size of the focusing circle of the WDS decreases, the energy resolution gets worse. EDS resolution depends on the energy resolution of Si(Li) detector and amplifier.

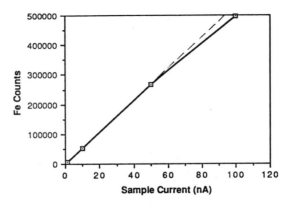

Figure A7.1. Iron x-ray counts versus sample current over a wide range of currents. Deviation from linearity above 50 nA (>25,000 counts/sec) indicates a dead time.

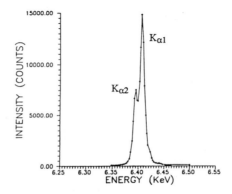

Figure A7.2. Resolution of the FeK_α doublet at 20 kV and 30-nA sample current.

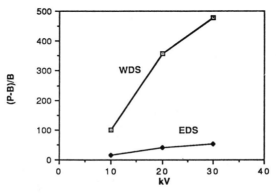

Figure A7.3. *P/B* versus voltage for WDS and EDS. Note that *P/B* for the WDS is about 10 times that for EDS.

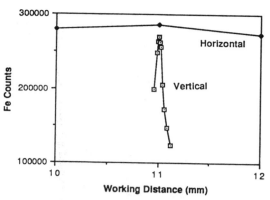

Figure A7.4. FeK_α x-ray intensity versus working distance for vertical and horizontal WDS systems

---

Experiment 7.5: Peak-to-Background Ratio for WDS. The P/B ratio increases with increasing voltage as shown in Figure A7.3. However, the penetration depth of the electron beam also increases so the spatial resolution gets worse with increasing voltage. The P/B at 20 kV for WDS is approximately 350 on iron, whereas the P/B at 20 kV for EDS is approximately 40 on iron.

Experiment 7.6: X-Ray Intensity versus Working Distance. The peak intensity for a vertical WDS is very sensitive to sample height. Flat polished specimens provide constant sample height across the specimen. A change in height of only 0.1 mm (100 µm) can reduce the x-ray peak intensity in a vertical spectrometer by 50% (see Figure A7.4). A horizontal or inclined WDS is much less sensitive to specimen height (working distance) and thus is most commonly employed on SEMs without light optical microscopes. Even EDS detectors can be sensitive to specimen height if the different parts of the detector are illuminated as the height changes.

7.3 Typical WDS Microanalysis Situations

Experiment 7.7: Characteristic X-Ray Peak Overlaps. Several peak overlaps in the EDS are so bad that an element may be completely missed or may be very difficult to measure quantitatively. One such overlap is MnK_α and CrK_β (Figure A7.5). These two peaks are easily resolved with WDS, as shown in Figure A7.6.

Experiment 7.8: Quantitative Analysis. The K ratio for Mn is formed by dividing the Mn counts obtained on the 316 stainless steel by the Mn counts obtained on pure Mn. The K ratio must be multiplied by the ZAF factors to obtain the Mn concentration. The following typical data are for the WDS and the EDS collected simultaneously at a take-off angle of 40° for 100 sec with 5 nA beam current. For the EDS the MnK_α had to be deconvoluted from the CrK_β.

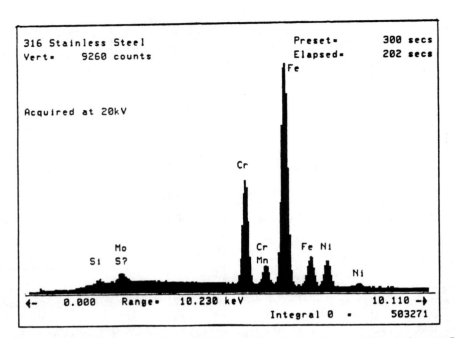

Figure A7.5. EDS spectrum of 316 stainless steel. Note serious peak overlaps at MoL/SK and MnK_α/CrK_β.

	WDS	EDS
K ratio for Mn	0.016	0.178
Z (atomic number)	0.973	0.974
A (absorption)	1.057	1.058
F (fluorescence)	0.966	0.966
(K ratio) x (ZAF)	1.6 ± 0.2 wt%	1.8 ± 0.5 wt%

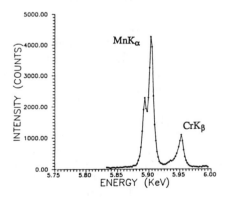

Figure A7.6. WDS spectrum acquired in the same time as the EDS spectrum in Figure A7.5. The MnK_α is clearly resolved from the CrK_β (plotted on an energy scale). In addition for WDS both P and P/B are much larger than with EDS, yielding about ten times lower (better) minimum detectable elemental concentrations.

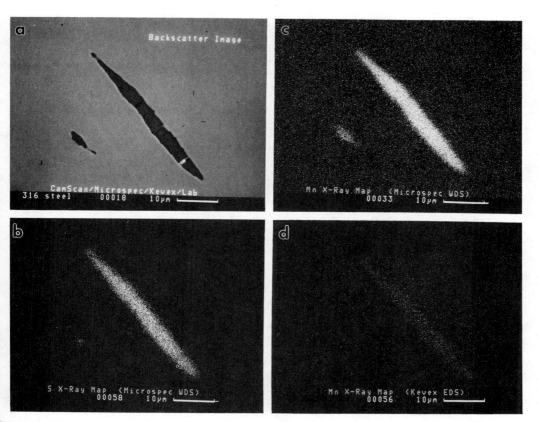

Figure A7.7. X-ray dot maps of manganese sulfide inclusion in 316 stainless steel at 20 kV. (a) Backscatter image. (b) Sulfur map by WDS. (c) Manganese map by WDS. (d) Manganese map by EDS.

Experiment 7.9: X-Ray Dot Maps. The x-ray dot maps in Figure A7.7 show that the contrast in the WDS manganese x-ray map is higher than that in the EDS map because of the greater number of counts obtained in the peak as compared to the background. The sulfur map probably could not be obtained by EDS since it is generally not possible to deconvolute SK from MoL in EDS dot maps.

Part II

ADVANCED SCANNING ELECTRON MICROSCOPY

SOLUTIONS

Laboratory 8
Backscattered Electron Imaging

8.2 Everhart-Thornley Detector as a BSE Detector

Experiment 8.1: E-T Detector Collection Efficiency. The solid angle and efficiency of a specific E-T detector for direct collection of BSEs is:
(a) Area, A, of scintillator (cm^2) = 1.5 cm^2.
(b) Distance, r, from specimen to scintillator (cm) = 4 cm.
(c) Solid angle, $\Omega = A/r^2 = 1.5/16 = 0.094$ steradians.
(e) For a specimen set normal to the beam (0° tilt), approximate take-off angle = 45°.

Experiment 8.2: Imaging with the E-T Detector. Figure A8.1 shows a positively-biased E-T image of the Raney nickel alloy. The image contains a mixture of atomic number contrast and topographic contrast (from SEs generated by the electron beam). Some BSEs generated by the electron beam are also collected if they have a direct line of sight to the detector. Note that the detector must be located at the top of the photo (the bright side of the features). The atomic number contrast is only weakly visible, and two phases can be made out only with difficulty. If the detector were negatively biased, the only electrons collected would

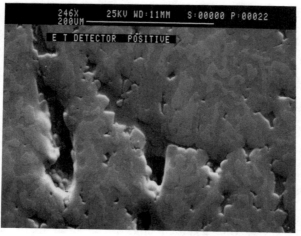

Figure A8.1. Electron image of Raney nickel with a positively biased Everhart-Thornley detector collecting both SEs and BSEs. Note the strong topographical contrast but weak atomic number contrast between the two phases.

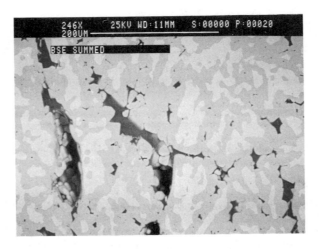

Figure A8.2. Image from BSE detector in sum mode showing strong atomic number contrast on Raney nickel specimen.

be BSEs that have a direct line-of-sight into the scintillator. Topographic contrast would be very strong owing to the strong asymmetry of the collection and the relatively low detector take-off angle. Atomic number contrast would be extremely weak (see the images of Laboratory 3).

8.3 Dedicated BSE Detector in Sum Mode

Experiment 8.3: Dedicated BSE Detector Collection Efficiency. The solid angle and efficiency for direct collection into the dedicated BSE detector in the sum mode (four segments added) is:
(a) Area, A, of scintillator or solid state diode (cm²) = 1.2 cm² x 4 = 4.8 cm².
(b) Distance, r, from specimen to detector (cm) = 1.5 cm.
(c) Solid angle, $\Omega = A/r^2 = 1.2$ cm²/1.5^2 = 0.53 steradians.
(d) Relative efficiency of a single detector $\varepsilon = \Omega/2\pi \times 100 = (0.53\ \text{sr}/2\pi) \times 100 = 8.5\%$. Relative geometric efficiency of entire detector: $\varepsilon_{total} = 4 \times \varepsilon = 34\%$.
(e) For a specimen set normal to the beam (0° tilt), approximate take-off angle = 70°.

Experiment 8.4: Imaging with a Dedicated BSE Detector in Sum Mode. Figure A8.2 shows a sum mode image of the Raney nickel alloy prepared with a solid state diode type of dedicated BSE detector placed above the specimen. The image contains a mixture of atomic number contrast and topographic contrast. The atomic number contrast dominates the image and the separate phases can be made out easily. The topographic contrast is now relatively weak because of the symmetry of the collection and the high detector take-off angle.

For a tilt of 60° the contrast is reduced because (1) the backscattered electron coefficients for all materials increase toward unity as the tilt angle increases to high values and (2) the directional nature of backscattering from tilted surfaces directs electrons away from the BSE detector placed above the specimen.

8.4 Dedicated BSE Detector in Difference Mode

Experiment 8.5: Imaging with the BSE Detector in Difference Mode. Figure A8.3 shows a set of difference mode images of the Raney nickel specimen prepared with a four-segment BSE detector. Various combinations of detector subtractions are depicted: Figure A8.3a, top-bottom; A8.3b, bottom-top; A8.3c, right-left; A8.3d, left-right. Several observations can be made:
1. The contrast is dominated by topographic effects, and atomic number contrast effects are suppressed.
2. The apparent lighting of the topography appears to come from the side of the image from which detector the signal is subtracted; i.e., when the bottom-top combination is used, the illumination appears to come from the bottom of the image. Thus, the sense of the topography depends strongly on the exact way in which the subtraction takes place.
3. The contrast depends on the orientation of a feature relative to the line which connects the two detector positions in an image. Features which run perpendicular to this line are seen in high contrast, while features which run parallel to this line virtually disappear.

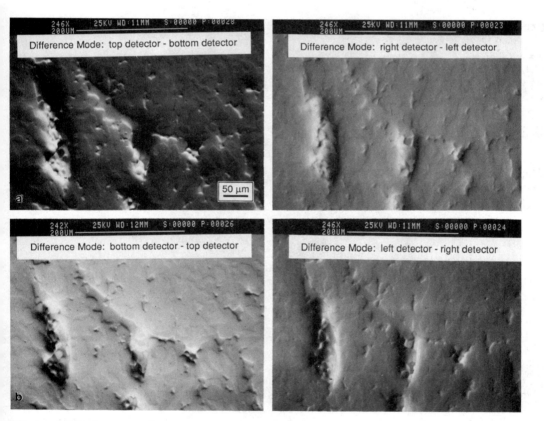

Figure A8.3. Images of Raney nickel with BSE detector in difference mode showing the effects of various possible subtraction. (a) Top-bottom; (b) bottom-top; (c) right-left; (d) left-right.

8.5 Specimen Current Signal as a BSE Detector

Experiment 8.7: BSE Imaging Using Specimen Current. Since electrons not emitted from the specimen are conducted to ground, the direct specimen current image should look like the negative of the direct BSE image. Consequently, the inverted specimen current image should be very similar to the direct BSE but with very little indication of specimen

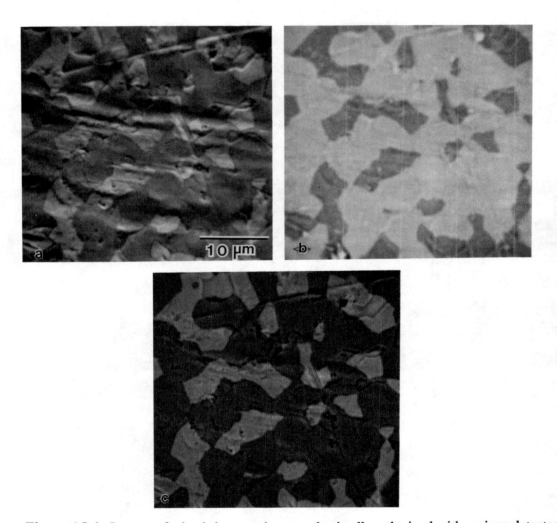

Figure A8.4. Images of a lead-tin eutectic superplastic alloy obtained with various detectors: (a) positively biased E-T detector image, revealing both atomic number contrast (lead-rich phase bright) and topography (shallow scratches); (b) direct specimen current image, showing predominantly atomic number contrast, with the sense of contrast reversed from that seen with the E-T detector (lead-rich phase dark); (c) inverted specimen image, with the sense of the atomic number contrast the same as that observed with the E-T detector. This image shows clear atomic number contrast without interference from the strong topographic contrast observed in (a).

topographical contrast. Contrast from topographic features arises primarily due to the directionality of electron emission which does not influence the specimen current signal. Specimen current is sensitive to number effects related to electron emission with the result that atomic number contrast dominates.

Figure A8.4a shows a positively biased E-T detector image with the correct sense of atomic number contrast. Figure A8.4b shows a direct specimen current image, and Figure A8.4c shows the same image after INVERT processing has been applied. The following observations can be made:
1. The atomic number contrast is easily seen in the inverted SC image (image is very similar to the BSE image).
2. Topographic contrast is very weak because the strong trajectory component of the contrast is completely suppressed in the SC image.

Laboratory 9
Scanning Transmission Imaging in the SEM

9.1 Scanning Transmission Electron Detectors

Experiment 9.1: Characteristics of the STEM Detector. The following calculation uses the characteristics of a particular scintillator STEM detector on a fiber optic:

Detector area = 0.79 cm^2
Specimen to detector distance = 5 cm
Solid angle Ω = area/(distance)^2 = 0.79 cm^2/5^2 = 0.031 steradians

9.2 STEM Images

Experiment 9.2: Comparison of SEM and STEM Images. Figure A9.1 shows a pair of images of the same field of view of mineral particles (antigorite) obtained with the E-T and STEM detectors with a frame time of 4 sec. Figure A9.2 shows the same images photographed with a longer frame time to improve the signal-to-noise ratio. The following observations can be made:

1. The E-T image shows much more of a noise component than the STEM detector because the signal generated from scattering off thin objects is small. The STEM detector is well placed for signal collection from weakly scattering small objects.
2. Many of the particles are sufficiently thick to completely block the transmission of the electron signal. Many particles can be seen to possess a complicated structure which cannot be observed in the corresponding SEM image.
3. The STEM detector image provides information on the internal structure of the thin particles.
4. The E-T detector image provides information on the surface structure of the particles, particularly the larger particles which are too thick for electron penetration in STEM.
5. Thus, it is of great value to have both the SEM and STEM images for complex particle fields.

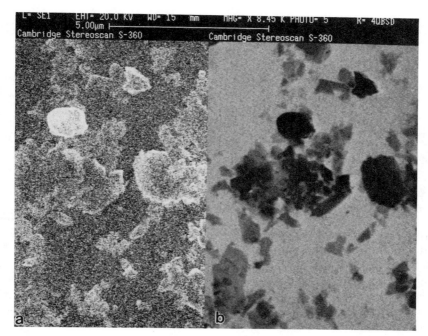

Figure A9.1. Comparison of SEM and STEM images of mineral particles on a carbon film. (a) Secondary electron image with the E-T detector above the specimen; (b) STEM image. Frame time = 4 sec.

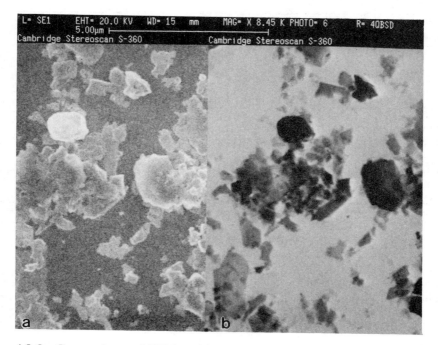

Figure A9.2. Comparison of SEM and STEM images. Same images as Figure A9.1 but taken with a longer frame time to improve the signal-to-noise ratio. Frame time = 40 sec.

Laboratory 10

Low-Voltage SEM

10.1 Topographic Contrast versus Beam Energy

Experiment 10.1: Estimating the Electron Range. The table below shows the dramatic decrease in the range of primary electrons as the beam energy is decreased.

Target: Steel (treat as Fe)	
Magnification: 1000x	
Pixel size: 0.1 mm	
Beam energy (keV)	Primary electron range
30.0	3.1 µm
15.0	0.99 µm
5.0	0.16 µm = 1.60 nm
2.5	0.05 µm = 50 nm
1.0	0.01 µm = 10 nm
0.5	0.0034 µm = 3.4 nm

Experiment 10.2: Contrast Effects at Low Beam Voltage. Figures A10.1a-f show the same region of a steel surface exposed by a hacksaw. At progressively lower beam energies surface details become more obvious and the charging characteristics of particles are modified.

10.2 Minimizing Charging Effects on Insulators

Experiment 10.3: Charging Effects. Figures A10.2a-d show part of a beam energy sequence on an uncoated surface of the mineral antigorite. The images at 15 keV (Figure A10.2a) and 5 keV (Figure A10.2b) are overwhelmed by gross charging effects, including strong effects on secondary electron collection (bright and dark areas), and in some locations by actual deflection of the primary beam. Images taken at 2 keV (Figure A10.2c) and 1.5 keV (Figure A10.2d) show much reduced charging artifacts, although locally there are still some areas of unusual brightness.

Having established a useful operating beam energy to reduce charging in the range $1.5 < E_0 < 2.0$ kV, a second series of images, shown in Figures A10.3a-d, was recorded at higher magnification in the center of the field-of-view. These photos were taken with the energy taken in 100-eV decrements from 1.9 keV to 1.5 keV. At first glance, these images appear identical, but upon careful examination, incipient charging artifacts are observed at 1.5 keV and 1.9 keV. The best images are those at 1.6 and 1.7 keV. Even those images appear to have some slight artifacts of charging.

10.3 Electron Beam Damage

Experiment 10.4: Electron Beam Damage of Polymers. When the polymer spheres shown in Figure A10.4a are imaged at high beam energies (>5 keV), the spheres are sometimes lost owing to the condition of mutual repulsion which results from charging. At lower energies, Figure A10.4b, the image of the particles may be stable, but each sphere is seen to have an apparent black spot on its surface which looks like a hole drilled into the surface. This is a stable charging artifact which can be reversibly induced and eliminated by operating at very low energy, < 1.5 keV, as shown in Figure A10.4c. Figure A10.5 shows high-resolution images of composite polymer spheres that collapse at 5 keV in a conventional SEM with a tungsten thermionic gun (Figure A10.5a) while the same type of specimen exhibits little or no damage at 1 keV with a field-emission gun instrument (Figure A10.5b).

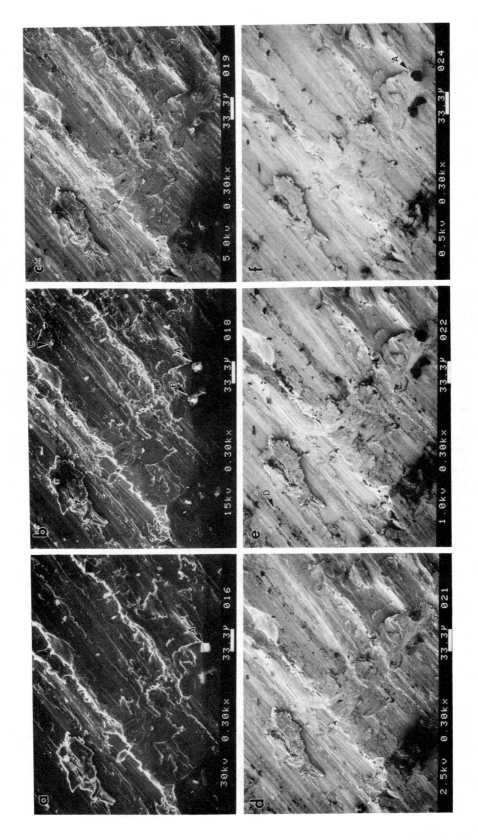

Figure A10.1. (a) Rough steel surface at 30 keV: The image is dominated by range effects at edges which produce extreme signal changes relative to regions away from an edge. Virtually all areas of the specimen not immediately adjacent to an edge show a constant level of signal. A region of negative charging (bright area, arrow labeled "A") is observed which is associated with a non-conducting surface particle or inclusion. The influence of the charged inclusion on the secondary electron collection field is evidenced by the extensive dark region near the charging particle. (b) 15 keV: General edge brightness similar to that at 30 keV is observed. The charging at the surface particle has become worse, and an adjacent particle (labeled "B") shows evidence of charging. Together the charging on these particles creates an even larger dark area of poor secondary electron collection. Two smaller particles (labeled "C") are also showing incipient charging effects because the reduced electron range causes more charge to be deposited in the particles. (c) 5 keV: General edge brightness similar to that at the higher beam energies is still observed. Most of the gross charging artifacts have been eliminated, but region "B" still shows a large dark area of poor secondary electron collection. (d) 2.5 keV: While some of the bright edge effect remains, the surface topographic contrast is no longer dominated by the edge effects. With the bright edge effect reduced, broad areas of the specimen surface now show brightness that is more representative of local surface tilt. Some of the objects which showed gross charging effects at higher beam energies are now revealed as distinct dark particles which have surface topography. (e) 1 keV: All of the bright edge effects have been eliminated, and virtually all of the charging artifacts are gone. The source of the charging is revealed to be a widely distributed class of particles, e.g., area D, which now appear anomalously dark relative to the rest of the surface. The origin of the very strong contrast of these particles against the surrounding metal (and native oxide surface), which is probably some form of compositional contrast, is not well understood. (f) 0.5 keV: In general, this image is similar to the 1-keV image, but there is less fine-scale detail visible. The range at 0.5 keV is so shallow that only the very outermost layers of the specimen are imaged. Many of the very fine black particles are lost, particularly those in area D. This may be due to a lack of penetration of any overlying contamination layer to reach the particles. Larger black particles, such as those at A remain visible.

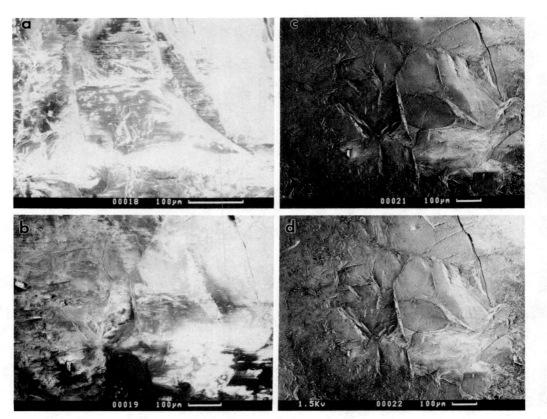

Figure A10.2. Charging effects versus beam energy. (a) 15 keV; (b) 5 keV; (c) 2 keV; (d) 1.5 keV.

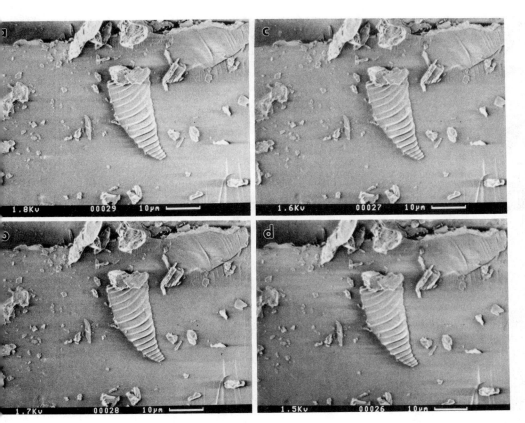

Figure A10.3. Finding the correct keV for uncoated specimens at higher magnification in the center of Figure 10A-2. (a) 1.8 keV; (b) 1.7 keV; (c) 1.6 keV; (d) 1.5 keV.

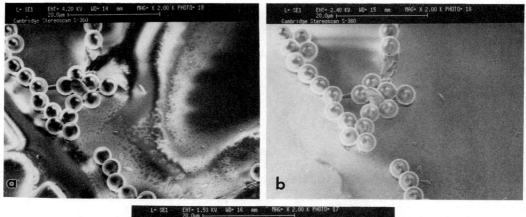
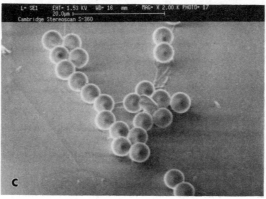

Figure A10.4. Polymer imaging. (a) At 4.2 keV the image shows high contrast but also local charging. (b) 2.4-keV particles show characteristic black spot on surface. (c) 1.53-keV image exhibits lower contrast but also beam damage and charging.

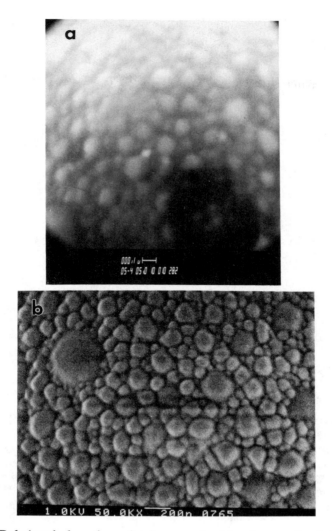

Figure A10.5. Poly(methyl methacrylate) covered with spheres of poly(styrene) latex (uncoated). (a) Sphere collapses (dark area) at 5 keV in conventional SEM. (b) 1-keV image of poly(styrene) surface spheres exhibiting no detectable beam damage. High resolution provided both by a field-emission gun and by through-the-lens extraction of the secondary electron signal.

Laboratory 11
High-Resolution SEM Imaging

11.2 Topographic SE-I Contrast

Type I imaging in SE signal mode provides contrast from signal electrons which have a small escape depth and an exit area diameter similar to the probe diameter. Such contrast provides the maximum topographic resolution for high-performance microscopes. Here, high resolution is defined as spatial resolution bulk surface features smaller than 10 nm. In the more conventional SE-II imaging mode such small details will not be imaged.

Experiment 11.1: Relief Contrast and Edge Brightness Contrast. Under normal operating conditions, the SE-I signal is buried by the BSE-dependent SE-II signal. At medium magnifications (< 50,000x) and high accelerating voltage (Figures A11.1a, c), SE-II electrons generate nontopographical contrast (atomic number contrast and mass-thickness contrast) and topographic contrast (relief contrast and edge brightness contrast) with the SE-IIa component. At low voltage (1 kV) relief contrast is lost and topographical contrast is reduced to edge brightness contrast (Figures A11.1b, d).

High-performance SEMs provide such small probe diameters (2 nm) that at high magnifications, good edge sharpness and distinct surface (SE-I contrast) features are easy to establish at high voltage (Figure A11.2a). At low voltages the beam size is larger and the fine details from SE-I contrast mechanisms are lost (Figure A11.2b). The SE-III component, if large, can reduce the SE-I S/N ratio to the point where the SE-II signal swamps it. Surfaces appear flat and small details of dimensions equal to the probe diameter are missing (Figure A11.2c).

If the background SE-III signal is reduced (using a BSE absorption plate, a converter plate with positive bias, an electromagnetic precollection field, or specimen biasing), SE-I contrast imaging may be possible. However, the SE-I component is so small that high beam currents are required to increase the S/N. If these conditions are met, SE-I contrast will be clearly visible and will dominate the image. At sufficient probe currents this SE-I signal presents a striking surface image with all expected topographic SE-I contrast (Figure A11.2d). Edge brightness contrast fringes measure 1-4 nm and the smallest bright particles are about 3-nm wide. SE-IIa contrast is also present, identifiable as edge brightness fringes and as pseudotopographic contrast, each 10-20 nm wide. However, thin layers of contamination often reduce the signal strength and obscure small particles. Microscope instabilities may further reduce image quality.

Experiment 11.2: Particle Contrast. The parameters for high-magnification imaging can be verified by imaging with SE-I particle contrast. This strong contrast is generated only

Figure A11.1. SE-I topographic contrast on bulk gold deposited on carbon taken with an unmodified SEM. Low magnification images (5000x) taken at (a) 30 kV and (b) 1 kV show similar atomic number contrast. Medium magnification images (50,000x) taken at (c) 30 kV and (d) 1 kV show different contrast effects. At high voltage (c) topographic details of the individual crystals can be seen (edge brightness and relief contrast) via SE-II components. Additionally, crystallographic contrast is observed. At low voltage (d) topographic information is reduced to edge fringes and atomic number contrast (no relief contrast).

on very small particles (particle diameter < probe size plus electron escape depth) at high magnifications. Locally enhanced surface potential may increase SE emission from small platinum particles and may account for the bright appearance of some of the particles (Figure A11.2b).

11.3 Nontopographic Contrast

Low (500x) and medium (20,000x) magnification images are dominated by Type II contrast. On inhomogeneous planar film specimens, the signal collected from thin surface layers depends on the atomic number of the layers. Images of thin carbon and platinum layers on silicon demonstrate the signal contribution of the different electron components. Similar layers are often used as surface coatings.

Experiment 11.3: Atomic Number Contrast and Mass-Thickness Contrast.
Only the 2-nm-thick platinum film is imaged by BSEs with a strong atomic number contrast. The thin carbon film does not generate a significantly different number of BSEs to be recognized (Figure A11.3a). The SE image (Figure A11.3b) shows both films, the platinum as bright islands and the carbon as dark islands. Part of the SE signal is generated by BSEs produced in the metal. Many of these BSEs must be nearly elastically backscattered and thus must have produced SE-IIa electrons. However, BSEs are also backscattered from the bulk silicon underneath the platinum film and can penetrate through the metal. Some SEs produced by these multiply scattered BSEs (SE-IIb) in the silicon will be absorbed by the metal. Therefore, the SE signal collected from the metal islands is composed of SE-IIa and b signals.

This mechanism of signal generation is confirmed by the carbon/silicon contrast. Only a few SE-IIa are expected to be produced in the carbon because the BSE yield is minimal. Therefore, the contrast is mostly produced by SE-IIb because most BSEs are expected to be produced in the bulk silicon. Since the SE yield for bulk silicon is higher than that of carbon

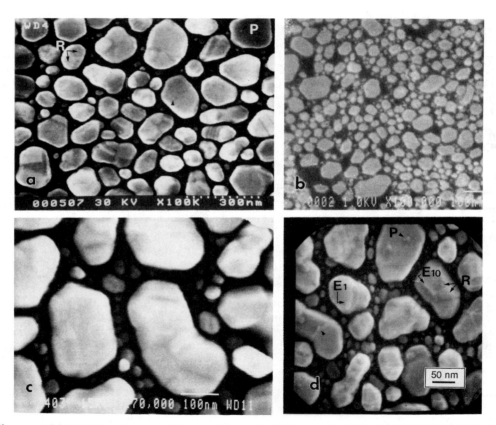

Figure A11.2. Microtopography specimen. At higher magnification (100,000x) the 30-kV image (a) reveals some topographic details in SE-I contrast (at P). The 1-kV image (b) exhibits very flat particle surfaces with no high-resolution detail. At the highest magnifications (approximately 200,000x) a 15-kV image from an unmodified SEM (c) exhibits a poor S/N ratio due to a large SE-III background signal and the SE-II signal component dominates the image. Image (d) shows enhanced SE-I topographic contrast due to an SEM modification (an SE-III absorption plate--see Laboratory 12) which reduces the SE-III component.

see *SEMXM*, Figure 3.29), most of the SEs produced in the silicon must have been absorbed in the 2-nm-thick carbon film. The remaining SEs, produced by the BSEs in the carbon, could not equal the number of SEs produced in the uncoated silicon surface. Therefore, the carbon islands are imaged dark. A similar signal generation mechanism is found at low voltage (Figure A11.3d). However, because most of the BSE emerge after multiple scattering from a depth of 20-50 nm, both thin films will be imaged by an SE-II atomic number contrast.

At low voltage, the signal exit area is strongly reduced through the reduced range of the primary electrons. The escape depth of the SEs remains unchanged, though the proportion of the thin film/bulk excitation volume of the SE signal is increased (Figures A11.3c,d). Lowering the magnification also gives an apparent increase in contrast on the platinum and carbon features (Figures A11.3e,f) and allows contrast comparisons to be made over a wider field of view. Note that the contamination layers also show a dark (mass-thickness) contrast which can obscure the contrast of the carbon on the silicon (Figure A11.3f).

These results indicate the role of thin contamination layers on Type II signal generation. At low magnifications atomic number contrast and mass-thickness contrasts dominate. However, as will be indicated in Experiment 11.4 (Type I imaging), despite the reduction in SE yield, topographic contrast generation is in principle not affected.

1.4 SE-I Imaging Requirements

Experiment 11.4: SE-I Yields from Thin Film Coatings. Thin films deposited on a bulk substrate produce strong atomic number and mass-thickness contrast at low magnifications. Several signal components contribute to the image contrast but cannot be discerned at low magnification (Figure A11.4a) because the exit areas of the different signals are imaged with only a few pixels. The range of 20-kV PEs in silicon is on the order of 2 µm. At 500x the exit area of the multiply scattered PE is described by a radius of similar dimensions. The signal collected from this exit area is used to produce the image. Assuming a 10-cm photographic image of 2000 lines at 500x, 2 µm will equal 20 lines (pixels). At higher magnification, the exit distance of the signal electrons can be better resolved, but image contrasts may significantly alter (Figure A11.4b). At 10,000x, the exit distance of the multiply scattered signal is contained within 400 pixels. However, it was shown in Section 11.2 that the main contrast features at this magnification are produced by the SE-IIa signals from the silicon and the platinum. On silicon the SE-IIa electrons exhibit an exit distance of 100-200 nm, but on the metal film, the exit distance may be on the order of the metric film thickness, i.e., 3-5 nm. At 10,000x these dimensions are represented by 20-40 pixels and by 1 pixel, respectively. Therefore, the 0.5-µm-wide fuzzy borders of the metal patches represent true mass-thickness contrast.

At higher magnifications, the contrast contribution to each pixel from the SE-IIa signal, generated in the silicon, is reduced. The SE-IIa component, produced in the platinum, provides the major contrast for small object elements. At 200,000x (Figure A11.4c) the platinum dependent signal exits from an area of ~20 pixels and enhances contrast of small details. Owing to the small excitation volume within the metal film, small variations in composition and thickness of materials are resolved, i.e., the 2-5-nm shadows between the particles and the platinum film. The SE-IIs, produced by BSE, contribute only a background signal since the silicon is of homogeneous composition. The SE-I signal component produced at the platinum surface cannot be identified but is expected to be smaller than the SE-IIa component because the S/N on the small platinum islands is so low.

On the carbon patches (Figure A11.4d), small particles can be identified and imaged with weak topographic contrast. Since very few SE-IIa are produced in the carbon film itself, this

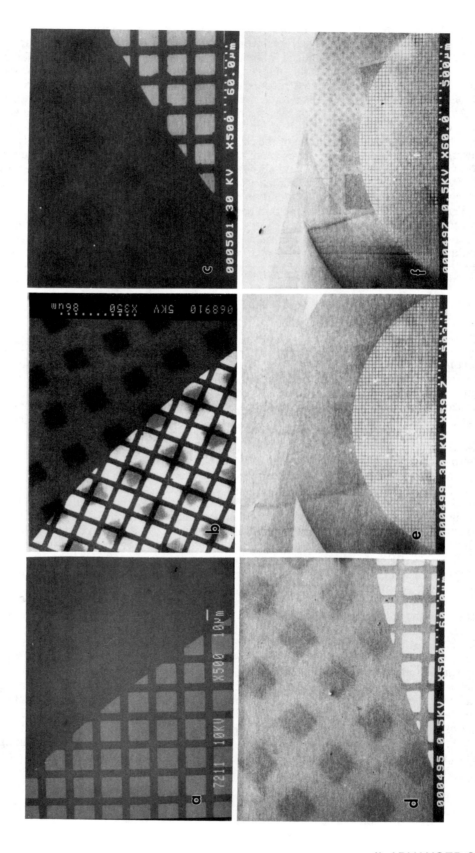

Figure A11.3. Thin-film nontopographic constrast. Inhomogeneous thin surface film specimen composed of 2 nm platinum and 2 nm carbon on bulk silicon. Whereas the BSE signal (a) images only the platinum areas as bright, the SE signal (b) also images the carbon film that was deposited over both silicon and platinum regions. A reduction of the signal excitation volume by lowering the voltage from 30 kv (c) to 0.5 kV (d) increases the atomic number contrast. In (d) the platinum regions appear bright while the carbon regions appear darker than the silicon substrate. A further increase of contrast is obtained by reduction of magnification. At 30 kV (e) only the platinum regions show up bright. At 0.5 kV (f) the platinum can be seen as bright regions and the carbon as dark regions. Although the S/N is reduced, the low-voltage image shows that very thin contamination layers (dark rectangles) can be recognized.

topographic contrast must represent pure SE-I signal. The nontopographic contrast between carbon and silicon is so small that it cannot be conclusively identified. The SE-II component generated by electrons backscattered from the silicon contributes only a background signal. Thus, the thin low-Z coating used here provides the best images of topographic features, and the "topographic plus Z contrast specimen" provides a good test for the quantitative relationship of SE-I to SE-II components.

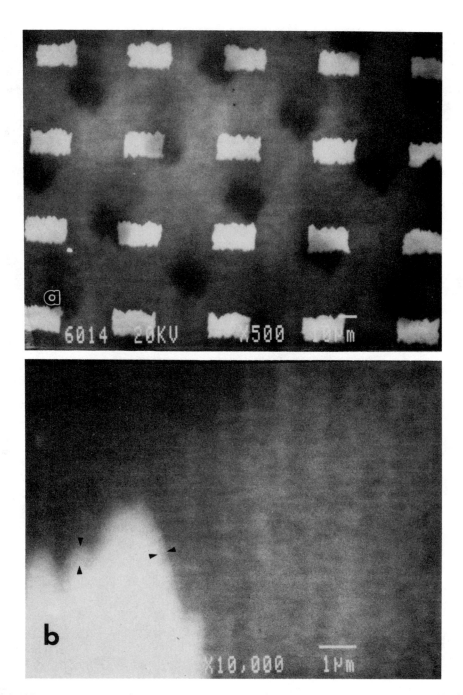

Figure A11.4. SE-I contrast on thin film coatings. Small low-Z particles on bulk silicon shadowed with 2 nm platinum or carbon. At low magnification (500x) the platinum and carbon film islands are imaged with strong SE-II atomic number contrast and mass-thickness contrast. At medium magnifications (10,000x) a thickness effect (mass-thickness contrast) is visible at the edges of the islands (arrows). No further structural details are resolved (pixel resolution limitation). At high magnifications (200,000x) fine structural details become visible.

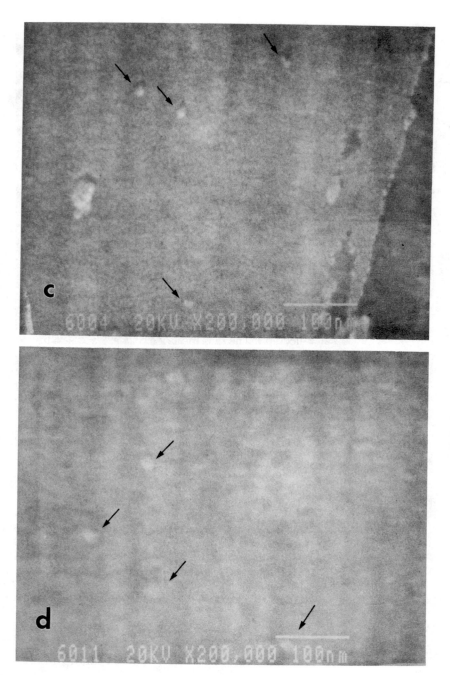

Figure A11.4 (cont'd). (c). Particles and their shadows are imaged by a SE-IIa signal with strong contrast. Both the platinum-decorated islands and the film discontinuities are resolved (arrows). On the carbon film at 200,000x (d), SE-I topographic contrast is high enough to image the particles with the same increased width (12-15 nm). This properly reflects the carbon deposition on the particles. Atomic number contrast is so low that the carbon/silicon shadow is not recognizable.

Laboratory 12

SE Signal Components

12.1 Enhancing SE Signal Collection Efficiency

The SE signal collection efficiency of the conventional E-T detector is limited by the asymmetry of the collection field resulting from the detector position and from the surface potential of rough specimens. Specimen biasing should be routinely applied to optimize signal collection for a given specimen and imaging situation. The voltage supply must be of extreme stability which can be provided by dry batteries such as 45-V farm batteries. Two batteries connected in series provide for an easy change of polarity. There is no rule to predict the effect of specimen biasing on signal collection. The bias modifies only the accelerating voltage of the probe and the collection field of the collector. Thus, SE components will still be collected. (*Note*: Only a grounded specimen grid which shields the specimen from all other biased surfaces allows establishment of a positive field between the grid and the specimen for the absorption of SE-I+II as described in experiments of Section 12.3.)

However, specimen biasing will distort the collection field of the E-T detector either improving or reducing the SE-I+II collection. Observation of the signal waveform while applying a bias and changing polarity (and field strength) provides a quick check of the usefulness of specimen biasing in a particular imaging situation.

Experiment 12.1: Biasing of Rough Specimens. On rough specimens and at higher magnifications either positive (or negative) specimen biasing may improve the SE-I+II collection. On silver crystals, a positive bias distorts the collection field to such an extent that the SE-I+II signal is not collected while the BSE-dependent SE-III signal is efficiently collected (see Figure A12.1a). Note edge brightness contrast associated with BSE electrons with about 300 nm exit radius. This contrast, also present in the unbiased specimen-specific signal, is produced by SE-IIb electrons. On this same area of the sample a negative specimen bias promotes effective SE-I+II collection (Figure A12.1b) allowing the imaging of fine surface details with the SE-I and SE-IIa signal components superimposed on an SE-IIb and SE-III background.

Experiment 12.2: Biasing of Flat Specimens. On flat surfaces the effectiveness of bias polarity is more predictable. A positive bias will favor SE-III collection. On inhomogeneous specimens such altered signal collection will enhance high-Z features in the image and reduce contrast in low-Z regions of the specimen (Figure A12.2a) as expected for BSE signal collection. Thin carbon films on platinum patches become barely visible and are imaged only with low S/N. However, negative specimen biasing (Figure A12.2b) enhances the contrast of carbon relative to both the platinum and the silicon support due to more efficient SE-I+II signal collection.

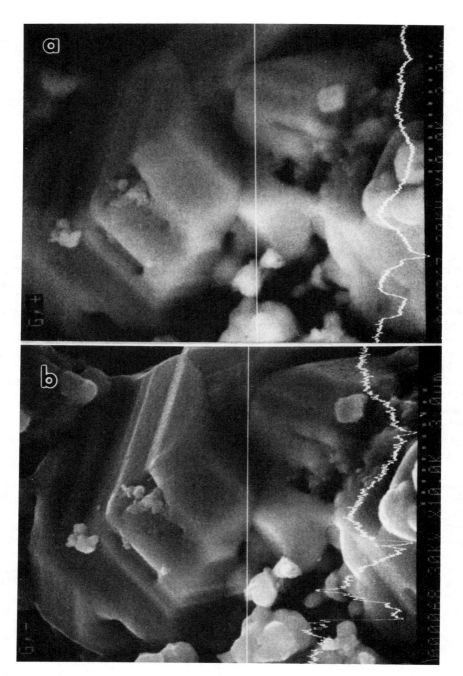

Figure A12.1. Specimen biasing of rough specimens. Silver crystals at 30 kV. (a) SE image with the specimen at +45 V showing suppressed SE-I+II signals from specimen relative to the lower resolution SE-III signal. (b) SE image with unchanged amplifier but the specimen at -45 V showing enhanced collection of SE-I+II components.

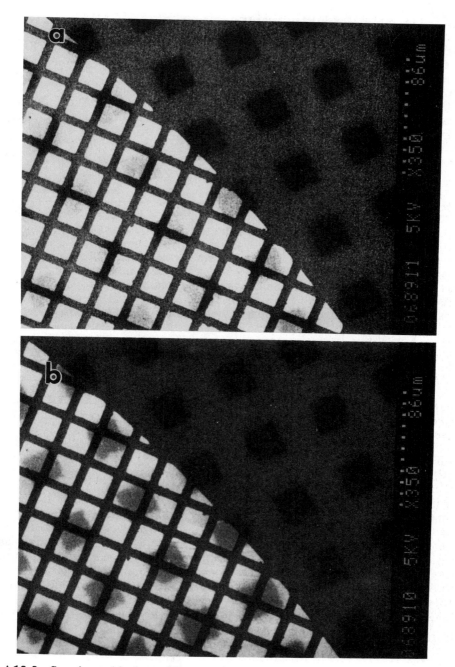

Figure A12.2. Specimen biasing of flat specimens. Inhomogeneous thin film specimen at 5 kV. (a) SE image with specimen at +45 V showing reduced specimen-specific SE-I+II signals. Carbon thin film (dark squares on right) is imaged with low S/N ratio. (b) SE image with specimen at -45 V showing enhanced SE-I+II collection providing better specimen-specific atomic number and mass-thickness contrast.

12.2 SE-III Signal Contributions

In a conventional SEM with the specimen position below the final lens, the SE-III signal component contributes a major fraction of the signal. The SE-III signal may be used as a "converted BSE" signal and is especially useful for low-voltage imaging if no special low-energy BSE detector is available. However, the SE-III component is also a background signal for the SE-I+II component and reduces specimen-specific SE contrast. At medium and high magnification, the SE component produced by multiply scattered PEs contributes a large exit radius signal background of SE-IIb. In a conventional microscope the SE-IIb signal is enhanced by the SE-III signal produced by BSEs at the pole piece and the specimen chamber. Therefore, control of the SE-III signal component becomes an essential procedure for SE imaging.

Experiment 12.3: Mirror Imaging of the Specimen Chamber. The origin of SE-III electrons can be visualized through mirror imaging of the specimen chamber (Figure A12.3). An electrostatic mirror lens of several keV is produced by scanning a small area of an insulator at high accelerating voltage. Then, a low-voltage probe is reflected on the concave electrostatic field and can easily be focused on the specimen chamber. SEs produced by the PEs of the probe are collected by the E-T detector and produce the contrast. This contrast is not produced by SE-IIIs which are generated by BSEs produced on an uncharged sample. However, the contrast seen in mirror images does indicate the collection efficiency for SEs produced at that particular area of the specimen chamber, particularly at the conical surfaces of the pole piece. The collector grid of the E-T detector also can potentially produce a high SE-III signal. The same mirror imaging can be used to produce EBIC contrast on the BSE detector (Figure A12.4).

Experiment 12.4: The Converter Plate. If the converter is placed in high take-off angle position underneath the pole piece, it can be used to convert BSEs to SE-IIIs. In the present experiment the plate controls the emission of SEs generated by reflected PEs. A negative biasing of the plate will emit an amplified SE signal (Figure 12.5a). A positive biasing will retain SEs produced at the converter. Thus, in normal imaging using the positive biased converter plate will suppress SE-IIIs caused by BSEs striking the polepiece.

A simpler but still effective way to modify SE-III generation may be provided by coated aluminum disks placed underneath the pole piece. The disks should be wide enough to optically shield the conical surfaces of the pole pieces from the specimen and should be coated either with carbon "DAG" (BSE absorption plate) or with gold (BSE amplification plate). The effect of such plates on image quality can be seen from a comparison of topographic contrast collected with the insertion of the gold plate (Figure A12.6a) or the carbon plate (Figure A12.6b). SE-IIa (and SE-I) signals are improved as the SE-III signal strength is reduced using the carbon-coated plate.

12.3 Quantitative Measurement of SE Signal Components

Experiment 12.6: Imaging of the Faraday Cup. The proportion of the different signal components relative to Type SE-IV can be measured by line scan procedures using a Faraday cup. Secondary electron imaging of the aperture surface provides a measurement of the total SE signal. The isolated SE-IV signal is collected in the hole allowing relative measurements to be made by scanning across the edge of the aperture. The collection of SE-as well as of SE-I+II is independent of the type of precollection field used for the E-T detector

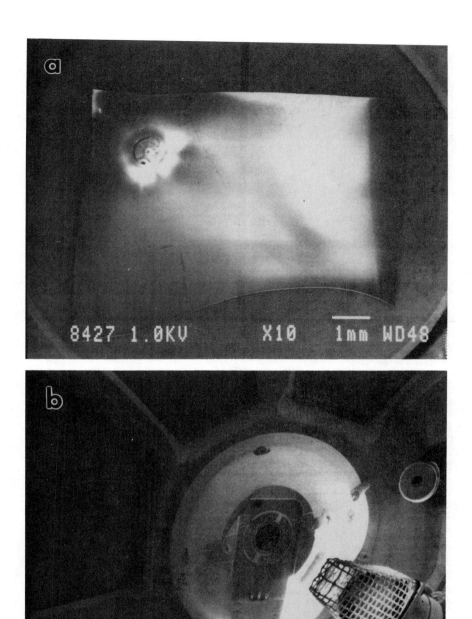

Figure A12.3. Mirror imaging with the SE detector. Charging of a glass cover slip at high voltage produces a local electrostatic mirror-lens. (a) At low magnification the deflecting part of the charge buildup can be seen at the insulating surface. (b) Mirror imaging on the charged area of glass reflects the primary electron beam to scan the specimen chamber. This technique indicates the collection efficiency for SEs produced at the surfaces of the specimen chamber (SE-IIIs during normal microscopy). A dedicated BSE detector can be seen mounted directly beneath the pole piece in addition to the E-T and EDS detectors.

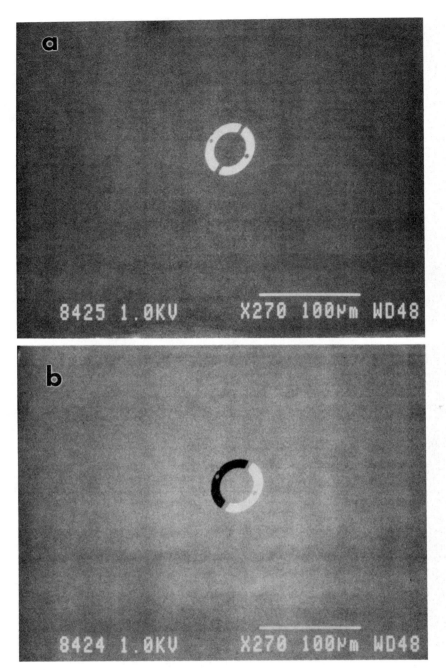

Figure A12.4. Mirror imaging with the BSE detector. The BSE detector can be imaged using the current induced by the primary electron beam in the BSE detector. (a) Image of the summed segments of the BSE detector. (b) Image of one segment only. The image background is independently adjusted through the amplifier settings.

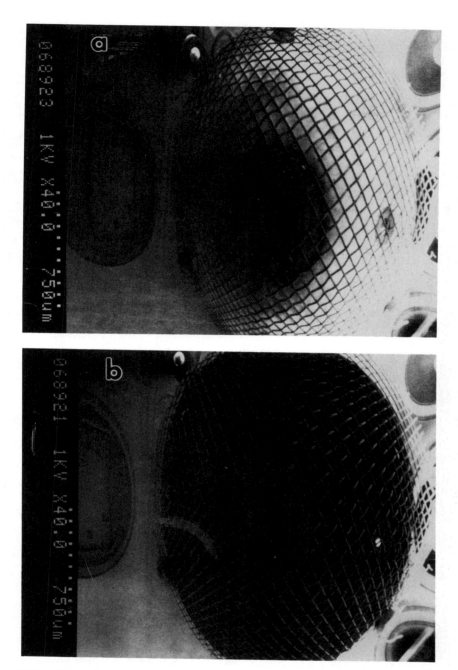

Figure A12.5. Function of the converter plate. SE emission from a converter mounted underneath the pole piece and imaged by an electrostatic mirror. (a) A negatively biased plate (-45 V) emits electrons. (b) Positively biased plate (+45 V) absorbs electrons.

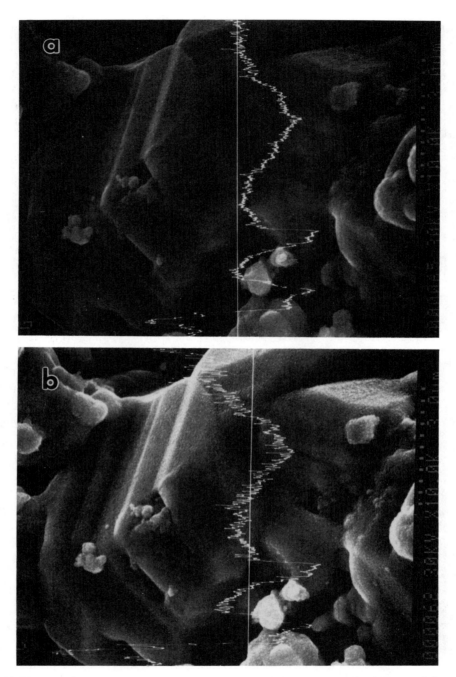

Figure A12.6. Effects of coated aluminum plates. Effects on the SE image of the two smaller plates (Figure 12.3) mounted on a retractable BSE detector for easy positioning underneath the pole piece. (a) The gold plate introduces additional SE-IIIs to the total SE signal and thus reduces the contrast of the signal proper. (b) Same image with a carbon-coated plate after adjusting the signal (in order to regain the same signal level). The contrast of the SE-I+II signal is increased and thus more topographic details are better visible.

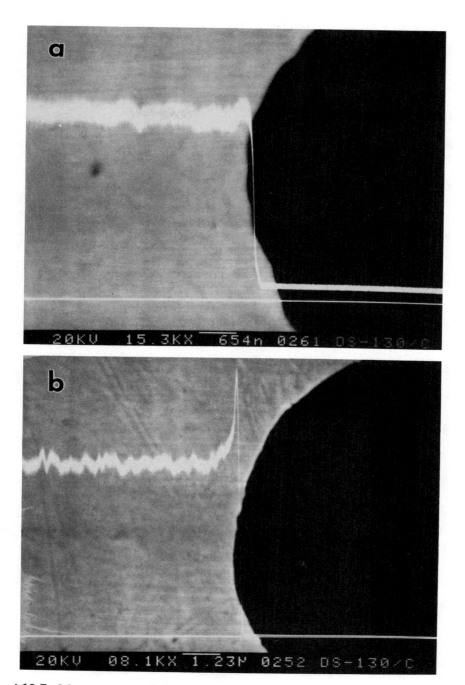

Figure A12.7. Measurement of signal components using a Faraday cup only. Image of the Faraday cup rim and of line scans over the center area using the SE signal and the amplifier baseline. (a) For an in-lens, specimen position, the electromagnetic precollection field collects only SE-I+II from the aperture, but collects a relatively high SE-IV signal over the cup. (b) For a below-lens specimen position, the electrical precollection field of the E-T detector collects SE-I+II+III on the aperture, but relatively few of SE-IVs over the cup.

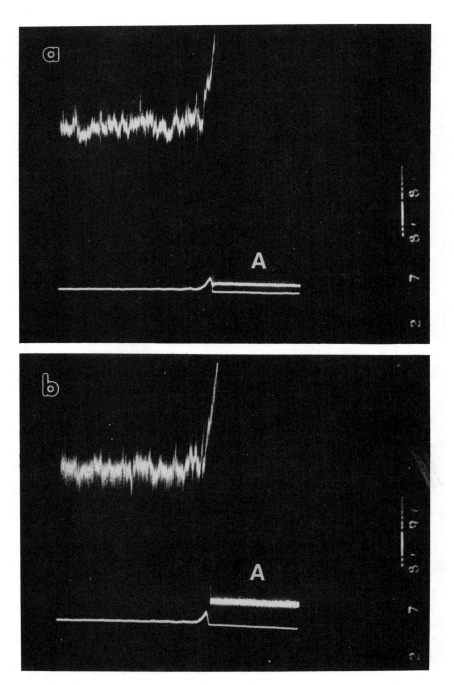

Figure A12.8. Effect of final aperture size on the SE signal. SE-IVs from the final aperture measured as a separate component over the Faraday cup. (a) Reduction of the aperture size from 120 μm (a) to 60 μm (b) causes a threefold increase in the SE-IV contribution to the total SE signal (compare the relative signals at A).

(a) Using the electromagnetic field of the in-lens specimen position, relatively high proportions of SE-IVs are measured: as high as 20% of the total SE signal (Figure A12.7a). However, for this collection mode, SE-III contributions are negligibly small. The electrical pre-collection field of the conventional below-lens arrangement sometimes collects up to 10% SE-IV from the final aperture; however, Figure A12.7b shows a case where SE-IVs were only 1% of the collected SE signal. A negative specimen bias would increase the signal on the aperture by enhancing the collection of the SE-I+II components (see Figure A12.16).

(b) Use of different size apertures, adjusting the signals through amplification to identical levels, demonstrates the effect of probe scattering at the final aperture (Figure A12.8). If the final aperture is located above the final lens (the virtual objective aperture) this effect will not be observed.

Experiment 12.7: Modifying the SE Signal with a Converter Plate. If a converter plate is mounted beneath the pole piece, the SE signal can be modified by biasing the plate during the scan on the aperture rim (Figure A12.9). If the plate is negatively biased, SE-IIIs are enhanced at the plate and are added (A) to the SE signal. Positive biasing reduces the signal to the desired SE-I+II component collected from the specimen plus the SE-IV component from the final aperture (B). Over the Faraday cup hole only the SE-IV component is collected (C).

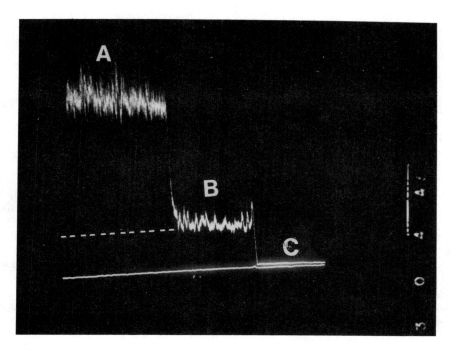

Figure A12.9. Signal components separated with a converter plate. If a converter plate is mounted beneath the pole piece, SE-IIIs may be added to (A) or subtracted from (B) the SE-I+II+IV signal collected when the beam is on the platinum aperture. The SE-IV component is separately collected over the Faraday cup hole (C).

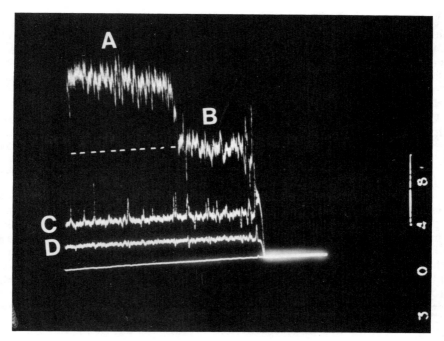

Figure A12.10. Signal components separated with a converter plate and a specimen grid. A combination of converter and specimen grids allow the separation and enhancement of the SE-III component but is not useful for the separation of the specimen-specific SEI+II signals. The negatively biased converter (A) adds additional SE-III to the signal collected when the converter is grounded but the specimen is negatively biased (B). A positively biased converter eliminates most of the SE-IIIs produced (C). If both the specimen and the converter are positively biased only SE-III produced at the specimen and the converter grid are collected (D). The latter component cannot be separated and must be collected together with the specimen-specific signals in similar proportions in an unmodified microscope.

Experiment 12.8: Modifying the SE Signal with a Grounded Specimen Grid.
A specimen grid may be used to separate the SE-III signal from the other components. This procedure allows BSE imaging (with converted BSEs) at a high take-off angle. It is especially useful for low and very low voltage imaging. In order to suppress SE-I+IIs, a specimen grid has to be installed to provide a bias reference for the specimen surface and not to alter the collection field of the E-T detector. However, as will be evident from appropriate measurements, the grid reduces signal collection efficiency even if the specimen is biased negatively. The specimen grid also provides an additional SE-III source. (a) The first top scan of Figure A12.10 indicates again the emission characteristics of the converter and includes the SE+I,IIs emitted from the negatively biased specimen and the SE-IVs. If the converter is biased negatively (A) the SE-III signal emitted is higher than from the grounded plate (B). (b) The next scan obtained with a positively biased converter reveals the SE-IV and all those SEs collected between the converter plate and the specimen (C). (c) SE-+I,IIs are additionally suppressed with both the specimen and the converter positively biased, and only those SE-III emitted from these grids will be recorded (D). The scans prove that a specimen grid is not advisable for specimen-specific signal collection because the signal is reduced and SE-IIIs are contributed from the specimen grid itself. However, it is very useful for BSE imaging.

Laboratory 13

Electron Channeling Contrast

13.2 Wide-Area Channeling Patterns

Experiment 13.1: Obtaining a Wide-Area Channeling Pattern. A wide-area electron channeling pattern (ECP) is illustrated in Figure A13.1. The resulting pattern consists of bands and lines which intersect and provide a "crystallographic road map." The crystal has been intentionally placed so that the edges of the crystal are visible in the field of view. Notice that both a positional scan, which forms a conventional image, and an angular scan, which forms the ECP, are taking place simultaneously.

Experiment 13.2: Properties of Channeling Patterns. Figure A13.2 shows the effect of translating the crystal laterally. The crystal moves, as evidenced by the displacement of the bright blobs, but the ECP remains fixed since the relative orientation of the beam and the crystal planes remains unchanged. Figure A13.3 shows the effect of rotating the crystal. The ECP also rotates since the relative orientation of the crystal planes and the beam has been changed by rotation. Figure A13.4 shows the effect of tilt. The ECP orientation changes because tilting changes the relative orientation of the crystal planes relative to the beam.

13.3 Measurements with Channeling Patterns

Experiment 13.3: Calibrating the Angular Scale. Consider the ECP of silicon (a_0 = 0.542 nm) in Figure A13.5 oriented with the [100] direction nearly parallel to the electron beam. The beam energy is 25 keV (λ_{rel} = 0.00765 nm). The main bands which form the [100] pole (the fourfold symmetry square) have Miller indices of (022). The Bragg angle calculated with equation (13.3) for this situation is thus 0.0199 rad = 1.14 degrees. Because of the symmetry of the ECP, this number must be doubled to give the full width of the band. The angular measure indicated by the double-pointed arrow (022) on the ECP is thus $2\theta_B$ = 0.0399 rad = 2.28°. From the measured width w of the [220] band in the image in centimeters (2.6 cm), an angular calibration can be calculated as $2\theta_B/w$ = 0.877°/cm. The SEM generated angular scale bar reading 4.3° is incorrect and needs to be recalibrated.

Experiment 13.4: Sensitivity to Lattice Parameter and Beam Energy Changes. The effect of a change in the beam energy with a fixed crystal spacing, which is equivalent to a change in the lattice parameter at fixed electron beam energy, is illustrated for cubic silicon carbide in Figure A13.6. Figure A13.6a shows the ECP recorded at 24 kV. After increasing to 25 kV (Figure A13.6b), the coarse structures of the pattern are essentially unchanged, but the fine structure in the center of the [111] pole has changed dramatically.

Figure A13.1. Wide area electron channeling pattern of silicon single crystal near [111] imaged with the specimen current signal.

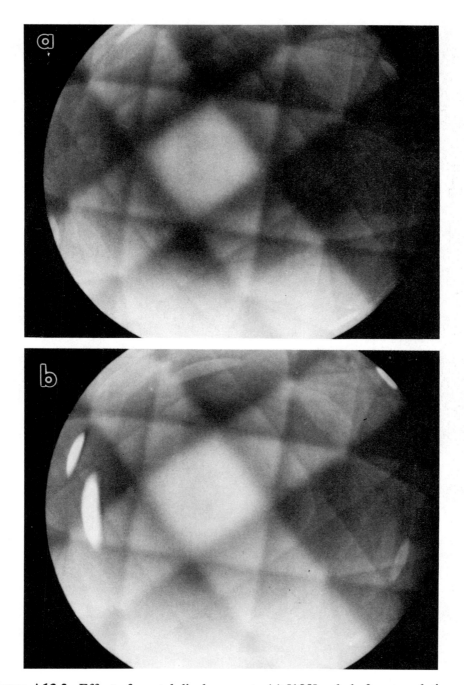

Figure A13.2. Effect of crystal displacement. (a) [100] pole before translation with specimen stage; (b) after translation. Channeling pattern remains stationary.

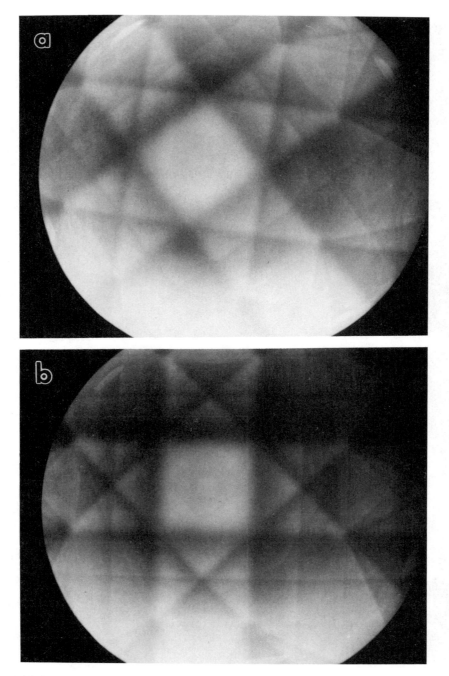

Figure A13.3. Effect of crystal rotation. (a) Before rotation; (b) after rotation. Channeling pattern rotates with crystal.

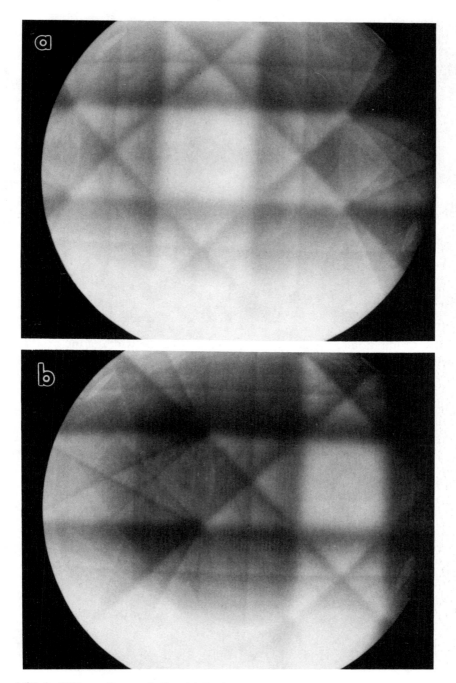

Figure A13.4. Effect of crystal tilt. (a) Before tilting; (b) after tilting about 3° with the tilt stage. A different crystal orientation is in the pattern center.

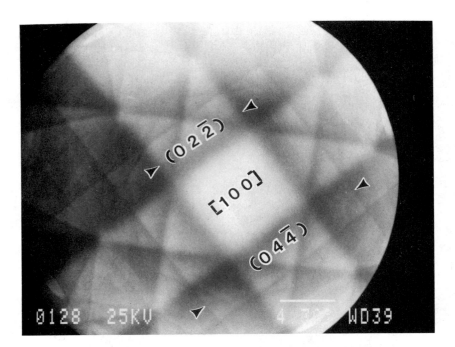

Figure A13.5. Calibration of the pattern. From measuring the width of the band a bar that converts distance into angle may be devised.

Figure A13.6. Effect of beam energy changes on fine structure. The detail in the center of a [1̄11] zone of cubic silicon carbide changes when the voltage is changed from (a) 24 kV to (b) 25 kV.

Figure A13.7. Channeling contrast in polycrystalline nickel. Grains exhibit various shadings because the entire grain takes on the shading of a particular point (orientation) of the channeling pattern.

13.4 Channeling Contrast Imaging of Microstructures

Experiment 13.5: Channeling Contrast of Microstructures. Figure A13.7 shows a typical example of channeling contrast from an annealed polycrystalline nickel sample. Grain boundaries can be readily seen. Note that the light and dark grains are produced by the entire grain taking the contrast of a portion of the ECP at its particular orientation.

Experiment 13.6: Properties of Channeling Contrast Images. The effect of tilting by 3° is shown in Figures A13.8a and b. Note the changes in the visibility of grains 1 and 2, depending on the exact nature of the local crystal orientation relative to the beam. When grain sizes are to be studied by this method, it is clear that a series of images at different tilts may be needed to ensure that all boundaries are found.

Experiment 13.7: Resolution of Channeling Contrast Images. A grain boundary of an annealing twin in polycrystalline nickel is shown in Figure A13.9. The channeling contrast image shows fine-scale spatial detail at a level of approximately 0.2 μm. The probe size calculated by the brightness equation for the indicated conditions is 0.16 μm, which corresponds reasonably well with the limiting detail.

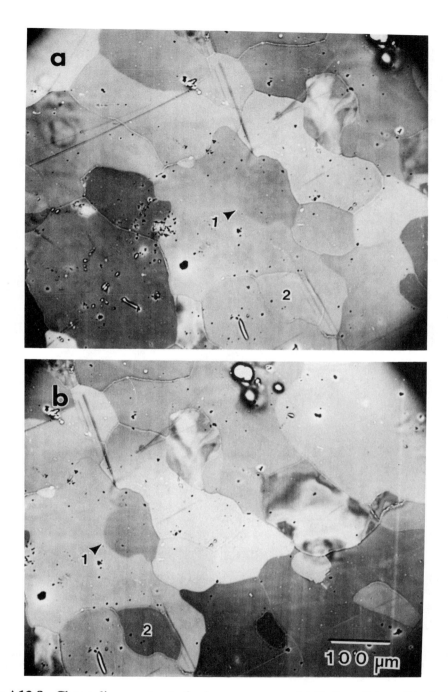

Figure A13.8. Channeling contrast of microstructures. (a) Before tilting; (b) after tilting. Note change of contrast for grains 1 and 2.

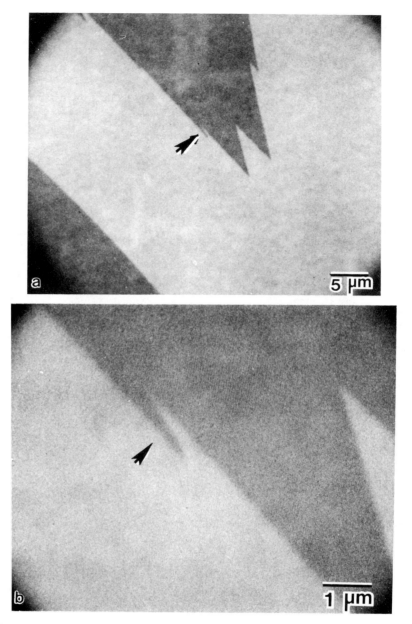

Figure A13.9. Resolution of channeling contrast at a grain boundary of nickel is about 0.1 μm at vertex near arrow.

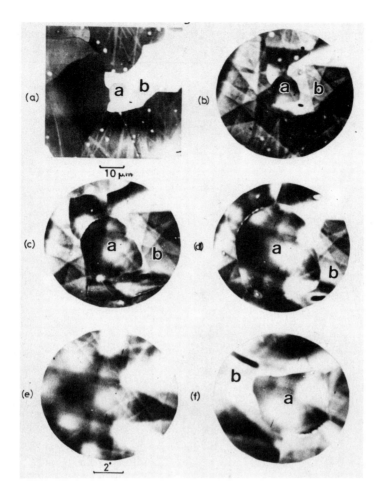

Figure A13.10. A through focal series on Fe-3% silicon; (a) is channeling contrast image. Channeling pattern occurs at (e). Note reversal in real image as focus changes from (d) to (e) to (f). A selected area electron channeling pattern from grain "a" alone is shown in (e).

13.5 Selected Area Electron Channeling Patterns

Experiment 13.8: Taking an SACP. The variety of SACP optics incorporated in SEMs makes it difficult to generalize a discussion of the specific questions asked in this experiment. An example of a "through-focus" series in the deflection focusing mode of SACP generation is shown in Figure A13.10. Figure A13.10a is a conventional SEM channeling contrast image of Fe-3% Si. Figures A13.10b-f show the deflection focusing mode of SACP generation with the objective lens strength progressively weakened from an initial crossover above the specimen surface (multiple ECPs) to a point coincident with the specimen surface (single ECP) and finally below the surface (note reversal of the image in A13.10f relative to Figures

A13.10c,d). For reference, note the relative positions of grains "a" and "b" in the conventional channeling micrograph and the SACP through focus series. A small hole is also present near grain "a" which aids in detecting the reversal.

Laboratory 14

Magnetic Contrast

14.1 Type I Magnetic Contrast

Experiment 14.1: Obtaining Type I Magnetic Contrast. Figure A14.1a shows a Type I magnetic contrast image from a square wave recorded on magnetic tape. The image is prepared with an Everhart-Thornley detector biased positively to collect secondary and backscattered electrons. Figure A14.1b shows the same area with the E-T detector biased negatively to exclude secondary electrons. The Type I contrast disappears. This particular square wave was recorded at 1 kHz with a tape transport speed of 4.77 cm/sec. This combination produces a spacing of 47.7 µm/cycle, which is close to the value measured from the micrograph, 43.3 µm/cycle.

Experiment 14.2: Effect of Specimen Rotation on Type I Magnetic Contrast. The effect of a 180° rotation is illustrated in Figure A14.2. The domains are observed to reverse in contrast. The deflection of secondaries in Figure A14.2a is away from the detector, so the domains appear dark, while in Figure A14.2b it is toward the detector.

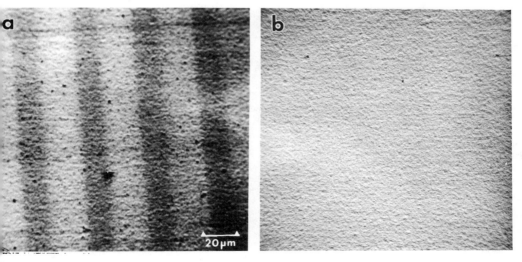

Figure A14.1. Type I magnetic contrast. (a) Positively biased E-T detector; (b) negatively biased E-T detector showing that contrast was due to secondary electrons.

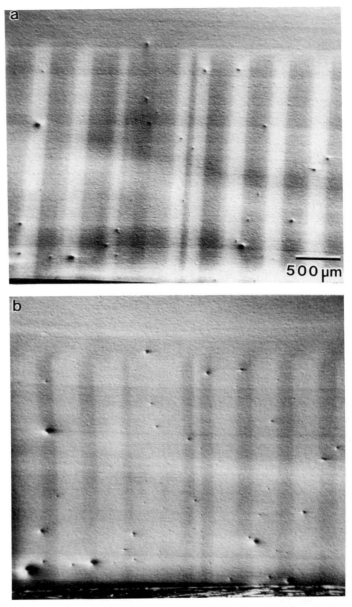

Figure A14.2. Effect of rotation on Type I magnetic contrast. (a) Specimen at 0° rotation (b) specimen rotated 180°

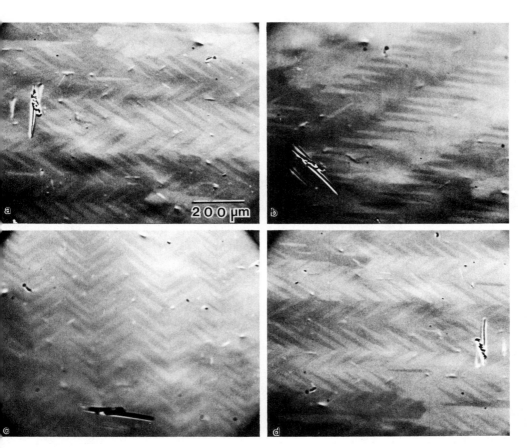

Figure A14.3. Type II magnetic contrast in iron at various specimen rotations. (a) 0°; (b) 5°; (c) 90°; and (d) 180°.

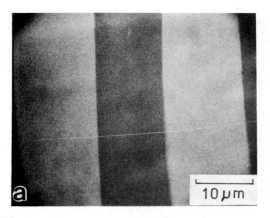 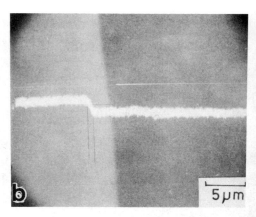

Figure A14.4. Spatial resolution of Type II magnetic contrast of domains in transformer steel.

Experiment 14.3: Spatial Resolution of Type I Magnetic Contrast. The edge of the domains is found to be indistinct, as shown in Figure A14.1. The magnetic field within the specimen varies abruptly at a domain wall, but the leakage field above the specimen is more diffuse. The spatial resolution is about 2 μm.

14.2 Type II Magnetic Contrast

Experiment 14.4: Obtaining Type II Magnetic Contrast. Figure A14.3 shows a tree domain pattern from an iron single crystal at several different relative rotations: (a) 0°, (b) 45°, (c) 90°, (d) 180°. The contrast reverses at 180° relative to the 0° image and disappears (middle gray) at 90°. A second domain pattern comes into contrast at 90° and disappears at 180°.

Experiment 14.5: Spatial Resolution of Type II Magnetic Contrast. Figure A14.4a shows an example of iron-silicon transformer steel containing several grains. Figure A14.4b shows a line scan across single grain of iron-silicon transformer steel. The domains show a limiting resolution of approximately 1 μm.

Laboratory 15

Voltage Contrast and EBIC

15.1 Voltage Contrast

Experiment 15.1: Static Voltage Contrast. (a) There are two reasons for the positive lead having dark contrast and the negative lead having light contrast. Secondary electrons emitted from the wire at a positive potential relative to ground are partially recollected by the wire itself and hence the secondary yield is lower than that from the wire carrying a negative potential which repels its own secondary electrons. In addition the collection field from the Everhart-Thornley detector is smaller at the positive wire than at the negative wire because the potential difference between the Faraday cage on the E-T detector and the positive wire is less than the potential difference between the cage and the negative wire. Hence the field is smaller and the collection efficiency for secondaries is lower.

(b) By changing the acceleration voltage in small steps we find that on a typical 741 chip the E2 energy is about 1.7 keV. This value may be higher if the chip has been kept for any period of time after being decapped.

(c) As the specimen tilt is increased, E2 rises because more electrons are emitted at each energy. As the table below demonstrates the rise is slow at first and then quite rapid. At large tilt angles, where E2 is becoming high, the measurement is not very accurate because the SEM energy can only be adjusted in many instruments in 1-keV steps in that range.

Tilt (degrees)	0	10	20	30	45	60
E2 (keV)	1.7	1.7	1.9	2.2	3.5	between 6 and 7

Clearly E2 can be moved upwards in energy a significant amount by tilting through angles of the order of 45 or 60 degrees. It can be shown that if the E2 energy for normal beam incidence is E2(0), then the value of E2 for an angle of incidence θ, E2(θ), is given by the expression

$$E2(\theta) = \frac{E2(0)}{\cos^2\theta}$$

(d) Note that as power supplies are connected or disconnected voltage contrast is visible, but will fade away after a few seconds or so. This is because the incident beam, the passivation layer, and the device surface together act like a parallel plate capacitor. When the potential on the device surface is changed, this potential immediately appears on the top surface of the passivation layer and so produces voltage contrast. If the applied potential was positive, then the surface becomes positive and starts to recollect secondaries. If the applied potential was negative, the surface becomes negative and repels electrons. In either case the resultant variation in electron emission changes the current balance at the surface of the passivation and the resultant current will charge up the capacitor in such a sense as to remove the contrast. Thus static dc potentials from a circuit covered by passivation are only visible when being switched on or off. Note that the best image is usually obtained at a high scan rate and that often no contrast at all will be visible on a slow speed or photoscan rate. Thus to be able to obtain a stable image to potentials it is usually necessary that the passivation be removed chemically before observation.

Experiment 15.2: Dynamic Voltage Contrast. Figure A15.1 shows barber pole contrast from a 741 chip being driven by an input signal at about 20 kHz. The input connecting lead on the right-hand side of the micrograph, and the large test pad in the center of the field of view, both show strong voltage contrast since they are directly connected to the ac signal. The features on the left-hand edge of the micrograph show mostly static voltage contrast (bright = negative potential) because these are test structures connected to the negative power rail, although the ac contrast is still weakly visible because electrons leaving these area are also affected by the fields from neighboring structures. Note that in the presence of an ac field, static voltage contrast is sometimes more persistent because of the periodic reversal of polarity within the passivation layer. The image was taken at 2.1 keV, which gave the best overall image even though this was just above the E2 energy (1.7 keV). At lower energies the image was too noisy while at higher energies the charging of the passivation layer was enough to eliminate useful contrast. While this mode of voltage contrast observation is only qualitative, it does illustrate important principles of the technique and can be a useful diagnostic tool in simple devices.

Figure A15.1. Dynamic voltage contrast. A stroboscopic "barber pole" effect is visible on the square pad in the center and the connecting lead on the right since these regions are being driven at 20 kHz.

Figure A15.2. EBIC image of 741 operational amplifier at 20 kV.

5.2 Electron Beam Induced Current (EBIC)

xperiment 15.3: EBIC Observation of Devices. Figure A15.2 shows the EBIC nage recorded from a 741 op-amp under the given experimental conditions. The specimen urrent amplifier was the GW Electronics Device set to a sensitivity of 10^{-8} A. The image ontains only three contrast levels--a gray background and then very dark and very bright egions. The gray areas are regions of the device from which no charge-collected signal is eing collected. When the beam is in the vicinity of a junction, such as at a transistor or diode, ien the depletion field at the junction will separate the electron-hole pairs and, because the CA is connected to the power rails of the chip and hence has a path to every junction in the evice, these charge carriers will flow through the external circuit producing the visible ontrast. Depending on whether the *p*- or the *n*-type side of the active device is connected to e rail the current flow will be positive or negative and the corresponding image area will be right or dark. For a measured incident beam current of about 3×10^{-9} A the average indicated urrent on the SCA was about 3×10^{-8} A. But if the beam is stopped on one of the junction egions (bright or dark), then the magnitude of the collected current increased to about 5×10^{-7} .. We are therefore seeing a current gain of several hundred times in the vicinity of the inction depletion fields. This value is lower than would be predicted theoretically (e.g., beam energy/electron-hole pair energy for silicon, i.e., $20,000/3.6 \approx 3000\text{x}$) because the ollected current flowing through the input impedance of the amplifier is creating an ohmic oltage drop which acts to forward bias the junctions under observation and so reduce their ollection efficiency.

If the scan rate is set too high, the image streaks and blurs in the direction of the scan iotion. This is because there is a capacitance associated with each junction in the device and le net effect of all of these capacitors placed in parallel with the input impedance of the mplifier creates an *"RC"* circuit which has a time constant which may approach a significant fraction of a second. EBIC images of devices must always therefore be recorded at slow visu rates if artifacts are to be avoided.

Figure A15.3 compares the secondary (left-hand) and EBIC (right-hand) images from a portion of the device area. The bright area in the EBIC image is seen to be the junction area below the contacts on the secondary image. Note also that the EBIC image contains what appears to be topographical contrast from the interconnects in the region where they lay over top of the junction. This is because the incident beam has to pass through these connects to reach the junction. If some of the beam is backscattered before reaching the junction then the signal is reduced. Note that the edge of the junction region is not sharp in the EBIC image even though the secondary image is well focused and resolved. This is because the spatial resolution of the EBIC image is governed by both the interaction volume of the beam and the minority carrier diffusion lengths of the material. This point is examined in further detail below.

Experiment 15.4: Effect of Electron Beam Energy. Figures A15.4a,b show the same area as that of Figure A15.2, but at beam energies of 15 and 13.1 keV, respectively. At 15 keV most of the contrast detail is still visible but is noticeably weaker. By 13.1 keV the image is very weak and noisy and only a few of the contrast features remain. At still lower beam energies no EBIC contrast is visible at all. In order for charge collection images to be formed the incident beam must deposit electrons into the depletion region of the device to be imaged. Therefore junctions which lie deeper than incident electron range will produce no contrast and will not be visible until the energy is increased to a sufficiently high value for electrons to penetrate down to them. A sequence of pictures of the same area recorded at increasing beam energies will thus reveal the three-dimensional arrangement of structures in the device. If the energy at which a junction first becomes visible is E, then its depth R can be

Figure A15.3. Comparison of secondary electron image (left) and EBIC image of the same area (right)

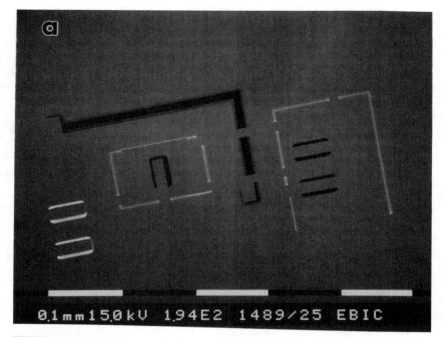
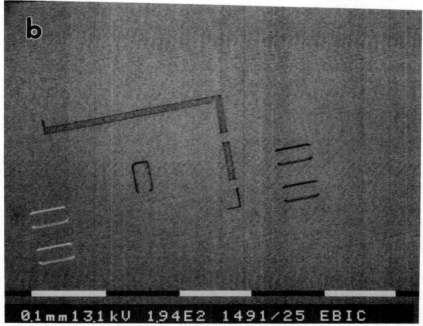

Figure A15.4. Effect of electron beam energy on the EBIC image of Figure 15.2. a) 15 keV; (b) 13.1 keV. At lower beam energy the EBIC image has no contrast.

estimated from the simple range equation:

$$R = \frac{70E^{1.66}}{\rho}$$

where with E in keV and ρ in g/cm^3 then the range is in nanometers. Thus for silicon, where ρ is 2.34 g/cm^3, a structure which first becomes visible at 15 keV lies at a depth of about 2700 nm (i.e., 2.7 μm) beneath the surface. Since, in this case, we have a passivation layer on top of the chip, the thickness of this must be taken into consideration when assessing the actual position relative to the surface of the silicon. By increasing the beam energy to 30 keV or even more it is therefore possible to examine structures which may lie 5 or 10 μm into the chip. As the penetration of the beam increases so does its lateral spread. At high beam energies, therefore, the spatial resolution of the image will become poorer than at low energies. A slow scan line profile (Figure A15.5) across the sharp bright line structure visible in Figure A15.2 shows that the width of the edge is about 4 or 5 μm even though the physical edge of the junction region is quite abrupt. This broadening is the visible evidence of the spreading of the beam and the magnitude noted here is consistent with the range of the electrons as predicted from the equation given above. The minority carrier diffusion length in the material can also contribute to the resolution, but this effect is only strongly evident when the collecting junction area is very large in extent compared to the size of the beam interaction volume.

Experiment 15.5: EBIC Observation of Materials. Figure A15.6 shows a typical example of an EBIC image from a solar cell. In this case the amplifier polarity was such that the active area, i.e., the area beneath the Schottky barrier was dark. Unlike the images of the previous device, where current could only be collected in the immediate vicinity of the junctions, the whole sample can be examined here because the entire surface is covered by the

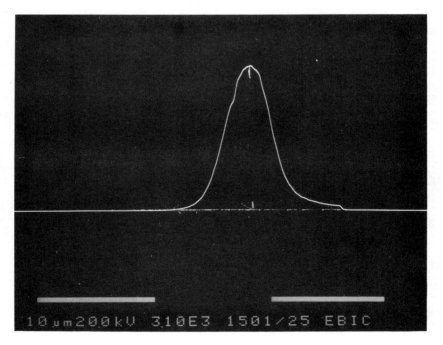

Figure A15.5. Slow scan line profile across bright line in Figure A15.2.

Figure A15.6. Example of EBIC image from a solar cell. The image contrast is reversed so the area beneath the Schottky barrier is dark.

Schottky barrier. A region extending below the Schottky barrier for a depth of typically 2 or 3 µm is depleted by the bias at the barrier and it is in this region that the charge carriers are separated. The average collected signal is of the order of 5 to 10 x 10^{-7} A, representing a current gain of about 1000 times. This value is lower than the theoretical estimate both because of the forward biasing of the Schottky barrier by the ohmic drop at the amplifier input, and because at 20 keV a significant fraction of the beam energy deposition (and hence of the electron-hole pair formation) occurs beneath the depleted region and hence charge collection is not complete. The lines of contrast, a few micrometers wide, which are visible running across the image, arise from electrically active crystallographic defects in the silicon. Because these defects have electrical fields associated with them they can trap both electrons and holes and cause them to recombine within the sample rather than flowing through the external amplifier circuit. In the region of a defect therefore the collected signal current will be reduced by typically 10%-15% compared to the local average background value. In all other respects this EBIC image is similar to the previous ones, and hence varying the incident beam energy will change both maximum imaging depth and the lateral spatial resolution of the micrograph.

Experiment 15.6: EBIV Measurements of a Schottky Diode. Measurements of open circuit voltage V_{oc} versus electron beam current allow calculation of the height of the Schottky barrier. The data collected in the experiment are shown in the table below.

Spot size	Beam current	V_{oc} (mV)
#6	0.53	70
#5	0.80	75
#4	1.40	88
#3	4.65	124
#2	14.00	150

These data were collected at 20 keV from a solar cell Schottky diode. The measured gain of the diode was 3570 and the total area of the diode was 280 µm x 290 µm. It can be shown that the output voltage of the diode V_{oc} is related to the incident beam current I_B by the equation

$$V_{oc} = n\left[\phi_B + \left(\frac{kT}{q}\right)\ln H\right] + n\frac{kT}{q}\ln I_B$$

where ϕ_B is the height of the Schottky barrier (in eV), n is the ideality factor, k is Boltzmann's constant, T is the temperature in degrees Kelvin, and q is the charge on an electron. The quantity $H = [(\text{gain of diode}/(\text{area} \times AT^2)]$ where A is Richardson's constant, which equals 121 A/cm^2/°K^2.

For this experiment, done at room temperature so that $T = 273°$K, the area of the diode 280 µm x 290 µm $\approx 6.5 \times 10^{-4}$ cm^2. AT^2 is 8.94×10^6 A/cm^2 so $H = 3571/(6.5 \times 10^{-4} \times 8.94 \times 10^6) = 0.617$. The value of (kT/q) for an electron at room temperature is 0.027 eV. So substituting numerical values in the equation we have:

$$V_{oc} = n[\phi_B - 0.013] + 0.027n \ln I_B$$

A plot of V_{oc} against $\ln I_B$ will be a straight line with a slope of $0.027n$, and an intercept of $n[\phi_B - 0.013]$. Substituting the measured slope and intercept from such a plot we find here that ϕ_B is 0.65 eV and that $n = 1.05$.

The barrier height depends on the semiconductor material and the metal which is used to form the Schottky barrier, and usually lies between about 0.5 and 0.8 eV. It is this potential which creates the depletion field beneath the barrier layer and so allows the separation of the electron-hole pairs. The ideality factor is a measure of how closely the diode approaches the theoretical expectation. Here $n = 1.05$ so the diode is behaving as predicted. Diodes which are leaky or faulty in some way may show n values as high as 2 or 3.

Laboratory 16

Environmental SEM

16.2 Signal Quality

Experiment 16.1: Collection of SE and BSE Signals. Several gas/electron interactions affect image quality in the environmental SEM. The scattering of probe electrons in the high-pressure zone reduces the probe current and contributes an electron tail around the focused probe. The latter degrades the spatial imaging character of the probe. Scattered primary electrons (PEs) as well as the signal electrons (SEs, BSEs) can ionize gas molecules and produce additional electrons (environmental SEs = ESEs). Also, the ESEs may be amplified by accelerating them with the collector field in a cascade fashion, generating more ESEs. The total number of ESEs produced by the PEs, SEs, BSEs and the cascade effects depends on the gas pressure, the working distance, the collection voltage, and the position of the collector. At low pressure the contribution of the PEs to the ESE signal background is negligible. Also, the number of BSEs collected by the ESE detector is too small to be recognized. However, the contributions of BSEs and SEs to the signal vary significantly.

On the thin-film specimen, independent of the gas pressure, the backscattered electron detector collects high-energy BSEs which image only the platinum patches by atomic number contrast (Figure A16.1a). The positively biased gaseous SE detector may collect (ESE detection) low-energy electrons in addition. At a certain gas pressure, the carbon patches will also be imaged (Figure A16.1b), indicating that SEs produced at the specimen surface contribute the image contrast in a way identical to the SE signal collected by the Everhart-Thornley detector under high vacuum conditions (see Laboratory 11).

The proportional contribution of SE and BSE to the ESE signal can be analyzed if the test specimen is imaged at different gas pressures but otherwise identical conditions (Figure A16.2a through d). At very low pressures (Figure A16.2a) the SEs from the specimen surface are collected as a weak signal. At increased gas pressures (Figure A16.2b) this SE signal is amplified through ESE cascades and images both the platinum and the carbon. At high gas pressures (Figure A16.2c) most of the SE-produced ESEs are absorbed, but those produced by the BSEs near the ESE detector will effectively be collected contributing a "converted BSE" signal which is similar to the signal collected with the BSE detector (Figure A16.2d). However, ESEs produced by PEs will reduce the BSE signal contrast. At a given pressure, conditions required for SE cascade amplification can be established through variation of collector bias and working distance. This possibility makes SE imaging independent of the gas pressure.

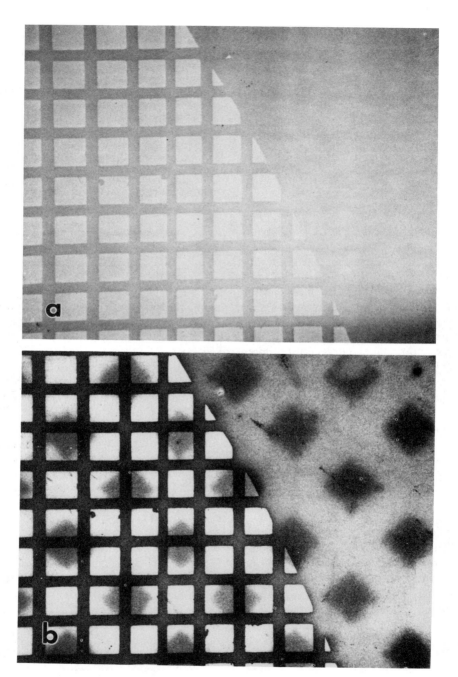

Figure A16.1. Collection of SE and BSE signals from the inhomogeneous thin film specimen. (a) BSE signal collected with a conventional scintillator detector images only the platinum patches; (b) SE signal collected as environmental SE electrons (ESEs) with the new gaseous detector reveals the carbon patches as well as the platinum patches. The imaging conditions for the carbon are identical to those found in high-vacuum SEM (compare with Laboratory 11, Figure A11.3).

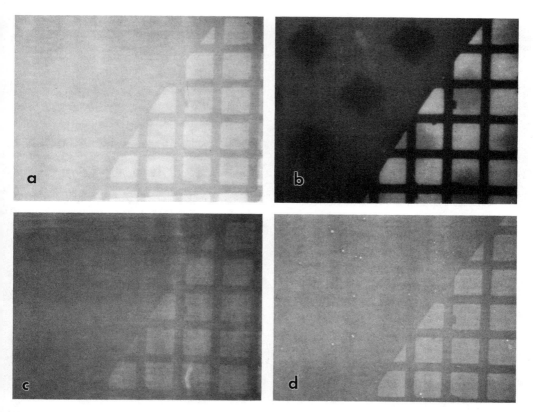

Figure A16.2. Change of ESE signal quality at different gas pressures (20 kV, 9.2 mm WD, 500 V collection voltage). (a) At low pressures (< 0.1 Torr) only the SEs generated at the specimen surface are collected. (b) At higher pressures (1.5 Torr) this SE signal is amplified through gas ionization (the cascade effect). (c) At even higher gas pressures (5.5 Torr) most of the ESEs are absorbed by the gas but those produced by BSEs in the immediate vicinity of the ESE detector are still collected and produce a "converted BSE" signal which is similar to the BSE signal (d) collected by the BSE detector.

16.3 Imaging of Liquid Water

Experiment 16.2: Control of Saturated Water Pressure. One of the important new features of environmental microscopy is the possibility to alter the type of gas and its pressure over a large range. The aqueous environment is only one example among many possibilities. The imaging of liquid water is a fascinating new aspect of electron microscopy. At saturated water vapor pressures, water is stable indefinitely. Increasing or decreasing the vapor pressure allows condensation or evaporation of water, and provides means to generate water films. As can been seen from Figure 16.1, the stabilizing pressure for water at 20°C is 17.5 Torr. Such a pressure would require a very short working distance for SE imaging. Therefore, it is preferred to reduce the specimen temperature and to establish vapor saturation at lower pressures. Liquid water can only be stabilized at pressures greater than about 5 Torr (see Figure 16.1).

Figure A16.3. Imaging of a thin water film produced at a saturated water pressure of 12 Torr in a condensation chamber. (a) Condensing water droplets, imaged in strong topographic contrast, coalesced and partially filled the groove. (b) After evaporation of some of the water (asterisk), a thin water film remained in the cover grid. (c) Unsupported water films are stable at saturated vapor pressure and can be imaged at higher magnifications. A thin water layer also covers the surface of the grid bars.

In SE imaging mode (and in real time) it can be observed that at the walls of the condensation chamber water accumulates first forming drops which then enlarge (Figure A16.3a). These drops coalesce and partially fill the chamber (Figure A16.3b). Evaporation of water at slightly higher temperatures will cause the water to form thin films between the bars of the TEM grid (Figure A16.3c). These thin films can be maintained at appropriate saturation conditions, and can be made thicker or thinner, producing concave or convex shapes as seen by topographic relief contrast (Figure A16.2b). The SE image arises from only the outer few nm of the water layer and is produced with low signal yield. Note the dark appearance of bulk water in Figure A16.3b (at the asterisk). Most of the signal electrons which form the image of the thin water films are generated by BSEs within the condensation chamber.

16.4 Imaging of Insulators

Another important aspect of environmental microscopy finds its basis in the ionization of gas molecules predominantly by the signal electrons. The gas ions neutralize surface charges in a self-regulated fashion allowing the imaging of flat as well as rough surfaces of insulators.

Experiment 16.3: Surfacing Imaging of a Polymer.
(a) The flat surface of tempered PVC is imaged at any chosen accelerating voltage. The neutralization efficiency of the ions is limited by ion diffusion. Such limitation may only be recognized on special specimens, as in the thin surface film specimen. Whereas the BSE detector images the metal patches with strong atomic number contrast (Figure A16.4a, left side), the ESE detector images the metal patches differently (Figure A16.4b, left side). The metal patches are seen as gray disks with a dark rim and a wide dark zone at the side which faced the incoming scanning beam. The metal film acts as a capacitor plate and the increase of the plate bias during primary beam bombardment causes increased attraction of ions. High ion and gas concentrations at the plates modify and reduce the SE signal.

(b) Surface topography, seen in shallow depressions of the granular plastic material, and mass-thickness contrast, seen between the thin film patches and the polymer surface, are imaged by SEs as well as BSEs (Figure A16.4, right side). Of course, surface imaging of polymers allows elimination of metal coatings for charge neutralization and contrast enhancement. However, the SE signal reveals additional surface information. Although the BSE and SE yields on low-Z, low-density materials are reduced in comparison to metal-coated surfaces, the SE signal allows the imaging of the outermost surface with a depth discrimination similar to that found in high-vacuum microscopy on conductors. Comparison of BSE and SE images at low (Figure A16.5a,b) and medium magnifications (Figure A16.5c,d) indicates that the BSE contrast is dominated by multiply scattered signal electrons BSE-IIb. This can be demonstrated by noting the 3-5-µm exit radius from edge brightness contrast (Figure A16.5a). These signal electrons mainly reveal information from a depth equal to their escape depth (roughly 1/2 the PE range) and produce strong contrast from average atomic number variations which are larger than the signal excitation volume. The SE signal is dominated by the SE-IIa, which are produced by BSE-IIa electrons (Figure A16.5b). Note the 100-200-nm exit radius of signal generating electrons from the thinnest detectable lines in edge brightness contrast. The BSE-IIa electrons produce strong SE particle contrast on small surface particles of dimensions smaller than the escape depth of the BSE-IIa electrons.

Experiment 16.4: Imaging of Uncoated Polymer Foam or Biological Tissue.
The advantages of charge neutralization in the environmental SEM can be utilized on inhomogeneous insulators of rough topography. Such samples are found in polymer foams and fixed and dried biological tissues. Conventional biological high vacuum SEM can utilize the high depth of field available for the imaging of large tissue pieces. However, to compensate for the uneven charging of such specimens, thick metal coatings are used but produce coating artifacts and obscure fine details. In the environmental SEM the charge neutralization at high gas pressures allows full advantage of the SE imaging capabilities without the use of metal coating (Figure A16.6a). Dried lung tissue has a highly inhomogeneous mass thickness distribution which results in a strongly varying charge distribution. However, only moderate pressures are necessary to provide for sufficient charge neutralization to allow excellent imaging of the tissue at high voltage.

In a conventional high-vacuum SEM, it is recommended to apply low voltages to prevent charge accumulations on insulators. Since the electron yield on biological tissue depends strongly on mass density and tilt angle, even at very low voltages of 1 kV or less, charge

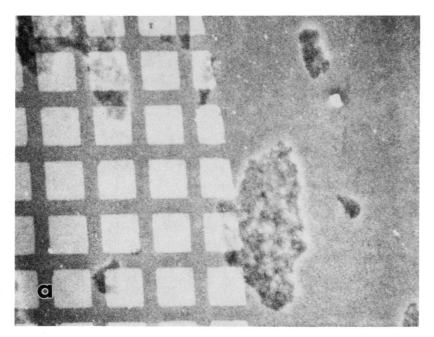

Figure A16.4. Contrast mechanisms on electrical insulators at 20 kV revealed by imaging thin surface films of 2 nm platinum and 2 nm carbon deposited onto tempered PVC. (a) The BSE signal images the metal patches with typical atomic number contrast (left side) and the surface depressions with topographic contrast (right side). (b) The ESE signal, produced at 7 Torr, images the specimen surface very differently. As indicated in the appearance of the carbon patches, "true surface SEs" are imaged. This signal reveals abundant details in mass-thickness contrast (recognizable as the small dark surface details on the uncoated polymer) and strong topographic contrast (seen in the increased microroughness contrast in the depressions). However, the metal islands exhibit a different contrast mechanism, depending on the electron beam induced biasing.

accumulation cannot be prevented. This results in strong charging artifacts (Figure A16.6b). Additionally, in low-voltage microscopy at high vacuum, strong beam damage is seen in volume loss and deposition of contaminations. All these effects are diminished in high-gas-pressure microscopy.

16.5 Dynamic Environmental Experiments

Experiment 16.5: Crystallization of Salts. Using TV rate imaging at saturated water vapor pressures, crystallization of water-soluble salts can be observed and recorded. Crystals are dissolved in condensing water and the salt can be crystallized by evaporating water from the solution. Water vapor pressures are adjusted by regulating the specimen temperature as described above. Concentrated solutions of NaCl can be crystallized at elevated support

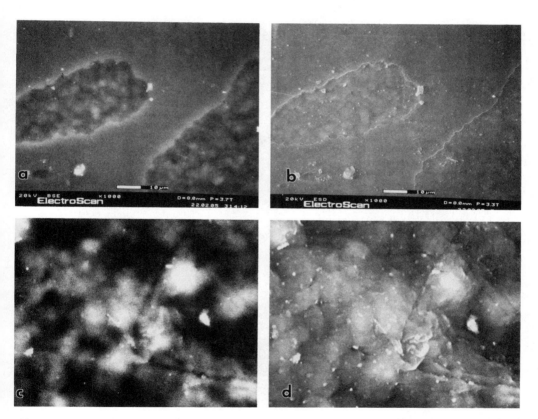

Figure A16.5. Surface imaging of uncoated polymer. BSE image (a) and ESE image (b) at low magnification. This surface imaging is not affected by charge neutralization and is identical to that seen on conductors in conventional high-vacuum SEM. The BSE signal (collected with a scintillator detector) images the surface as well as mass-density variations underneath the surface. The SE signal, collected as an amplified ESE signal, images mainly the surface detail and small particles. Some background signal contrast, produced by BSEs, is recognizable in the depression. The ESE image (c) and BSE image (d) at higher magnification (3,000x) shows the difference in surface imaging capability between these two signals.

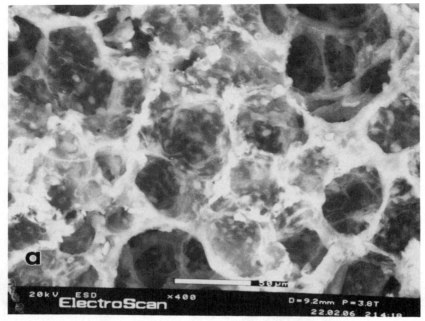
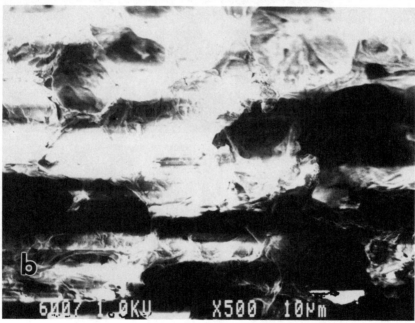

Figure A16.6. Imaging of dried uncoated biological tissue which has a low electrical conductivity and a highly uneven mass-density distribution. (a) High-gas-pressure environmental SEM at 20 kV. Environmental SEM can image the tissue without the need of metal coating applied for the purpose of charge suppression. (b) High-vacuum SEM at 1 kV. Low-voltage microscopy of the same sample in high vacuum did not prevent serious charging artifacts. Heavy metal coating would have been necessary for charge suppression.

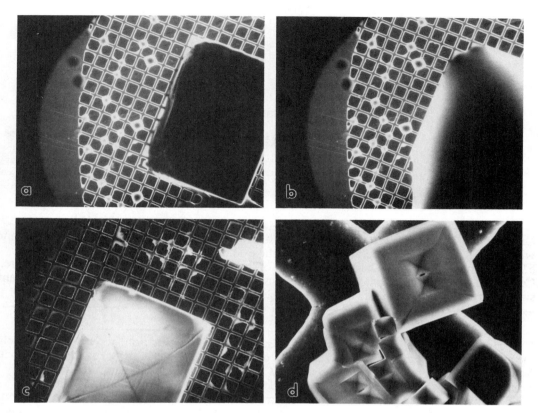

Figure A16.7. Crystallization of sodium chloride. TV rate imaging in SE imaging mode and control of partial water vapor pressures through specimen temperature make possible the visualization of (a) crystallization, (b) dissolution, and (c) recrystallization of sodium chloride. The rapid crystallization produces inward directed crystal growth and often generates (d) twinned crystals (20 kV, 9 Torr, 8.5 mm WD).

temperature (Figure A16.7a), dissolved at reduced temperatures (Figure A16.7b), and again crystallized by repeating the cycle (Figure A16.7c). Under these conditions crystallization takes place rapidly and proceeds from the outside to the inside of the droplet producing typical twin crystals (Figure A16.7d). Comparison of images of the salt solution and the crystals indicates a very low signal yield obtained from the liquid due to reduced mass density. However, topographic relief contrast is strongest on the solution droplet.

Laboratory 17
Computer-Aided Imaging

17.1 Digital Image Acquisition

Experiment 17.1: Digital Scanning. For a 128x128 image there are 16,384 pixels, and with a dwell time of 100 μsec/pixel the image takes 1.6 sec to acquire (frame time). A 512x512 image contains 262,144 pixels and, for the same dwell time, takes 26 sec to acquire. Compared to a 1000-line analog image (the usual SEM Polaroid), a 512x512 image of the same size is very similar but exhibits barely percetible lines and pixels.

Experiment 17.2: Digital Intensity Levels. In analog acquisition, we typically match the amplified analog image signal to a specified voltage range (e.g., 0-10 V) as measured on an oscilloscope which has been set up so that 0 V corresponds to black (on the CRT or the eventual photographic print) and 10 V corresponds to full white. The continuous range of CRT intensity or photographic density is referred to as the "gray scale." The human observer, in a lighted room under good viewing conditions, can typically distinguish only about 16-20 distinct gray levels on a photographic print. However, in a darkened room with a back-lit image, as is the case for a CRT, the human visual system can perceive several hundred gray levels. This fact is often the source of disappointment when we attempt to record a CRT image onto positive film.

In the digital case, intensity levels are discrete and there is no longer a gray scale continuum. Based upon the analog case, we might believe that the number of digital levels to record in a positive print would simply be 16 or 4 bits (4 bits corresponds to $2^4 = 16$). This is not really enough levels to provide a smooth image. Figure A17.1 shows one of the consequences of digitizing to an inadequate number of levels. This common defect is known as plateauing and is manifested as a series of apparent intensity terraces which follow the contours of the image.

However, the purpose of image digitization is to allow for subsequent numerical manipulation. The mathematical operations which we wish to perform are continuous in nature, by which we mean that there is an infinity of values between any two specified values. Unless we have a sufficient number of discrete intensity values in our digitized representation of the image, the mathematical functions will be degraded by what is referred to as round-off error. For mathematical manipulations of images, the digitization should take place at a level of 16 bits (65,536 levels) or even 32 bits (4, 294, 967, 296 levels). Note that no matter how many discrete levels are present in the stored digital image, the final displayed image is usually converted to an 8 bit intensity image (256 levels) for presentation to the digital-to-analog converter which controls the brightness of the CRT. This is achieved by scaling all or a portion of the stored digital intensity range into 8-bits. Thus, it is important to note that we are dealing with two distinctly different concepts, a digital acquisition space and a digital display space.

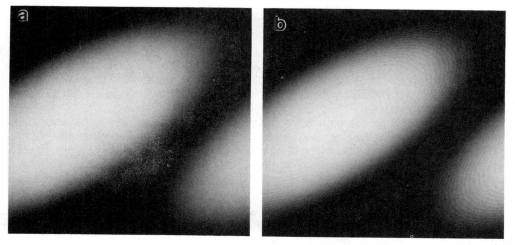

Figure A17.1. Display of image intensity. (a) 8-bit image of 256 levels (appears similar to a continuous gray scale); (b) 4-bit image of 16 levels. Note the plateauing artifact.

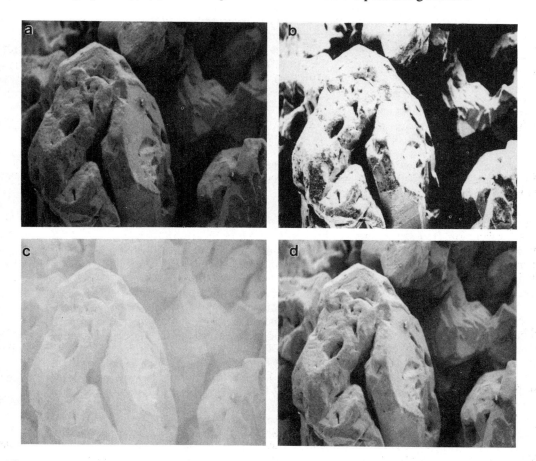

Figure A17.2. Single pixel contrast enhancement. (a) Image collected with normal contrast range; (b) contrast enhancement of image (a) which drives the white areas to saturation and the dark areas to full black; (c) low-contrast image collected with insufficient beam current; (d) contrast enhancement of image (c) producing an image with a normal contrast range similar to (a).

17 COMPUTER-AIDED IMAGING

17.2 Image Contrast

One of the most important modes of signal processing in the conventional analog SEM is the contrast enhancement operation variously called "black level," "differential amplification," or "contrast expansion." In analog operation, this consists of subtracting a fixed (dc) level from the video signal and then amplifying the difference signal to span the complete range of gray levels of the display.

In the digital case, the same differential amplification operation may be carried out mathematically with much more ease, speed, and fine control by manipulating individual pixels. The algorithm applies the same straight-line function to the intensity value of each pixel in the digital matrix. The slope and intercept of this straight-line function can be varied to alter the contrast and brightness.

Experiment 17.3: Single Pixel Contrast Enhancement. If the collected digital image already has signal excursions which span the entire gray scale range from black to white (Figure A17.2a), then the application of contrast enhancement drives the lighter areas into full white and the darker areas into full black (Figure A17.2b). This leads to a loss of information in these regions but an increase in contrast of the relatively flat middle gray regions.

If the original as-collected image has very low contrast, as in the case of images with very low probe currents (see Laboratory 3), then contrast enhancement will provide more information at all gray levels. Figure A17.2c is an image of low contrast that appears flat with much of its detail difficult to see. After contrast expansion (Figure A17.2d), detail is much more visible because it now spans the full gray scale range.

17.3 Quantitative Intensity Measurements from Images

Because of the perceptual limitations and artifacts of human vision we need a way to transfer information to the brain in a manner that by-passes the part of the "information channel" responsible for nonlinear effects.

Experiment 17.4: Line Scans. Let us take an SEM image and draw a straight horizontal line through any part of it. The resulting line will consist of a row of pixels all having the same y value. We could now plot on a piece of graph paper the intensity in each pixel along the line as a function of the x coordinate of that pixel. For historical reasons we will call such a plot a "line scan." We have transformed, for one line of the image, information which otherwise would have entered our mind as "brightness" information into "vertical displacement" information. It is, of course, unnecessary to restrict the direction of the line scan to the horizontal. The "line" can have any angle through the micrograph. The set of pixels which define the "line" we will call the "locus" of the line scan. The actual superposition of the locus on the micrograph tells us where the data for the plot came from. See Figure A17.3.

There are several significant advantages which accrue from this transformation. (1) The information reaches our mind via a linear path. The eye does not take the logarithm of the information and there are no "edge enhancement" artifacts at sharp transitions. (2) The information is quantitative, especially if we label the axes of the plot with meaningful numbers. (3) We may discern more than the 20 or so intensity levels which we would ordinarily perceive in the image. Indeed, if the plot is spread out sufficiently in the vertical direction, it is possible to distinguish, and identify numerically, hundreds of levels. It is usual practice to superpose the plot onto the micrograph itself. This procedure permits the mind, using the information it knows from one line in the image, to extend this knowledge to neighboring lines and even other regions of the image.

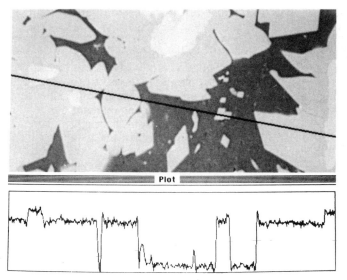

Figure A17.3. Image of flat specimen exhibiting atomic number contrast with a line scan of intensity versus distance superimposed. The locus of points displayed in the line scan is shown as a black line in the image.

Experiment 17.5: Image-Intensity Histograms. Image histograms are an especially effective way to condense an important aspect of the information content of an image: the relative number of pixels of a certain gray level. The abscissa of the image histogram is usually presented in units of the "display" intensity scale. For example, 0 to 255 levels represent full black to full white of a standard digital display monitor. The ordinate indicates the number of pixels possessing that gray level. See Figure A17.4. The areal fraction of a particular phase can be estimated from the area under one of the peaks in the histogram compared to the area of all the peaks. For example, the area of peak B in Figure A17.4 is somewhat greater than the sum of the two smaller peaks showing that the areal fraction of the phase represented by peak B is about 0.6 (or 60%).

Experiment 17.6: Histogram Modification. An image histogram of an excessively bright image will indicate this fact by having a large peak in the higher gray level range. If the image data have been incorrectly acquired or inappropriately mapped into the display range, these facts will manifest themselves as excessive numbers of gray levels being unused. Figure A17.5a is an example of an image histogram indicating inadequate usage of the display range because most of the available gray levels are unused.

  (a) *Histogram Stretching.* Histogram stretching consists of assigning the lowest intensity (darkest) pixels in an image to pure black and the highest intensity (brightest) pixels to full white. Intermediate intensities are linearly scaled between these extremes. The result of this particular operation is to fully utilize the available intensity range of the display system. An example of histogram stretching is illustrated in Figure A17.5b where the histogram of Figure A17.5a has been stretched to fill the entire 0-255 display range.

  (b) *Histogram Normalization.* When an image has a preponderance of pixels having nearly the same intensity, as manifested by a histogram with a particularly large and narrow peak, it is sometimes beneficial to "normalize" the histogram.

The result of this operation is a new image and histogram in which pixel intensities are as evenly distributed as possible over the gray scale range. An example of histogram normalization is illustrated in Figure A17.5c.

(c) Density Slicing. Density slicing is a display option where only those pixel intensities between certain values are displayed. Intensity values that fall outside of the specified range are set to full black. This operation is used to highlight structures in an image that have certain intensities and eliminate other details which might interfere with interpretation. A variant of this option sets the pixels within the chosen intensity range to full intensity. Density slicing has an obvious application to compositional maps, where a selected concentration range, e.g., 45% to 55%, may be selected for display. An example of density slicing is illustrated in Figure A17.6.

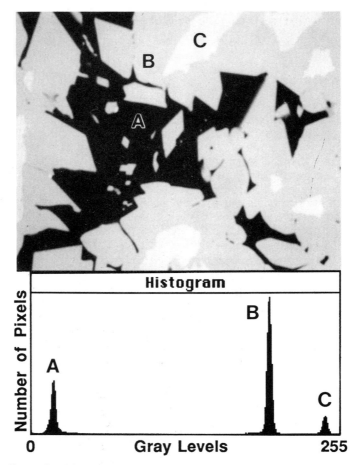

Figure A17.4. Intensity histogram of an image exhibiting three regions (phases) of distinct brightness level. The histogram shows the number of pixels (vertical scale) at each intensity level (horizontal scale). Letters relate peaks in histogram to intensities in the image.

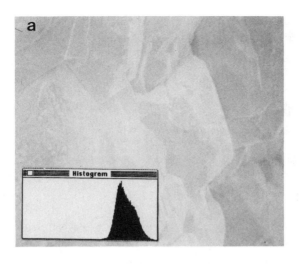

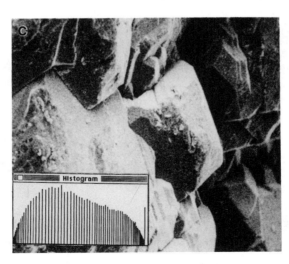

Figure A17.5. Example of histogram modification. (a) Image with inadequate usage of available display range (image is represented by gray levels ~150-255); (b) stretched image now represented by gray levels 0-255; (c) equalized image (normalized image).

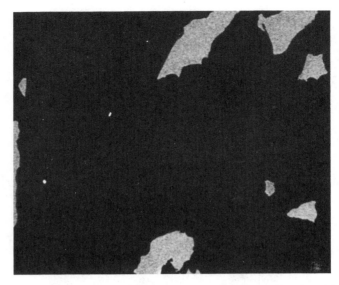

Figure A17.6. A density slice of image from Figure A17.4 using histogram peak C.

17.4 Image-Processing Kernels

A kernel is a square array of adjacent elements. In order to have a central element to operate on a particular pixel, the array must have an odd number of elements (e.g., 3x3, 5x5, etc.). To operate, a kernel is centered on a pixel at the upper left of the image such that the kernel does not hang over the edge of the image. The pixel intensity value under each kernel element is multiplied by the value of the kernel element. All such products are summed and the resulting number is placed in the position of the central pixel of the new "processed" image. The kernel is then moved to the next pixel where the process is repeated and so on until the kernel has passed over the entire image. For reasons of computational speed, the mathematics is performed with integer arithmetic rather than real arithmetic. "Ones" require no arithmetic to be performed at all, and sign changes (negative ones) are extremely fast to implement.

Experiment 17.7: Smoothing Noisy Images. When a scanning electron micrograph is obtained under high-resolution conditions the image is usually modulated by noise. Similarly, x-ray micrographs are universally noisy owing to the low quantum efficiency of the process. These SEM images often can benefit from the application of a "smoothing" kernel. The central concept here is that of spatial averaging involving the nearest neighbors to the pixel of interest. The simplest of these is the so-called "block" kernel in which the center pixel under the kernel is replaced by an unweighted average of all the pixels under the kernel, e.g., a 3x3 block kernel:

$$\text{Block kernel} \quad \begin{matrix} 1 & 1 & 1 \\ 1 & 1 & 1 \\ 1 & 1 & 1 \end{matrix}$$

A more useful set of smoothing kernels is the weighted variety in which the center value has the greatest effect. These are usually called "tent" kernels for obvious reasons. A 3x3 example is:

Tent kernel	1 1 1 1 2 1 1 1 1

The useful size of tent kernels in SEM applications can exceed 27x27. Examples of smoothing with tent kernels are illustrated in Figure A17.7.

The Gaussian smoothing kernels provide, in general, a "softer" smooth than either the tent or block kernels.

Gaussian smoothing kernel	1 1 2 2 2 1 1 1 2 2 4 2 2 1 2 2 4 8 4 2 2 2 4 8 16 8 4 2 2 2 4 8 4 2 2 1 2 2 4 2 2 1 1 1 2 2 2 1 1

The smoothing of images is the two-dimensional equivalent of smoothing an x-ray spectrum. And, just as that tool has often been abused, we can only expect the same with image smoothing. *Smoothing produces artifacts.* Under certain circumstances (unfortunately very easily achievable in the SEM), smoothing can convert an image with correct, but difficult to see, information into an image with easy to see nonsense. However, under appropriate circumstances, smoothing can produce spectacular improvements in an image. It must be realized that artifacts are always lurking at some level. *Smoothing artifacts usually manifest*

Figure A17.7. Smoothing kernels. (a) Unsmoothed x-ray image; (b) image after smoothing with a 3x3 tent kernel; (c) after smoothing with an 7x7 tent kernel; (d) after smoothing with a 11x11 tent kernel; and (e) after smoothing with a 17x17 tent kernel. Note that wormlike artifacts appear most obvious in images (c), (d), and (e). Note that the real signal-to-noise is greater in (e). Therefore if the brightness were decreased, the worms in image (e) would be suppressed into background.

themselves as a "wormy" pattern in the image with the width of the worms about the size of smoothing kernel. Smoothing kernels should be chosen to be as large as the period of the lowest spatial frequency component that represents noise. Consequently, 3x3 smoothing kernels are often completely inadequate for noisy x-ray images where smoothing could be th most beneficial. This may be seen in Figure A17.7. Note that in Figure A17.7 (b and c) the are wormlike features that are not noticeable in the original image (Figure A17.7a). These worms are somewhat suppressed when a larger kernel is used as in Figure A17.7e.

Experiment 17.8: Edge Enhancement. Laplacian operators closely approximate the visual process of the human observer. Laplacian operators are useful for finding "edges" in specimen since the Laplacian unambiguously maps out all the inflection points in the image. Inflection points are those points in the image where the first derivative of the function is not zero and the second derivative exists. Put another way, we assume that the edges of image features are correlated with the regions of maximum intensity gradient. The following "form Laplacians are those for which the integer weights are as close as possible to the rigorous rea values derived from the mathematics:

```
        Formal Laplacian (5x5)
   +2     +2     +1     +2     +2
   +2     +0     -4     +0     +2
   +1     -4    -12     -4     +1
   +2     +0     -4     +0     +2
   +2     +2     +1     +2     +2

        Formal Laplacian (3x3)
          +0     +1     +0
          +1     -4     +1
          +0     +1     +0
```

Figure A17.8. Effect of 5x5 formal Laplacian kernel operator on image of Figure A17.4. Note that the intensity of the outline is greatest where the change in gray levels (the intensity gradient) is greatest.

Figure A17.9. High pass 1 kernel applied to the image of Figure A17.2a.

There are many other edge detection kernels, e.g., the "hat operator" and the Sobel operator. The Laplacian and other derivative-type kernels are often used directly as outliners. An example of the effect of a Laplacian kernel is illustrated in Figure A17.8.

Kernels may be combined into more sophisticated operators which perform two or more mathematical processes simultaneously. These operators can be expressed in a single kernel which is a convolution of the individual operators. An example is to add a Laplacian operator in a fixed proportion to the original image all in one kernel operation. These kernels are called high pass kernels and are of particular utility.

High Pass 1			High Pass 2		
-1	-1	-1	+0	-1	+0
-1	+9	-1	-1	+5	-1
-1	-1	-1	+0	-1	+0

An example of the high pass 1 kernel is illustrated in Figure A17.9. Compare this image with the original shown in Figure A17.2a.

7 COMPUTER-AIDED IMAGING

Part III

ADVANCED X-RAY MICROANALYSIS

SOLUTIONS

Laboratory 18

Quantitative Wavelength-Dispersive X-Ray Microanalysis

18.1 Operating Conditions for WDS Analysis

Experiment 18.1: Measurement of X-Ray Intensities. (a-e) The tables on the following pages (A18.1-A18.3) summarize the data obtained from the nickel, chromium, and aluminum standards as well as from four or five analyses of the NiCrAl sample at each of three operating voltages of 10, 15, and 30 kV. Counting times of 100 sec per analysis were used. The beam currents used were 20 nA at 10 kV, 11 nA at 15 kV, and 5 nA at 30 kV. The beam currents were selected to obtain no more than 20,000 counts per sec of AlK_α from the Al standard. Values of N_B, N_T, and N_P are listed in the tables as well as the calculated values of the relative counting error (RCE). The RCE is well below 1%. Errors in the analysis of >1% relative are due to factors other than the counting uncertainties of each measurement.

18.2 Calculation of Composition

Experiment 18.2: ZAF Correction of Intensities. For each analysis the K ratio, N_P (sample)/N_{P0} (standard), was calculated and is listed in the attached tables.

(a) A ZAF quantitative analysis procedure was used to convert the K ratios for nickel, chromium, and aluminum in Tables A18.1-A18.3 to concentrations. These ZAF corrections include the absorption (A) correction of Philibert, the atomic number (Z) correction of Duncumb/Reed which utilizes the mean ionization potentials of Berger/Seltzer and the backscatter correction of Heinrich, and the fluorescence (F) correction of Reed.

A ZAF calculation performed at 15 kV and a 40° takeoff angle for the first data point at 15 kV in the NiCrAl alloy is shown in Table A18.4.

For the 15-kV ZAF correction, the chromium concentration is adjusted only -2.7% relative to the K ratio and the nickel concentration is adjusted only +1.9% relative to the K ratio. However the absorption correction for Al is a factor 2.12 and the total ZAF correction of 1.94 leads to a major adjustment in the K ratio of Al by almost +100%. Clearly any major ZAF calculation error would occur in the Al concentration due to the large absorption correction. At 10 kV the ZAF correction for Al is less, 1.40, and at 30 kV the ZAF correction for Al is much greater, 4.19, than at 15 kV.

Tables A18.1-A18.3 also include the calculated compositions for each data point at each voltage as well as a listing of the percent relative error [equation (18.5)]. Figure A18.1 shows the variation of P/B of chromium, aluminum, and nickel with voltage for the three pure element standards. The companion Figure A18.2 shows the variation of P/B with voltage for the NiCrAl alloy.

Table A18.5 summarizes the average composition of the NiCrAl alloy obtained by multiple analyses at each voltage. The table also includes the calculated percent relative error at

Table A18.1 10-kV NiCrAl Results by WDS

10 kV, 100 sec, 20 nA

Standards	N_{B-}	N	N_{B+}	N_P				
NiK_α	953	44684	915	43750				
CrK_α	491	128190	540	127674				
AlK_α	11668	2025377	7896	2015595				

X-ray line	N_{B-}	N	N_{B+}	N_P	RCE (%)	K Ratio	Meas. wt(%)	% rel error
NiK_α	897	24913	827	24060	0.623	0.550	55.4	-5.5
CrK_α	448	48095	434	47654	0.454	0.373	37.1	-3.0
AlK_α	2410	45131	2105	42874	0.459	0.021	2.99	0.1
NiK_α	826	25916	816	25095	0.611	0.574	57.8	-1.2
CrK_α	435	49195	427	48764	0.449	0.382	38.0	-1.2
AlK_α	2452	45636	2166	43327	0.456	0.022	3.03	-0.9
NiK_α	882	25946	806	25102	0.611	0.574	57.8	-1.1
CrK_α	467	49088	400	48655	0.449	0.381	37.9	-1.5
AlK_α	2448	45379	2097	43106	0.458	0.021	3.01	-0.3
NiK_α	779	25899	777	25121	0.612	0.574	57.9	-1.1
CrK_α	431	48903	417	48479	0.450	0.380	37.8	-1.9
AlK_α	2410	45122	2210	42812	0.459	0.021	2.99	0.3
NiK_α	854	26145	803	25316	0.609	0.579	58.2	-0.4
CrK_α	402	49368	412	48961	0.448	0.383	38.1	-1.1
AlK_α	2492	45428	2042	43161	0.457	0.021	3.01	-0.3
Ave Ni						0.570	57.4	-1.8
Ave Cr						0.380	37.8	-1.9
Ave Al						0.021	3.01	-0.2

Table A18.2 15-kV NiCrAl Results by WDS

kV, 100 sec, 11 nA								
Standards	N_{B-}	N	N_{B+}	N_P				
NiK_α	1874	294695	1897	292810				
CrK_α	697	259505	818	258747				
AlK_α	10695	1938150	6765	1929420				
X-ray line	N_{B-}	N	N_{B+}	N_P	RCE (%)	K Ratio	Meas. wt(%)	%Rel. error
NiK_α	1549	167009	1284	165592	0.244	0.566	57.6	-1.61
CrK_α	571	101341	571	100770	0.313	0.389	37.9	-1.66
AlK_α	1491	31076	1206	29727	0.555	0.0154	3.00	0
NiK_α	1585	168330	1532	166771	0.243	0.570	58.0	-0.91
CrK_α	570	99631	619	99037	0.316	0.383	37.2	-3.52
AlK_α	1559	30752	1262	29342	0.557	0.0152	2.97	1.01
NiK_α	1602	167192	1480	165651	0.243	0.566	57.6	-1.58
CrK_α	529	99701	595	99139	0.316	0.383	37.2	-3.38
AlK_α	1558	31117	1414	29631	0.553	0.0154	3.00	0
NiK_α	1568	165449	1485	163922	0.245	0.560	57.0	-0.95
CrK_α	566	98376	566	97810	0.318	0.378	36.7	-4.79
AlK_α	1561	30435	1278	29016	0.560	0.0150	3.04	-1.32
Ave Ni						0.565	57.5	-1.27
Ave Cr						0.383	37.3	-3.34
Ave Al						0.015	3.00	-0.08

Table A18.3 30-kV NiCrAl Results by WDS

30 kV, 100 sec, 5 nA

Standards	N_{B-}	N	N_{B+}	N_P				
NiK_α	3170	870362	3479	8676038				
CrK_α	1103	509989	1401	508737				
AlK_α	9645	1799120	6368	179113				

X-ray line	N_{B-}	N	N_{B+}	N_P	RCE (%)	K Ratio	Comp. wt(%)	%Re error
NiK_α	2429	481418	2487	478960	0.144	0.552	59.4	1.5
CrK_α	826	198151	854	197311	0.224	0.387	37.4	-3.0
AlK_α	731	12678	673	11976	0.863	0.0067	2.81	6.
NiK_α	2350	480760	2445	479463	0.144	0.553	59.5	1.6
CrK_α	855	200106	899	199229	0.223	0.392	37.7	-2.
AlK_α	757	12811	677	12094	0.858	0.0068	2.83	6.0
NiK_α	2364	431335	2458	478924	0.144	0.552	59.4	1.
CrK_α	855	200106	899	199229	0.224	0.392	37.7	-2.0
AlK_α	762	12897	643	12194	0.856	0.0068	2.86	4
NiK_α	2465	481067	2342	478663	0.144	0.552	59.6	1.8
CrK_α	829	198925	875	198073	0.224	0.389	37.6	-2.
AlK_α	853	12767	630	12025	0.859	0.0667	2.88	6.
NiK_α	2410	480425	2404	478018	0.144	0.551	59.6	1.8
CrK_α	830	198242	879	197387	0.224	0.388	37.5	-2.
AlK_α	725	12937	629	12260	0.856	0.0668	2.88	4.
Ave Ni						0.552	59.5	1.0
Ave Cr						0.390	37.6	-2.4
Ave Al						0.0068	2.84	5.0

Table A18.4 ZAF Analysis for 15 kV and 40° Take-off Angle

X-ray line	K ratio	Z	A	F	ZAF	Conc. (wt%)
Ni K_α	0.5630	0.998	1.021	1.0	1.019	57.4
Cr K_α	0.4100	1.010	0.953	1.01	0.973	39.9
Al K_α	0.0156	0.917	2.120	1.00	1.940	3.02
						100.32

ach voltage by using the measured k ratios and the ZAF correction technique. The measured oncentrations of the NiCrAl add up closely to 100%.

uestions:

1. Using Table A18.5 one can observe that the relative error [equation (18.5)] in the etermination of aluminum in the NiCrAl alloy is much higher at 30 kV. The major reason is e much higher absorption of AlK_α radiation than at 10 or 15 kV. This large amount of osorption requires a very large absorption correction to the K ratio of a factor of 2. The 5.6% error obtained at 30 kV is relatively small considering the very large correction that is ade and indicates the effectiveness of the absorption correction used.

2. The relative errors for chromium and nickel do not vary with voltage. The relative rors range from +1.67% to -3.34% and vary irregularly with voltage (see Table 18.5). Most the relative errors are probably due to experimental problems such as the reproducibility of 'DS settings, variation in specimen height, and beam current drift. The ZAF corrections to e K ratios are less than 10% even at 30 kV and are not a major cause of the analytical errors.

3. The relative counting errors for Cr and Ni decrease with increasing voltage, from 0.45 0.2 for chromium and from 0.6 to 0.15 for nickel. The improvement of RCE (Equation 3.2) with voltage occurs because the chromium and nickel counting rate increase with voltage. his increase in chromium and nickel counting rate occurs despite the fact that the beam current as decreased from 20 nA at 10 kV to 5 nA at 30 kV in order not to exceed a count rate of),000 counts per second on the aluminum standard. The major factor which causes the

Table A18.5 ZAF Summary of NiCrAl Data

Element	Known composition (wt%)	10 k		15 kV		30 kV	
		Conc. (wt%)	Rel.error (%)	Conc. (wt%)	Rel.error (%)	Conc. (wt%)	Rel. error (%)
Ni	58.5	57.4	-1.9	57.5	-1.27	59.5	+1.67
Cr	38.5	37.8	-1.9	37.3	-3.34	37.6	-2.43
Al	3.0	3.01	-0.2	3.0	-0.08	2.84	+5.60
Total	100.0	98.2		99.7		99.9	

Table A18.6 Homogeneity Variation

	Average composition (wt%)	Composition variation (wt%)	Composition variation (%)	3 x RCE (%)
10 kV				
Ni	57.4	55.4-58.2	±3.5	±1.8
Cr	37.8	37.1-38.1	±1.85	±1.35
Al	3.0	2.99-3.03	±1.0	±1.35
15 kV				
Ni	56.8	56.2-57.4	±1.05	±0.72
Cr	39.9	0	0	±0.9
Al	3.0	2.95-3.02	±1.70	±1.65
30 kV				
Ni	59.5	59.4-59.6	±0.17	±0.42
Cr	37.6	37.4-37.8	±0.53	±1.11
Al	2.84	2.81-2.88	±1.40	±2.55

increase in the chromium and nickel counting rate is that the overvoltage for the production of K_α radiation increases greatly with voltage.

4. Using Tables A18.1-A18.3, one can determine how the measured nickel, chromium and aluminum concentrations vary as a function of position for each voltage measured. The compositional variations in weight percent and in relative percent are given in Table A18.6.

A simplified criterion that has been used to establish the homogeneity of a sample is that if all the data points fall within ±3 x RCE(%) of the average composition, the sample is considered homogeneous (*SEMXM*, p. 432). The ±3 x RCE(%) homogeneity criteria for each element and voltage were calculated and are listed in Table A18.6. At 30 kV the range of

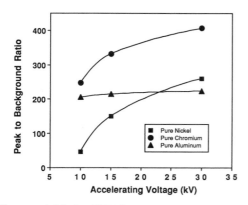

Figure A18.1. WDS peak-to-background ratios (N_P/N_B) for pure elements versus accelerating voltage.

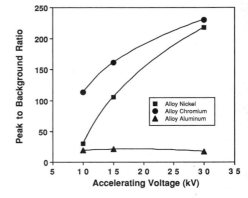

Figure A18.2. WDS peak-to-background ratios (N_P/N_B) for NiCrAl alloy versus accelerating voltage.

composition lies well within the homogeneity criteria of ±3 x RCE(%) and the sample is considered homogeneous. At 15 kV the range of compositional variation lies just at or slightly outside the homogeneity criteria. At 10 kV, however, the nickel composition variation clearly lies outside the homogeneity criteria. The nickel and chromium intensities are relatively low at 10 kV (see question 3), as little as 5% of the intensity at 30 kV. The lower counting statistics may explain the relatively large variation in nickel and chromium composition at the low, 10-kV, operating voltage. However the measurements at lower voltages may be more surface sensitive and the apparent variations in composition may be caused by polishing effects. More than four measurements may be required to establish the average composition at low voltage.

5. As shown in Figure A18.1, the P/B ratio increases with voltage as expected from the predicted variation of x-ray generation with voltage (*SEMXM*, pp. 106-107). However, the P/B for AlK_α decreases slightly with increasing voltage in the NiCrAl specimen (Figure A18.2). It is clear that some improvement in peak to background occurs for nickel and chromium, but it is questionable for aluminum in NiCrAl as overvoltages exceed 20. Furthermore, the increased absorption effects are clearly undesirable for quantitative calculations. The peak background ratios for chromium, nickel, and aluminum in the standards and the NiCrAl alloy are much superior using WDS than EDS (compare to Laboratory 19).

Laboratory 19

Quantitative Energy-Dispersive X-Ray Microanalysis

19.1 Operating Conditions for EDS Analysis

Experiment 19.1: Measurement of X-Ray Intensities. Figures A19.1 through A19.3 show EDS spectra from the NiCrAl specimen at 10, 15, and 30 kV, respectively. The take-off angle is 42.0° with the specimen tilted at 30.0° and the x-ray detector at a 12.0° take-off angle (when the specimen tilt was 0°).

Tables A19.1-A19.3 summarize the data obtained from the nickel, chromium, and aluminum standards as well as the analyses of the NiCrAl sample at each of the operating voltages, 10, 15, and 30 kV. Values of N, N_B, and N_P are listed in the tables. The net intensity values N_P were obtained by using a fit of the continuum background (for N_B) and subtracting this background from the peak intensities N.

Calculated values of the relative counting error (RCE) are also given in Tables A19.1-A19.3. The RCE is below 1.1% for nickel and chromium at all voltages. However, the relative counting error for Al increases with voltage from ~0.81% at 10 kV to ~1.46% at 15 k to ~2.2% at 30 kV. Errors in the compositional analysis greater than the RCE are due to factors other than the counting uncertainties of each measurement.

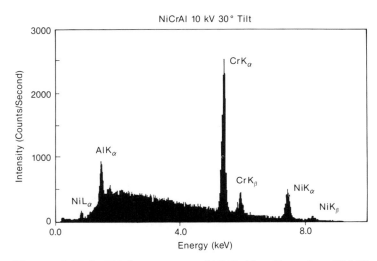

Figure A19.1. EDS spectrum of NiCrAl collected at 10 kV.

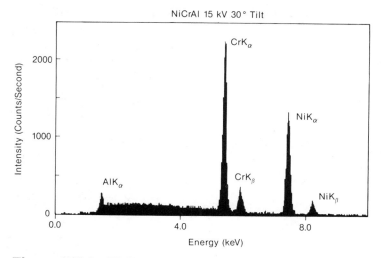

Figure A19.2. EDS spectrum of NiCrAl collected at 15 kV.

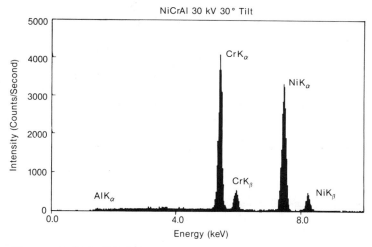

Figure A19.3. EDS spectrum of NiCrAl collected at 30 kV.

Table A19.1 10-kV NiCrAl Results by EDS

10 kV, 100 sec live time							
Standards	N	N_B	N_P				
NiK_α	11,116	1,482	9,634				
CrK_α	71,804	3,170	68,634				
AlK_α	229,880	4,122	225,758				
X-ray line	N	N_B	N_P	RCE (%)	K ratio	Meas. wt(%)	% Rel error
NiK_α	6,882	1,263	5,619	1.09	0.5670	57.10	-2.4
CrK_α	29,879	3,142	26,737	0.547	0.3920	39.00	1.2
AlK_α	9,569	3,514	6,055	0.813	0.0233	3.21	6.5
NiK_α	6,827	1,427	5,400	1.08	0.5610	56.50	-3.5
CrK_α	29,813	3,081	26,732	0.548	0.3870	38.50	
AlK_α	9,615	3,466	6,149	0.816	0.0228	3.14	4.4
NiK_α	6,772	1,401	5,371	1.08	0.5460	55.00	-6.3
CrK_α	30,362	2,998	27,364	0.548	0.3950	39.40	2.2
AlK_α	9,724	3,514	6,210	0.810	0.0237	3.25	7.6
NiK_α	6,992	1,375	5,617	1.07	0.5870	59.10	1.0
CrK_α	30,156	3,141	27,015	0.545	0.3960	39.30	2.0
AlK_α	9,556	3,508	6,048	0.814	0.0226	3.11	3.5
Ave Ni			5,502	1.08	0.5710	57.50	-1.7
Ave Cr			26,962	0.547	0.3930	39.10	1.5
Ave Al			6,116	0.815	0.0230	3.17	5.3

Table A19.2 15-kV NiCrAl Results by EDS

15 kV, 100 sec live time

Standards	N	N_B	N_P
$\text{Ni}K_\alpha$	30,901	1,017	29,884
$\text{Cr}K_\alpha$	68,264	1,773	66,491
$\text{Al}K_\alpha$	109,367	1,588	107,779

X-ray line	N	N_B	N_P	RCE (%)	K ratio	Meas. wt(%)	% Rel. error
$\text{Ni}K_\alpha$	17,635	938	16,697	0.733	0.5610	57.30	-2.09
$\text{Cr}K_\alpha$	27,721	1,446	26,275	0.585	0.3930	38.40	-0.26
$\text{Al}K_\alpha$	2,864	1,114	1,750	1.460	0.0150	3.19	5.96
$\text{Ni}K_\alpha$	17,791	910	16,881	0.730	0.5690	58.10	-0.69
$\text{Cr}K_\alpha$	27,606	1,567	26,039	0.585	0.3890	38.00	-1.32
$\text{Al}K_\alpha$	2,899	1,171	1,728	1.430	0.0155	3.30	9.09
$\text{Ni}K_\alpha$	17,848	953	16,894	0.728	0.5630	57.50	-1.74
$\text{Cr}K_\alpha$	27,459	1,478	25,981	0.587	0.3880	37.90	-1.58
$\text{Al}K_\alpha$	2,804	1,088	1,716	1.480	0.0147	3.13	4.15
Ave Ni			16,824	0.733	0.5630	57.60	-1.56
Ave Cr			26,098	0.585	0.3930	38.20	-0.79
Ave Al			1,731	1.460	0.0160	3.20	6.30

Table A19.3 30-kV NiCrAl Results by EDS

30 kV, 100 sec live time

Standards	N	N_B	N_P
NiK_α	79,795	1,214	78,581
CrK_α	117,252	1,888	115,364
AlK_α	93,236	1,188	92,048

X-ray line	N	N_B	N_P	RCE (%)	K ratio	Meas. wt(%)	% Rel error
NiK_α	43,903	1,026	42,877	0.47	0.5430	59.30	1.3
CrK_α	47,915	1,279	46,636	0.45	0.4000	38.90	1.0
AlK_α	1,078	464	614	2.30	0.0062	2.93	-2.3
NiK_α	43,910	1,099	42,811	0.47	0.5490	60.00	2.5
CrK_α	48,012	1,345	46,667	0.45	0.4000	38.90	1.0
AlK_α	1,157	530	627	2.16	0.0067	3.16	5.1
NiK_α	43,893	1,066	42,827	0.47	0.5460	59.70	2.0
CrK_α	48,349	1,353	46,996	0.45	0.4070	39.50	2.5
AlK_α	1,126	571	555	2.09	0.0064	3.02	0.6
NiK_α	43,697	1,053	42,643	0.47	0.5400	59.00	0.8
CrK_α	48,374	1,304	47,070	0.45	0.4050	39.40	2.5
AlK_α	1,157	504	653	2.21	0.0065	3.06	1.9
Ave Ni			42,681	0.47	0.5430	59.30	1.3
Ave Cr			46,766	0.45	0.4050	39.20	2.5
Ave Al			604	2.20	0.0066	3.10	3.2

Table A19.4 ZAF Analysis for 15 kV and a 42° Take-Off Angle

X-ray line	K ratio	Z	A	F	ZAF	Conc (wt%)
NiK_α	0.561	0.998	1.02	1.000	1.02	57.3
CrK_α	0.393	1.010	1.01	0.954	0.97	638.4
AlK_α	0.015	0.918	2.37	1.000	2.13	3.20
						98.9

19.2 Calculation of Composition

Experiment 19.2: ZAF Correction of Intensities. (a) A ZAF quantitative analysis procedure was used to convert the intensity ratio (K ratio) for nickel, chromium, and aluminum at each point to concentration values. The ZAF correction procedure includes the absorption (A) correction of Philibert, the atomic number (Z) correction of Duncumb/Reed, which utilizes the mean ionization potentials of Berger/Seltzer and the backscatter correction of Heinrich, and the fluorescence (F) correction of Reed.

A ZAF calculation performed at 15 kV and at a 42.0° take-off angle for the first data point at 15 kV from the NiCrAl alloy is shown in Table A19.4. For the 15-kV ZAF correction, the chromium concentration is adjusted only +1.3% relative to the K ratio and the nickel concentration is adjusted only +2.1% relative to the K ratio. However, the absorption correction factor for aluminum is 2.37 and the total ZAF correction of 2.13 leads to a major adjustment in the K ratio of aluminum by almost +100%. Clearly any major ZAF calculation error would occur in the aluminum concentration due to the large absorption correction. At 10 kV the ZAF correction for aluminum is less, 1.40, and at 30 kV the ZAF correction for aluminum is much greater, 4.7.

(b) Tables A19.1-A19.3 also include the calculated composition for each data point at each voltage as well as a listing of the percent of relative error [Equation (19.5)].

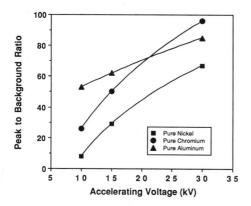

Figure A19.4. EDS peak-to-background (N_T/N_B) variation with voltage for pure elements.

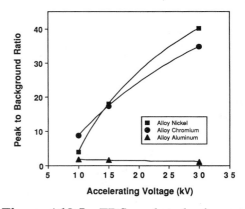

Figure A19.5. EDS peak-to-background (N_T/N_B) variation with voltage for NiCrAl alloy.

Table A19.5 ZAF Summary of NiCrAl Data

Element	Known composition (wt%)	10 kV Conc. (wt%)	10 kV Rel.error (%)	15 kV Conc. (wt%)	15 kV Rel.error (%)	30 kV Conc. (wt%)	30 kV Rel.error (%)
Ni	58.5	57.5	-1.74	57.6	-1.56	59.3	+1.35
Cr	38.5	39.1	+1.53	38.2	-0.79	39.5	-2.53
Al	3.0	3.17	+5.36	3.2	+6.30	3.1	+3.20
Total	100.0	99.8		99.0		101.9	

(c) Figure A19.4 shows the variation of P/B of nickel, chromium, and aluminum with voltage for the three pure element standards. The companion Figure A19.5 shows the variation of P/B with voltage for the same elements in the NiCrAl alloy.

Table A19.5 summarizes the average composition of the NiCrAl alloy obtained by multiple analyses at each voltage. The table also includes the calculated percent relative error for each composition measurement. The measured concentrations of the NiCrAl add up closely to 100%.

Questions:

1. The relative error in the determination of aluminum in the NiCrAl alloy is 3.5%-7.7% rel at 10 kV, 4.1%-9.1% rel at 15 kV, and -2.4%-5.1% rel at 30 kV as given in Tables A19.1-A19.3. The 10- and 15-kV data are always slightly over-corrected by the ZAF technique. The maximum range of the percent relative error is encountered at 30 kV where, because of very high absorption, the largest correction in the K ratio is made.

2. The relative error (percent) for chromium does not vary in any regular way with voltage and rarely exceed 2.5%. The analytical errors for nickel are somewhat higher at the low 10-kV operating voltage than at 15 and 30 kV (see Tables A19.1-A19.3). This higher relative error is most likely due to counting statistics from the low number of nickel counts measured (~5,500 at 10 kV versus 42,500 at 30 kV) in 100 sec. The ZAF corrections to the K ratios are much less than 10% even at 30 kV and are not a major cause of analytical errors. Most of the relative errors are probably due to experimental problems such as the reproducibility of the specimen height, the low NiK_α count rates when $U < 2.0$ and beam current drift. The latter problem is most serious in SEM instruments where stabilization of the beam current is not usually available.

3. The relative counting errors for chromium and nickel decrease with increasing voltage, from 0.55 to 0.45 for chromium and from 1.08 to 0.47 to nickel. This improvement of RCE with voltage occurs because the chromium, and most significantly the nickel, counting rate increases with voltage. The major factor which causes the increase in the chromium and nickel counting rate is that the production of K_α radiation increases with voltage relative to the background (see Figures A19.1-A19.3).

4. X-ray intensities varied over 10% relative if the specimen was not brought to the same height. This level of variation precludes quantitative analysis in the SEM. The major reason for the lack of reproducibility is that the EDS detector behind the collimator intercepts a different fraction of x-rays emitted from the sample when the height of the sample is changed. In the SEM instrument used here, the only way to obtain a reproducible sample height, which minimized variations in measured x-ray intensity, was to lock the objective lens focus at one

point and move the specimen vertically with the z-control until the specimen image was in focus. A magnification of 1000x or higher was ideal to use for focusing purposes.

5. Other factors which influence quantitative results are constancy of take-off angle and beam current. In many SEM instruments there is no long-term stability of the electron probe current. Reproducibility is worsened if different samples and standards cannot be analyzed without turning off the current to admit the specimen to the vacuum chamber of the instrument.

6. Using Tables A19.1-A19.3, one can determine how the measured nickel, chromium, and aluminum concentrations vary as a function of position for each voltage measured. The compositional variation and the relative variation in percent are given in Table A19.6. A simplified criterion that has been used to establish the homogeneity of a sample is that if all the data points fall within ±3 x RCE(%) of the average composition, the sample may be considered homogeneous (*SEMXM*, p. 432). The ±3 x RCE(%) homogeneity criteria for each element and voltage were calculated and are listed in Table A19.6. At 30 kV the range of composition lies within or just outside the homogeneity criteria of 3 x RCE and the sample would be considered homogeneous. At 10 and 15 kV the range of compositional variation lies within the homogeneity criteria except for the nickel which is clearly outside. The nickel intensities are relatively low particularly at 10 kV (see question 3), which degrades the counting statistics. In addition variations in peak heights from position to position are more difficult to deal with without light optics (question 4) and lead to variations in counting rates. Also, as discussed in question 5, there are the problems of constancy of take-off angle and beam current. It is probable that the 3 x RCE(%) criterion is too strict to establish the homogeneity of a sample using EDS. It is clear the NiCrAl standard is homogeneous if one examines the WDS data obtained in Laboratory 18 for the same sample.

7. As shown in Figure A19.4, the *P/B* ratio increases with voltage as expected from the predicted variation of x-ray generation with kV (*SEMXM*, pp. 106-107). However, the *P/B* for AlK_α decreases with increasing voltage in the NiCrAl specimen (Figure A19.5). It is clear that some improvement in peak-to-background occurs for nickel and chromium, but it is questionable for aluminum in NiCrAl when the overvoltage exceeds 20. Furthermore, the increased absorption effects are clearly undesirable for quantitative calculations. The peak-to-background ratios for nickel, chromium, and aluminum in the standards and in the NiCrAl alloy are much superior using WDS (Figures A18.1-A18.2) than EDS (Figures A19.4-A19.5).

Table A19.6 Homogeneity Variation

	Average composition (wt%)	Composition variation (wt%)	Relative variation (%)	3 x RCE (%)
10 kV				
Ni	57.5	55.0-59.1	±4.30	±3.24
Cr	39.1	38.5-39.4	±1.53	±1.64
Al	3.17	3.11-3.25	±2.40	±2.45
15 kV				
Ni	57.6	57.3-58.1	±8.70	±2.20
Cr	38.2	37.9-38.4	±0.79	±1.75
Al	3.20	3.13-3.30	±3.10	±4.40
30 kV				
Ni	59.3	58.6-60.0	±1.18	±1.41
Cr	39.2	38.6-39.5	±1.53	±1.35
Al	3.10	2.93-3.16	±5.50	±6.60

Table A19.7 Comparison of ZAF and $\phi(\rho z)$ techniques

Element	(ZAF) factor	Comp. (wt%)	$\phi(\rho z)$ factor	Comp. (wt%)
Ni	0.995	39.3	0.999	39.5
Cr	1.008	59.1	1.004	58.9
Al	1.378	3.11	1.347	3.04

Optional Experiment. A comparison is made between the ZAF and the $\phi(\rho z)$ calculation scheme for the 10 kV measured K ratio results, point 4.

The calculated results were very similar, within 2.2% rel for aluminum, 1.4% rel for chromium, and 0.5% rel for nickel.

The use of a "standardless" calculation technique led to inaccurate compositional values. For example, at 10 kV the results were 3.65 wt% Al, 64.1 wt% Cr, and 32.3% Ni, and at 30 kV the results were 4.23 wt% Al, 37.9 wt% Cr, and 57.9 wt% Ni (compare with Table A19.5). Thus, errors of over 50% rel are possible with the "standardless" procedure. For specimen elements that have x-ray lines which differ widely in energy, such "standardless" techniques often provide incorrect compositions.

19.3 Comparing "Standardless Analysis" with Quantitative Analysis Using Standards

Experiment 19.3: Silicate Analysis. The raw spectra for the silicate analyses (performed at 15 kV with a take-off angle of 40°) are shown in Figures A19.6-A19.13. The irregular shapes (non-Gaussian) of the Mg, Al, and SiK_α peaks and the complicated, nonlinear background are obvious as is the overlap of these peaks. In this experiment, the energy of the MgK_α peak was measured at 1 eV higher than its real value and the measured AlK_α and SiK_α were 2 eV too high. The gain was *not* recalibrated.

Analysis results using (1) a full quantitative program employing the MgO, Al$_2$O$_3$, and SiO$_2$ standard spectra, (2) a semi-"standardless" program employing standard spectra collected by the manufacturer on his SEM, and (3) a "standardless" program employing Gaussian peak shape estimates for the various elements are shown in Table A19.8. As can be seen, the results differ considerably, with the best results obtained for the full quantitative procedure (error < 0.1% for MgSiO$_3$) and the worst result for the completely standardless procedure. Thus, using the full quantitative procedure employing standards, one can obtain accuracy as good as that obtained with WDS techniques.

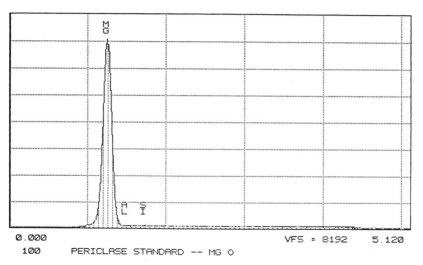

Figure A19.6. Magnesium reference peak from MgO standard.

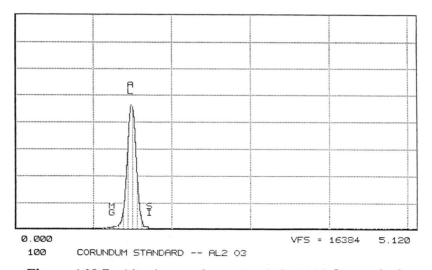

Figure A19.7. Aluminum reference peak from Al_2O_3 standard.

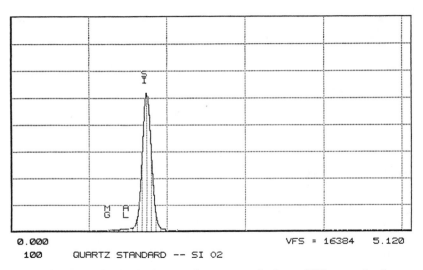

Figure A19.8. Silicon reference peak from SiO$_2$ standard.

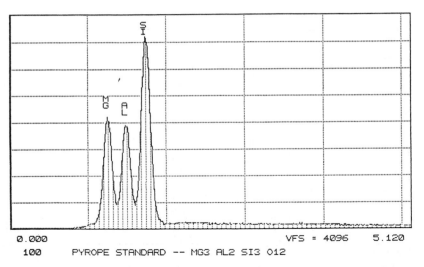

Figure A19.9. Mg-Al-Si reference spectrum from the mineral pyrope Mg$_3$Al$_2$Si$_3$O$_{12}$.

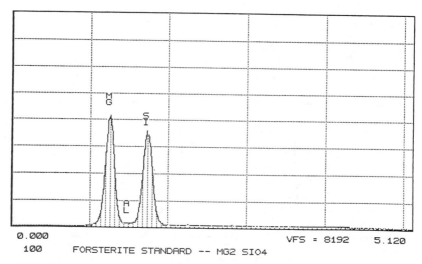

Figure A19.10. Mg-Si reference spectrum from the mineral forsterite Mg_2SiO_4.

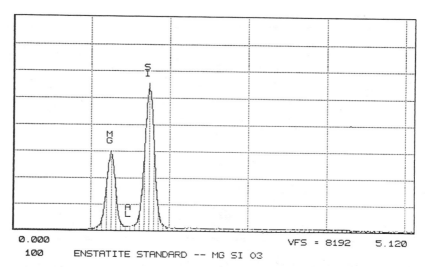

Figure A19.11. Mg-Si reference spectrum from the mineral enstatite $MgSiO_3$.

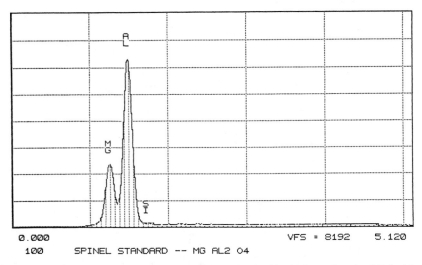

Figure A19.12. Mg-Al reference spectrum from the mineral spinel $MgAl_2O_4$.

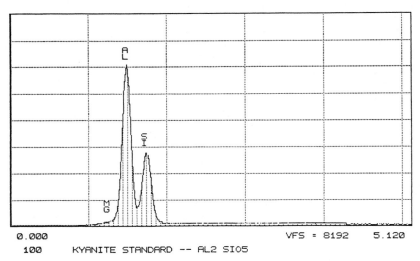

Figure A19.13. Al-Si reference spectrum from the mineral kyanite Al_2SiO_5.

Table A19.8 Silicate Analyses using EDS at 15 kV

Sample		Actual oxide wt%	EDS analysis results		
			Standards	Semistd	No stds
Mg_2SiO_4					
	MgO	57.30	57.50	52.59	57.56
	Al_2O_3		0.01	0.00	0.66
	SiO_2	47.70	42.47	47.18	51.63
	CaO		0.05	0.06	0.10
	FeO		0.01	0.17	0.00
	Total	100.00	100.04	100.00	99.96
$MgSiO_3$					
	MgO	40.15	40.12	35.71	30.44
	Al_2O_3		0.01	0.00	0.74
	SiO_2	59.85	59.80	64.11	68.71
	CaO		0.01	0.09	0.07
	FeO		0.01	0.09	0.00
	Total	100.00	100.04	100.00	99.96
$MgAl_2O_4$					
	MgO	28.33	28.17	25.94	22.97
	Al_2O_3	71.67	71.70	73.98	76.51
	SiO_2		0.10	0.00	0.38
	CaO		0.01	0.03	0.03
	FeO		0.05	0.10	0.08
	Total	100.00	100.03	100.05	99.98
Al_2SiO_5					
	MgO		0.33	0.00	0.56
	Al_2O_3	62.92	63.16	61.19	62.70
	SiO_2	37.08	36.43	38.57	36.43
	CaO		0.01	0.03	0.07
	FeO		0.01	0.21	0.09
	Total	100.00	99.95	100.00	99.86
$Mg_3Al_2Si_3O_{12}$					
	MgO	30.00	30.08	26.82	22.90
	Al_2O_3	25.29	25.00	23.88	25.92
	SiO_2	44.71	45.02	49.09	51.08
	CaO		0.06	0.09	0.05
	FeO		0.03	0.15	0.01
	Total	100.00	100.20	100.03	99.95

Laboratory 20
Light Element Microanalysis

20.1 Data Collection for Light Element Analysis

Experiment 20.1: Carbon K Peak Shape.
  Pure Carbon
  (a) Operating conditions: accelerating voltage = 10 voltage, take-off angle = 40.0°, beam current = 110 nA. For carbon, a lead stearate crystal was used. All specimens were aluminum coated at the same time to minimize x-ray absorption.
  (b) A WDS scan for graphite is shown in Figure A20.1. The scan is from 0.229 keV to 0.346 keV. The CK_α peak is at 0.276 keV.
  (c) Integrated peak intensity is 81,741 counts in 5 sec. $\sqrt{N}/N \times 100 = 0.35\%$.
  (d) Average background intensity is 716 counts in 5 sec.
  (e) No EDS measurements were made
  Iron Carbide, Fe_3C
  (a) A WDS scan for Fe_3C is shown in Figure A20.2. The scan is again from 0.229 keV to 0.346 keV.
  (b) The CK_α peak is at 0.275 keV. Two other strong peaks are shown, FeLl second order ($n=2$) at about 0.315 keV and FeL_α third order at 0.235 keV. The presence of these iron L peaks makes measurement of carbon background difficult. The graphite scan (Figure A20. does not include the FeL peaks. In addition, the carbon peak in Fe_3C is at a slightly lower energy (~1 eV lower).
  (c) Peak intensity is 46,643 counts in 58 sec $\sqrt{N}/N \times 100 = 0.46\%$. Average background intensity is 3,882 in 58 sec.
  (d) No EDS measurements were made.

Questions
  1. For the light elements the electrons that fill the vacancy in the K-shell often come from the valence band; thus, the K-spectra often show evidence of chemical bonding of the element wavelength shifts and peak shape changes.
  2. The Fe_3C standard provides a reproducible standard much more like steel since carbon atoms are surrounded by iron atoms as in the steel. Accurate quantitative analysis of light elements requires the use of integated intensities if the peak shape varies with chemical bonding. Maximum peak height measurements can only be used if there is no difference in peak shape between sample and standard.
  3. The energy resolution of the EDS detector is not sufficient to detect chemical shifts or changes in peak shape.

Experiment 20.2: Background Intensity.

Method 1: The carbon K_α intensity in Fe_3C is 45,740 counts in 61 sec. The carbon off-peak background intensity on the Fe_3C is 2,777 counts in 61 sec.

Method 2: The "carbon background" measured on pure iron at the carbon peak position is 4,380 counts in 61 sec.

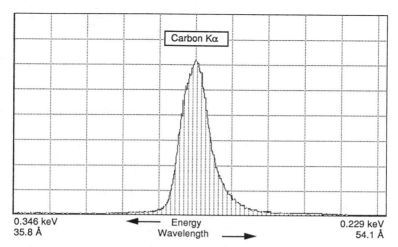

Figure A20.1. WDS 2θ-scan across the carbon K_α peak on graphite. Peak maximum is at 276 eV (44.8A). Counting time = 5 sec. $E_0 = 10$ keV. Maximum intensity, vertical full scale (VFS) = 2048.

Questions
1. The carbon background on pure iron at the carbon peak position (Method 2) is clearly higher than that measured from the Fe_3C off-peak intensities. If off-peak positions for background are measured on pure iron, the carbon background is 2,716 counts in 61 sec. This background value compares favorably with that obtained from Fe_3C (Method 1). The apparent carbon intensity measured at the carbon peak position on pure iron is real. Apparently a layer of carbon was deposited on the surface of the pure iron during either sample preparation, sample coating, the transfer process into the instrument, evacuation of the instrument, or analysis.

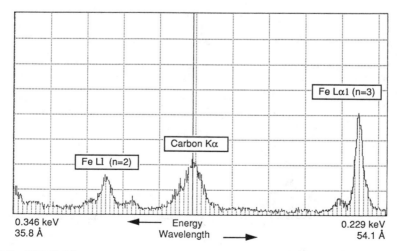

Figure A20.2. WDS 2θ-scan across the carbon K_α peak on Fe_3C. Peak maximum is at 275 eV (45.0A). Counting time 58 sec. $E_0 = 10$ keV. FeL peaks are also visible. Maximum intensity, vertical full scale (VFS) = 256.

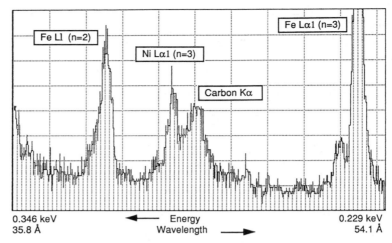

Figure A20.3. WDS 2θ-scan across carbon peak with pulse height analyzer open. Maximum intensity, vertical full scale (VFS) = 64.

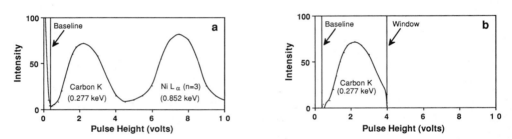

Figure A20.4. Schematic pulse height distribution curves. (a) PHD curve for Figure A20.3 showing acceptance of overlapping carbon and nickel signals. (b) PHD curve for Figure A20.5 showing rejection of iron and nickel signals by restricting the energy of the pulses that are accepted.

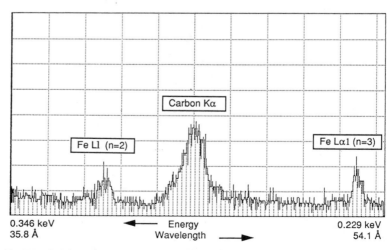

Figure A20.5. WDS 2θ-scan across carbon K_α peak with PHA adjusted to allow only the carbon peak to be accepted. Maximum intensity, vertical full scale (VFS) = 64.

2. The most accurate method for determining continuum background is to measure the off-peak background directly on the sample.
3. WDS has several advantages over EDS for background measurements of the light elements. Among these are the ability to avoid overlapping L and M peaks, to avoid the high EDS background at low energies, and to utilize high count rates for better detectability and precision ($\sqrt{N}/N$ effects).

Experiment 20.3: Peak Overlaps.
   (a) A wavelength scan was run through the carbon peak using a Fe-9.9 wt% Ni-1.04 wt% C standard, as shown in Figure A20.3. A detector voltage of 1850 V was used and the pulse height analyzer (PHA) was set at 0.4 V base line and 10 V window (wide open). The carbon peak position is indicated at 0.277 voltage. The FeL peaks, as observed in the Fe$_3$C scan of Figure A20.2, are also present in this WDS scan. The NiL_α has an energy of 0.8502 voltage and a wavelength of 14.56A. The third order NiL_α line appears at a $n\lambda$ value of 43.68 Å or an equivalent energy of 0.284 voltage, very close (within 7 ev) to the CK_α peak in the Fe-Ni-C standard. Figure A20.4a shows a schematic pulse height distribution curve for the carbon peak position of Figure A20.3. The third order NiL_α may diffract at the same angle as CK_α but the x-rays are of higher energy and thus give rise to a separate pulse height.
   (b) The pulse height analyzer was adjusted to exclude most of the high energy (0.852 voltage) x-ray peak from NiL_α as shown in Figure A20.4b. The new PHA settings were 0.4 V base line, a 4.0-V window. Figure A20.5 shows a wavelength scan over the same range as Figure A20.3. The interfering NiL_α third-order peak is not present and the FeL higher-order peaks are much smaller.

20.2 Measurement of Light Element Concentrations

Experiment 20.4: Quantitation by the ZAF Method. The measurements from the Fe$_3$C standard are

CK_α peak intensity I^{std}_{carbon} on Fe$_3$C = 49,628 counts in 56 sec

CK_α background intensity I^{std}_{carbon} on Fe$_3$C = 3,828 counts in 56 sec

$$I_{std} = I^{std}_{carbon} - I^{std\ bg}_{carbon} = 49{,}628 - 3{,}828 = 45{,}800 \text{ counts}$$

The measurements from the homogeneous steel sample are

CK_α peak intensity I_{carbon} from the steel sample = 11,090 counts in 56 sec

CK_α background I^{bg}_{carbon} measured directly on the steel sample = 3,174 counts in 56 sec

$$I_{SAMPLE} = I_{carbon} - I^{bg}_{carbon} = 11{,}090 - 3{,}174 = 7{,}916$$

The ratio of carbon x-ray intensity from the steel sample to that from the Fe$_3$C carbon standard, K_{carbon}, is given by

$$K_{carbon} = I_{SAMPLE}/I_{STD} = 7{,}916/45{,}800 = 0.173$$

If the correction factors in ZAF or equal 1.0 and the carbon content of Fe$_3$C is 6.67 wt%, then the uncorrected carbon content of the homogeneous steel sample is 1.15 wt%. With correction for carbon absorption in the steel, and using the measured data for iron (standard and background intensities), the ZAF program gives the measured carbon content in the steel as 0.8 wt%.

Table A20.1. Calibration Curve Data

Alloy	C Peak	C Backgnd	Peak-backgnd
Fe	8,544	5,747	2,797
0.30 C	12,785	5,617	7,168
0.62 C	16,205	5,766	10,439
1.01 C	22,366	5,865	16,501
1.29 C	27,863	5,811	22,052
Steel unknown	19,430	5,661	13,796

Experiment 20.5: Quantitation by the Calibration Curve Method. To obtain the calibration curve, four Fe-C standards were used plus pure iron. The standards contained 0.30, 0.62, 1.01, and 1.29 wt% C. Data were taken at 10 kV, 147 nA, 100 sec counting time and 40.0° take-off angle. The average data for each alloy are given in Table A20.1. The carbon x-ray intensity (peak-backgnd) versus carbon concentration curve is plotted in Figure A20.6. The carbon intensity does not go to zero when the carbon content is zero for the reasons discussed in Question 1, Experiment 20.2. The calibration curve is almost linear as shown by the best straight line drawn through the data. The value of $I_{\text{Peak-Background}}$ for the steel unknown is 13,769 as shown in the data above. Using the calibration curve (Figure A20.6) the value of the carbon concentration in the steel sample is 0.81 wt%. This value compares favorably with the known carbon content of 0.80 wt%.

Questions
1. The two methods of quantitation yielded very similar values of carbon content in the steel sample. The calibration method is more accurate since no calculations of absorption or atomic number effects are necessary.
2. If the operating voltage were increased, the absorption correction would become larger. However, as in Question 1, the calibration curve method would still be more accurate.

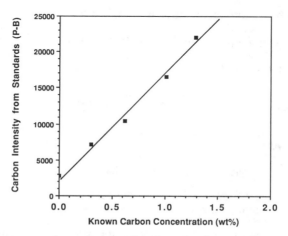

Figure A20.6. Carbon x-ray intensity (peak minus background) versus carbon concentration for the carbon in iron standards.

Laboratory 21
Trace Element Microanalysis

21.1 Data Collection for Trace Element Analysis

Experiment 21.1: Data from a High-Phosphorus Standard.
(a) WDS analysis of a meteoritic schreibersite standard $(FeNi)_3P$ with 15.5 wt% phosphorus. Accelerating voltage = 15 kV, take-off angle = 40.0°, beam current = 145 nA. For iron and nickel, a LiF crystal was used. For phosphorus, a PET crystal was used. The data are shown in Table A21.1.

Table A21.1. WDS Data from High-Phosphorus Standard

Element	Peak position (keV)	Background positions (keV)		N_S (counts)	N_{SB} (counts)	Time (sec)
FeK_α	6.406	6.222	6.603	384,711	970	10
NiK_α	7.484	7.233	7.753	61,000	1,311	10
PK_α	2.013	1.973	2.054	400,061	1,297	100

(b) EDS. Accelerating voltage = 15 kV, take-off angle = 40.0°, beam current = 145 nA. The EDS collimator entrance size was made smaller using an adjustable aperture until the dead time was <20%. Counting time = 100 sec.

Figure A21.1 illustrates the EDS spectrum from the phosphide standard, 15.5 wt% P and Figure A21.2 shows the phosphorus peak and continuum background in more detail. Note the energy regions containing the full-width-at-half-maximum (FWHM). The iron and nickel peaks and backgrounds were also measured this way. The data are summarized in Table A21.2.

Table A21.2. EDS Data from High-Phosphorus Standard

Element	Peak position (keV)	FWHM (ev)	Background region (keV) −	Background region (keV) +	N_S (counts)	N_{SB} (counts) −	N_{SB} (counts) +	N_{SB} (counts) Ave.
FeK_α	6.40	140	6.02-6.16	6.66-6.80	28,814	864	704	784
NiK_α	7.48	180	6.66-6.80	7.82-8.00	3,581	704	520	612
PK_α	2.02	120	1.74-1.86	2.24-2.36	16,740	2,706	1,641	2,174

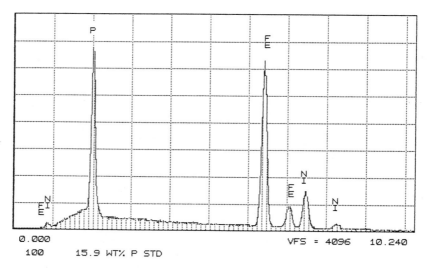

Figure A21.1. EDS spectrum from the phosphide standard (15.5 wt% P). X-ray energy range is 0 to 10.24 keV.

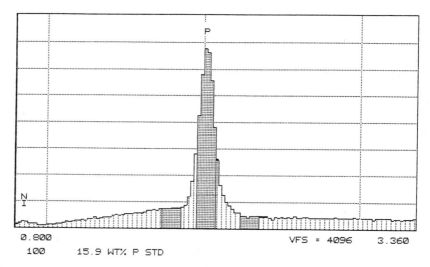

Figure A21.2. EDS spectrum around the phosphorus peak in Figure A20.1. Background windows are shown above and below peak.

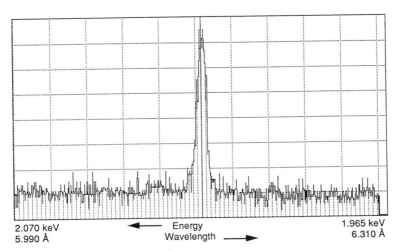

Figure A21.3. WDS 2θ scan around the phosphorus peak position in the Fe-Ni-P alloy containing 0.35 wt%P. X-ray energy range is 1.965 to 2.070 keV. Minimum intensity vertical scale. Vertical fall scale = 64.

Experiment 21.2: Establishing the Background at Low Phosphorus Concentration. (a) Figure A21.3 shows the WDS scan around the phosphorus peak position on the Fe-Ni-P alloy containing 0.35 wt% P. The scan was taken over a 100-eV range from 1.965 to 2.070 keV. The continuum background positions are chosen ±30 eV around the phosphorus peak at 2.043 keV and 1.983 keV. The data for N_B determination are shown in Table A21.3.

(b) Figures A21.4 and A21.5 show EDS spectra on the 0.35 wt% Fe-Ni-P alloy taken for 100 sec and 400 sec, respectively. The PK_α peak at 2.103 keV is hardly visible. Figure A21.6 shows an expanded region around the phosphorus peak. Note that the peak is visible above background after 400 sec but is indistinguishable from background after only 100 sec. The data for N_B are shown in Table A21.4.

Table A21.3. WDS Data from Low-Phosphorus Standard (WDS, Counting Time = 100 sec)

Element	Background Positions (keV) and Intensity (counts)				N (counts)	N_B (counts)
	-(keV)	(counts)	+(keV)	(counts)		
Phosphorus	1.983	1,022	2.043	1,912	11,469	1,467

Table A21.4. EDS Data from Low-Phosphorus Standard (EDS, Counting Time = 100 sec)

Element	FWHM (keV)	Background regions (keV)		N (counts)	N_B (counts)		
		-	+		-	+	Ave.
Phosphorus	120	1.74-1.86	2.24-2.36	1,810	1,889	1,769	1,829

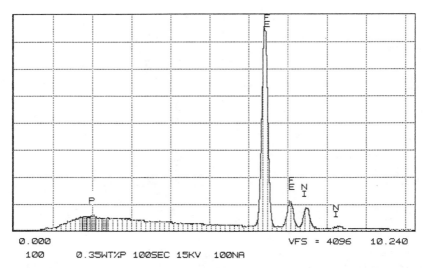

Figure A21.4. EDS spectrum from the Fe-Ni-0.35P alloy taken with a counting time of 10 sec. X-ray energy range is 0 to 10.24 keV. The position of the phosphorus K_α line is 2.014 keV.

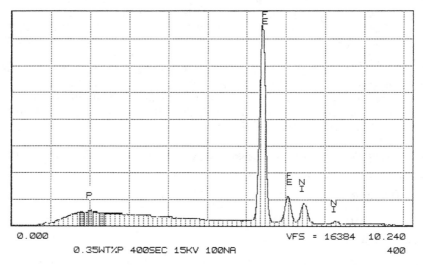

Figure A21.5. EDS spectrum from the Fe-Ni-0.35P alloy taken with a counting time of 40 sec.

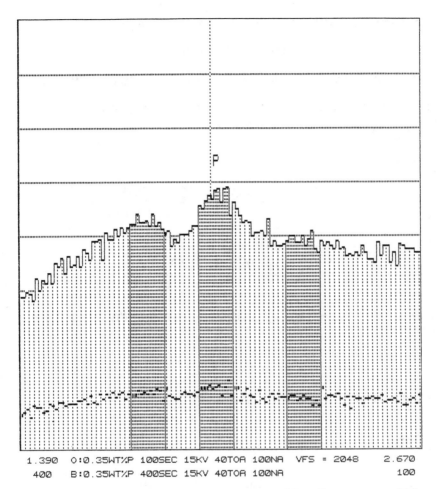

Figure A21.6. Expanded EDS region from the Fe-Ni-0.35P alloy spectra of Figures A21.4 and A21.5 collected with 400 sec counting time (top) and 100 sec counting time (bottom). X-ray energy range is 1.390-2.670 keV.

Note in Figure A21.6 that the 100-sec data N is essentially the same as N_B. Even for a 400-sec counting time, $N = 7,308$ and $N_B = 7,348$. Thus, using a set of regions above and below the phosphorus peak, we are not able to distinguish the presence of phosphorus although the phosphorus peak is apparently present. A background fitting technique may be more useful in determining N_B, since the background around the phosphorus peak is not symmetrical (see Figures A21.1 and A21.2).

21.2 Minimum Detectability Limits

Experiment 21.3: Estimation of Minimum Detectability from Alloy Data. To estimate a minimum detectability limit, C_{DL}, we assume $N-N_B$ in equation (21.1) just equals $3(N_B)^{1/2}$. Thus,

N_B(WDS) on 0.35 wt% P alloy = 1,467 counts, $3(N_B)^{1/2}$ = 114.9

N_B(EDS) on 0.35 wt% P alloy = 1,829 counts, $3(N_B)^{1/2}$ = 128.3

N_S(WDS) on 15.5 wt% phosphide = 400,061 counts, N_{SB} = 1,297

N_S(EDS) on 15.5 wt% phosphide = 16,740 counts, N_{SB} = 2,174

C_S = 15.5 wt%

Using equation (21.1),

$$C_{DL} = \left(\frac{N-N_B}{N_S-N_{SB}}\right)C_S = \left[\frac{3(N_B)^{1/2}}{N_S-N_{SB}}\right]C_S \quad (21.1)$$

For WDS, C_{DL} = 0.0045 wt%, 45 ppm
For EDS, C_{DL} = 0.135 wt%, 1350 ppm

Experiment 21.4: Estimation of Minimum Detectability from Pure Element Standards.

For the Ziebold equation,

$$C_{DL} = (3.29A)/(tP^2/B)^{1/2} \quad (21.2)$$

we use the following data to calculate C_{DL}.

Factor	WDS	EDS
A	1.0	1.0
t	100 sec	100 sec
$P = N-N_B$	24,000 cps	1,000 cps
$B = N_B$	78 cps	130 cps
C_{DL} (wt%)	0.012 wt%	0.375 wt%

The C_{DL} values determined by the Ziebold equation are more conservative, that is, about three times higher than those measured in Experiment 21.3.

Questions
1. The C_{DL} values are 30 times larger using EDS rather than WDS. Clearly the WDS technique yields a much improved detectability limit, well below 100 ppm. The WDS technique is better because the P and P/B values are so much higher for the same counting time.
2. The meteorite contains a P concentration 0.08wt%. The detectability limit for P using EDS is 0.135wt%. Therefore the EDS cannot detect the presence of P in the meteorite.

Table A21.5. Phosphorus Measurement Using WDS in a Meteorite

Point No.	Ni(counts) N	N_B	Fe(counts) N	N_B	P(counts) N	N_B
1	196,717	12,700	4,177,151	9094	2163	1014
2	205,735	12,563	4,210,170	9032	2210	1143
3	201,352	12,737	4,197,295	9061	2176	1100
4	201,760	12,444	4,208,298	9174	2143	1114
5	204,221	12,833	4,191,416	9071	2036	1100
6	204,477	12,743	4,167,921	9329	2159	1099

1.3 Measurement of Trace Element Concentrations

Experiment 21.5: Measurement of P Content in a Meteorite.
(a) Since C_{DL} for P was too high to measure using EDS, only WDS measurments were made in the low-nickel kamacite phase of the meteorite. The data from several positions in the meteorite are shown in Table A21.5. The analysis conditions for WDS were 15 kV, 145 nA beam current, and 100 sec counting time. The WDS scan around the phosphorus peak on the kamacite phase in the iron meteorite at 2.103 kV is shown in Figure A21.7. The scan ranges from 1.965 kV to 2.070 kV. The PK_α peak can be clearly seen above the background.

(b) The ZAF technique was chosen for the analysis of the metal phase. For point 1 in Table A21.5, the phosphorus concentration in the meteorite was 0.08 wt%. The ZAF factors shown in Table A21.6 were used to compute the wt% for each element from the K ratios:

$$K \text{ ratio} = \frac{N - N_B}{N_S - N_{SB}}$$

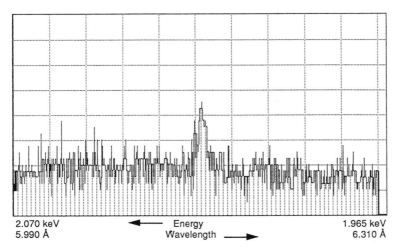

Figure A21.7. WDS 2θ scan around the phosphorus peak on the meteorite kamacite phase (low-nickel bcc phase) containing approximately 0.08wt%P. The scan covers an energy range of 1.965-2.070 keV. Maximum intensity vertical scale. Vertical fall scale = 32.

Table A21.6 Chemical Analysis Using ZAF Technique

Element	K ratio	Z	A	F	ZAF	wt%
Fe	0.946	1.001	1.0	0.995	0.996	94.3
Ni	0.044	0.985	1.06	1.0	1.044	4.6
P	0.0006	0.927	1.418	0.998	1.313	0.08

(c) Calculation of C_{DL} for Point 1, Table A21.5:

$$C_{DL} = \left(\frac{N-N_B}{N_S-N_{SB}}\right) C_S$$

where $N-N_B = 3(N_B)^{1/2}$ (Equation 21.1). From Table A21.5, Point 1, $N_B = 1014$ counts, $3(N_B)^{1/2} = 95.5$. From Table A21.1, $N_S = 400{,}061$ counts, $N_{SB} = 1{,}297$ counts, and $C_S = 15.5$ wt%. Therefore,

$$C_{DL} = \frac{95.5}{(400{,}061-1{,}297)}(15.5)\text{wt\%} = 0.0037 \text{ wt\%} \ (37 \text{ ppm})$$

Thus, C_{DL} is well below the phosphorus content in kamacite of 0.08 wt%.

(d) Phosphorus content in kamacite = 0.08 ± 0.0037 wt%.

Laboratory 22

Particle and Rough Surface Microanalysis

22.2 Spherical Particles or Fracture Surfaces

Experiment 22.1: The Microscope-Detector Relationship. (a) While many modern SEMs have EDS detectors mounted such they they look down on the specimen with a known take-off angle (at one particular working distance), most SEMs have EDS detectors that have an arbitrary relationship to the specimen. If the specimen lies below the detector and is tilted toward the detector, the take-off angle ψ may be calculated from

$$\psi = \theta + \arctan(\Delta z/x)$$

where θ is the specimen tilt angle, Δz is the distance that the beam impact point on the specimen is below the detector chip center line, and x is the distance between the detector chip and the electron beam impact point on the specimen. More details of this measurement may be found in SEMXM, p. 337.

(b) The relative position of the detector on the image may be found by viewing the specimen chamber from the top in the same orientation as the micrograph and noting the detector orientation.

(c) As shown in (a) the specimen should be kept at the same height, tilt angle, and x-y translation for each quantitative microanalysis.

Experiment 22.2: Rastered Beam and Point Analyses (Method 1). One can observe in Table A22.1 that conventional EDS methods produce very poor results. Results for the point analyses are particularly bad. The measured AlK_α peak intensity for the point analysis on the particle on the side away from the detector is only 6% of that measured on the flat specimen, while the NiK_α intensity measured on the same point is over 85% of that measured on the flat specimen. If the flat specimen were used as a standard as in Method 1 (Section 22.1), the ZAF processed EDS composition results would be as shown in Table A22.2. The relative errors for the normalized results with both the rastered beam and the point analysis toward the detector are about 10%, whereas the relative error in aluminum for the point analysis away from the detector is more than 85%. The AlK_α intensity is down, compared to the NiK_α, because of that x-ray line's much greater absorption by the sample and the longer path length, relative to the flat specimen, that x-rays must travel to leave the specimen.

Experiment 22.3: Peak-to-Background Ratios (Method 2). The peak-to-background ratios for AlK_α and NiK_α in the point analyses of the NiAl particle, as shown in Table A22.2, are certainly more constant than the peak intensities; however, there is still significant variation. The peak-to-background ratios for AlK_α for the point analyses (12.2 and 08) as given in Table A22.2 differ by as much as 20% relative to the value for the flat

Table A22.1. Example Data from Experiments 22.2-22.4

Take-off angle: 40°

20 keV		Flat specimen	Whole ptc specimen	Side of particle — Toward detector	Side of particle — Away from detector
Intensity:	NiL_α	16,029	9,207	21,543	1,568
	AlK_α	127,526	86,647	180,359	8,317
	NiK_α	153,208	89,783	155,316	132,202
Back-ground:	NiL_α	3,237	2,895	4,720	258
	AlK_α	12,621	13,371	14,820	916
	NiK_α	9,114	9,180	9,142	8,485
Relative intensity:	AlK_α/NiK_α	0.834	0.965	1.16	0.063
	AlK_α/NiL_α	7.960	9.410	7.21	5.300
Peak/back-ground:	$P/B\ NiL_\alpha$	4.95	3.18	4.56	6.09
	$P/B\ AlK_\alpha$	10.10	6.48	12.20	9.08
	$P/B\ NiK_\alpha$	16.80	9.78	17.00	15.60

10 keV					
Intensity:	NiL_α	13,881	4,829	16,179	3,429
	AlK_α	85,263	39,027	99,372	16,151
	NiK_α	4,015	1,508	4,022	3,922
Back-ground:	NiL_α	3,248	1,604	3,994	562
	AlK_α	11,287	6,717	12,864	2,259
	NiK_α	1,819	1,257	1,655	1,659
Relative intensity:	AlK_α/NiK_α	21.20	25.90	24.70	4.12
	AlK_α/NiL_α	6.14	8.08	6.14	4.71
Peak/back-ground:	$P/B\ NiL_\alpha$	4.27	3.01	4.05	6.10
	$P/B\ AlK_\alpha$	7.55	5.81	7.72	7.15
	$P/B\ NiK_\alpha$	2.20	1.20	2.43	2.36

specimen (10.1). The peak-to-background values for the less-absorbed NiK_a are closer to th less-absorbed NiK_α are closer to that for the flat specimen (17.0 and 15.6 versus 16.8), differing by less than 7.5% relative to the flat sample. On the other hand, the peak-to-background ratios for the whole particle scan are 35% to 45% lower than the peak-to-background ratio for the flat specimen. This is because rastering the beam over the whole particle surface results in some of the beam electrons penetrating into the substrate on which particle sits. As a result, the background continuum contains contributions from both particl and substrate while the characteristic x-ray peaks come only from the particle since the substrate does not contain nickel or aluminum. If the example data were processed using Equation 22.1 and the ZAF method, the results would look like those of Table A22.3.

Table A22.2. 20 kV EDS/ZAF Compositions of NiAl Particles from Peak Intensities (Method 1)

Unnormalized	Actual (wt%)	Rastered beam	Point analyses Toward detector	Away from detector
Al	31.49	22.94	40.58	2.67
Ni	68.51	41.14	69.42	59.50
TOTAL	100.00	64.08	110.00	62.17
Normalized				
Al	31.49	35.80	36.89	4.29
Ni	68.51	64.20	63.11	95.71
TOTAL	100.00	100.00	100.00	100.00
Atom fractions				
Al	0.500	0.548	0.560	0.089
Ni	0.500	0.452	0.440	0.911

Table A22.3. 20 kV EDS/ZAF Compositions of NiAl Particles from Peak-to-Background Ratios (Method 2)

Unnormalized	Actual (wt%)	Rastered beam	Point analyses Toward detector	Away from detector
Al	31.49	20.20	38.04	28.31
Ni	68.51	39.88	69.33	63.62
TOTAL	100.00	60.09	107.36	91.93
Normalized				
Al	31.49	33.62	35.43	30.80
Ni	68.51	66.38	64.57	69.20
TOTAL	100.00	100.00	100.00	100.00
Atom fractions				
Al	0.500	0.524	0.544	0.492
Ni	0.500	0.476	0.456	0.508

Table A22.4. 10 kV EDS/ZAF Compositions of NiAl Particles from Peak Intensities (Method 1)

Unnormalized	Actual (wt%)	Rastered beam	Point analyses Toward detector	Away from detector
Al	31.49	15.35	36.01	6.59
Ni	68.51	24.53	79.29	17.49
TOTAL	100.00	39.80	115.30	24.07
Normalized				
Al	31.49	38.49	31.23	27.36
Ni	68.51	61.51	68.77	72.64
TOTAL	100.00	100.00	100.00	100.00
Atom fractions				
Al	0.500	0.577	0.497	0.450
Ni	0.500	0.423	0.503	0.550

Experiment 22.4: Low Voltage. Table A22.1 shows that performing these analyses at 10 keV (using the AlK_α and NiL_α lines) somewhat improves results using peak intensities relative to those at 20 keV, but somewhat degrades results using peak-to-background ratios. The results using peak intensities are improved because (1) there is less absorption of the low-energy x-ray lines at 10 keV than there is at 20 keV, thus lessening the effect of differences in ray path length between particle and flat specimen, and (2) the NiL_α is closer in energy and absorption properties to the AlK_α line than is the NiK_α line. However, while the results using peak intensities are improved, they still are not good. Processing these intensity data to give compositions would result in analyses similar to those of Tables 22.4 and 22.5.

Table A22.5. 10 kV EDS/ZAF Compositions of NiAl Particles from Peak-to-Ratios (Method 2)

Unnormalized	Actual (wt%)	Rastered beam	Point analyses Toward detector	Away from detector
Al	31.49	24.23	32.20	29.82
Ni	68.51	48.79	64.98	97.87
TOTAL	100.00	72.53	97.18	127.69
Normalized				
Al	31.49	33.41	33.13	23.35
Ni	68.51	66.59	66.87	76.65
TOTAL	100.00	100.00	100.00	100.00
Atom fractions				
Al	0.500	0.522	0.519	0.399
Ni	0.500	0.481	0.481	0.601

Table A22.6. Experiment 22.5

Rastered beam analysis
Accelerating potential: 15 keV

Diameter: Shape:		Thick Flat	12 μm Trig.pr.	4 μm Tetr.pr.		
K-ratio:	MgK_α	1	0.1846	0.2792	___	___
	SiK_α	1	0.1816	0.2450	___	___
	CaK_α	1	0.2104	0.2291	___	___
	FeK_α	1	0.2278	0.2524	___	___
R-factor:	Mg/Si	1	1.017	1.140	___	___
	Ca/Si	1	1.159	0.935	___	___
	Fe/Si	1	1.250	1.030	___	___

From Figures A22.1-A22.3 (calculated):

R-factor:	Mg/Si	1	1.03	1.10	___	___
	Ca/Si	1	1.16	0.92	___	___
	Fe/Si	1	1.27	1.01	___	___

Point beam analyses
Accelerating potential: 15 keV

Diameter: Shape:		Thick Flat	4 μm Trig.pr	2 μm Tetr.pr		
K-ratio:	MgK_α	1	0.9518	0.3400	___	___
	SiK_α	1	1.2260	0.2640	___	___
	CaK_α	1	0.7298	0.2504	___	___
	FeK_α	1	0.7080	0.2880	___	___
R-factor:	Mg/Si	1	0.7763	1.288	___	___
	Ca/Si	1	0.5953	0.948	___	___
	Fe/Si	1	0.5775	1.091	___	___
P/B:	MgK_α	360.4	270.7	115.4	___	___
	SiK_α	1026.1	1055.0	476.9	___	___
	CaK_α	470.6	408.0	145.2	___	___
	FeK_α	157.8	143.2	83.4	___	___

(Contains substrate component)

Obviously, the unnormalized results for the particle differ greatly from those for the flat specimen. The normalized results are closer than those obtained at 20 keV, but still can vary by more than 10% relative. The variability in the normalized results is about twice that observed at 20 kV, although some of the results are closer to the flat sample values at 10 kV than at 20 kV. The variability at 10 kV is probably mostly due to the low values of peak-to-background being measured. Statistical fluctuations in the background intensity become much more important as the peak-to-background ratio approaches unity.

22.3 Irregular or Rough Particles

Experiment 22.5: Rastered Beam Analysis Utilizing Particle Shape (Method 3)

(a) and (b) Processing the example data (see data, Table A22.6) through a conventional correction program and normalizing the results to 100% would produce results similar to those in Table A22.7. (The results are normalized to 99.77% because MnO, which is present in the sample at a level of 0.23 wt%, was not analyzed.) Although the results are better than those obtained in Section 22.2 (in a case where the amount of absorption was considerably greater), they still are not very good. The apparent compositions of magnesium and iron, for example, differ between 10% and 20% from their correct values. For comparison, see the results in the next section.

(c) Correcting for particle ZAF effects dramatically improves the quality of the results. Compare the $R_{x/Si}$ factors in Table A22.6 to those calculated as a function of particle size in Figures A22.1-A22.3. For example $R_{Mg/Si}$ for the 12 µm particle was measured to be 1.02 whereas Figure A22.1 gives 1.03. The $R_{x/Si}$ values indicate how much particle size and shape affect the quantitative analysis. For large flat-topped particles $R_{x/Si}$ should approach 1. However, small particles and irregularly shaped particles will cause $R_{x/Si}$ to deviate from 1. $R_{x/Si}$ values can be used to correct the data for particle and shape effects. Using the sample data and sample geometry (Table A22.6), and the $R_{x/Si}$ factors given in Figures A22.1-A22.3 for various particle shapes, analyses similar to those of Table A22.8 may be obtained. The calculated compositions of the particles utilizing particle corrections produce results that are almost as good as those obtainable when analyzing thick polished specimens. The differences between measured and actual concentrations are less than 3% relative.

Table A22.7. Rastered Beam Analyses Uncorrected for Particle Effects

Normalized results	Actual (wt%)	12 µm ptc	4 µm ptc
Mg	9.17	8.77	10.29
Si	25.52	23.99	25.13
Ca	11.02	12.01	10.14
Fe	11.31	13.29	11.47
O	42.75	41.71	42.74
TOTAL	99.77	99.77	99.77

Table A22.8. Rastered Beam Analyses Corrected for Particle Effects

Normalized results	Actual (wt%)	12 μm ptc	4 μm ptc
Mg	9.17	9.17	9.40
Si	25.52	25.59	25.24
Ca	11.02	11.04	11.08
Fe	11.31	11.16	11.41
O	42.75	42.81	42.64
TOTAL	99.77	99.77	99.77

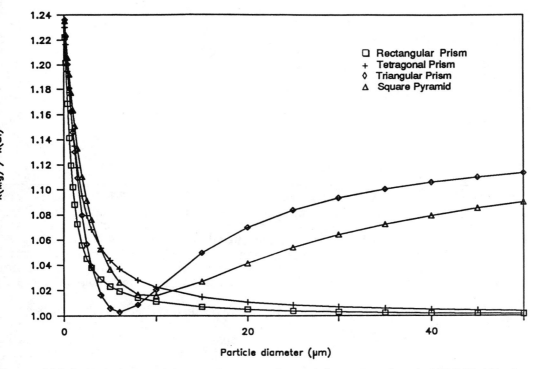

Figure A22.1. Relative particle K-ratios ($R_{x/Si}$ factors) for magnesium in NBS K-411 glass particles. 15 kV and 40° take-off angle.

Table A22.9. Point Beam Particle Analyses Using Peak-to-Background Ratios

Unnormalized results	Actual (wt%)	4 μm ptc	2 μm ptc
Mg	9.17	6.89	2.94
Si	25.52	26.24	11.86
Ca	11.02	9.55	3.40
Fe	11.31	10.26	5.98
O	42.75	41.18	18.51
TOTAL	99.77	94.12	42.69
Normalized results	Actual (wt%)	4 μm ptc	2 μm ptc
Mg	9.17	7.30	6.87
Si	25.52	27.82	27.71
Ca	11.02	10.12	7.95
Fe	11.31	10.87	13.98
O	42.75	43.65	43.26
TOTAL	99.77	99.77	99.77

Experiment 22.6: Peak-to-Background Ratios (Method 2). WDS produces better peak-to-background ratios and higher intensities for particles larger than about 1 μm. EDS provides better detection limits for particles smaller than this. Use of WDS requires beam currents of at least 1 nA for reasonable detection limits, while the EDS can be effectively used at currents as low as about 50 pA. A 1-nA electron beam is considerably larger in diameter than a 50-pA SEM beam. Below a minimum critical diameter the beam current required for WDS produces a minimum spot size that is larger than the particle diameter and the detectibility becomes worse.

The peak-to-background method can be used to obtain good particle analyses, but it has its limitations, as seen in the example data (Table A22.6). These data, when corrected by conventional techniques, would produce results similar to those of Table A22.9. The results for the 4-μm particle are not too bad, particularly when normalized. With the exception of magnesium, the various elements differ from the actual value by 10% relative or less. However, the results for the 2-μm particle are quite bad. Even the normalized results differ from the correct composition by as much as 25% relative. In both cases, but particularly in the case of the 2-μm particle, part of the measured background radiation probably was produced by electrons interacting with the substrate rather than with the particle. When using massive substrates, the peak-to-background method is limited to analyzing those particles that are significantly larger than the electron range. To analyze μm-sized or smaller particles using the peak-to-background method, the particles need to be mounted on something like a thin carbon film on a beryllium TEM grid that cannot produce any significant background radiation under electon bombardment.

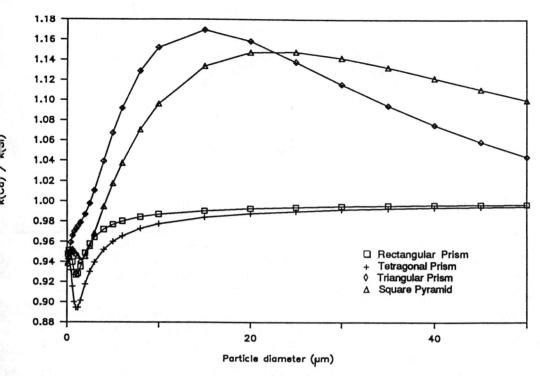

Figure A22.2. Relative particle K-ratios ($R_{x/Si}$ factors) for calcium in NBS K-411 glass particles. 15 kV and 40° take-off angle.

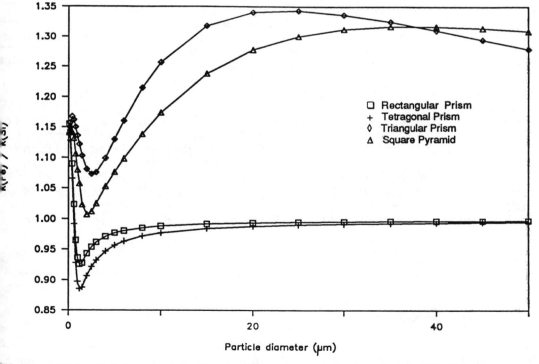

Figure A22.3. Relative particle K-ratios ($R_{x/Si}$ factors) for iron in NBS K-411 glass particles. 15 kV and 40° take-off angle.

Laboratory 23
X-Ray Images

23.1 Analog X-Ray Dot Maps

Experiment 23.1: Recording Dot Maps. (a) Adjustment of CRT recording dot. Various WDS iron x-ray images obtained with different settings of the CRT dot brightness reveal the subjective nature of the analog dot mapping procedure. A dim recording dot (Figure A23.1b) provides a very weak image even where the iron concentration is high and no information at all where the iron concentration is low. The optimum dot brightness image (Figure A23.1c) carries more information but the image is weak and noisy in the iron-poor regions because the total number of x-rays recorded in the 100-sec scan is inadequate to provide the necessary image statistics. Note that both characteristic and bremsstrahlung x-ray of the same energy are recorded since no background correction is applied in the analog WDS scan. The highest intensity dot produces an image (Figure A23.1d) in which the x-ray pulse locations are most easily evident, but the image is not satisfactory because blooming of the dot image causes a loss of resolution and contrast, obscuring fine scale details.

(b) Effect of number of x-ray counts. By increasing the scanning time (and the number of x-ray counts collected) at the optimum dot brightness, the image contrast improves dramatically (Figure A23.1e,f). For the 1000 sec scan (Figure A23.1f) the information content improves in the iron-poor regions but may be lost in the iron-rich regions because the dot density is so great that entire regions are completely white (saturated).

(c) Wavelength-dispersive spectrometer defocusing. The wavelength-dispersive x-ray spectrometer is a focusing device. When the spectrometer is adjusted to be on a characteristic ray peak, the focusing criterion is satisfied exactly only for an electron beam generating x-ray on the optic axis of the instrument. The act of scanning the beam to form an image takes the beam away from the optic axis, with the consequence that the collected x-ray intensity falls of Because the diffraction crystal has a significant width, typically 1 cm, the spectrometer has a line of focus parallel to the crystal width. This focus line is projected as a line of uniform x-r transmission in the plane of the specimen. At low magnifications (Figure 23.2c), the focus li is visible in the x-ray map and tends to dominate the image contrast, suppressing the true elemental contrast on either side of the line. Figure A23.2d shows a digital EDS image of the same area that exhibits no defocusing. An analog solution to the defocusing problem has bee implemented on some instruments. This analog technique effectively adjusts the spectromete position or the stage z-motion in synchronism with the scan on the specimen to maintain the spectrometer focus line to be coincident with the scan line it progresses through the image frame.

23.2 Digital X-Ray Images

Experiment 23.2: Choice of Digital Image Parameters. (a) Given $\rho = 5.5$ g/cm^3, $E_0 = 15$ kev, and $E_c = 7.1$ keV for iron, the x-ray range is found from Equation 23.1:

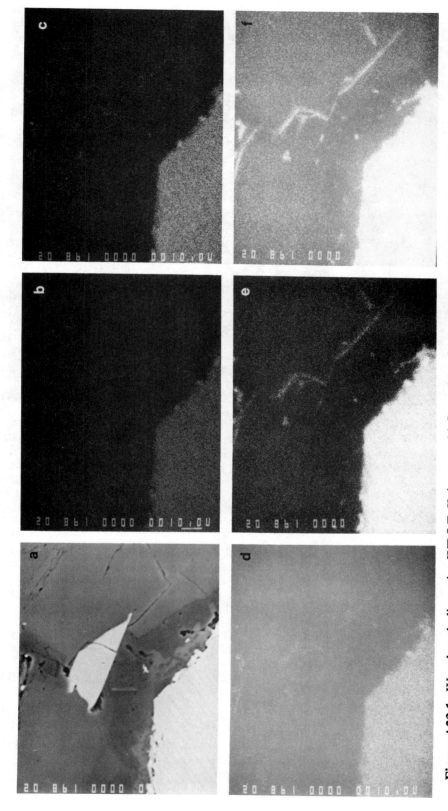

Figure A23.1. Wavelength-dispersive WDS FeK_α images. (a) Secondary electron image of multiphase basalt; (b) analog image dim recording dot (100 sec); (c) optimum recording dot (100 sec); (d) blooming recording dot (100 sec); (e) optimum recording dot (300 sec); (f) optimum recording dot (1000 sec = 16.7 min).

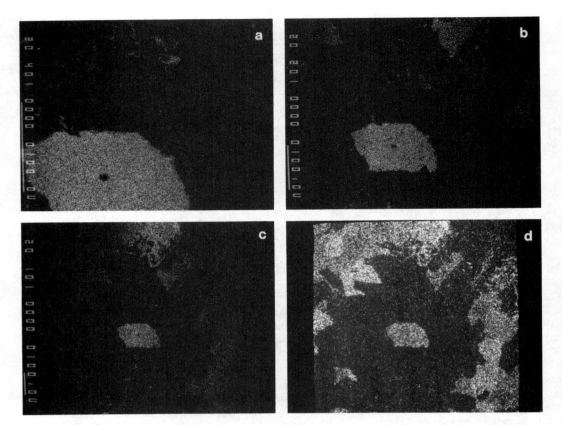

Figure A23.2. Wavelength-dispersive spectrometer defocusing. (a) WDS FeK_α image at 400x original magnification; (b) WDS FeK_α image at 200x original magnification; (c) WDS FeK_α image at 100x showing defocusing effects (weaker intensity on the left and right); (d) EDS image of the same area showing no defocusing (even x-ray intensity across the field).

$$R = \frac{0.064}{\rho}\left(E_0^{1.68} - E_c^{1.68}\right) \mu m$$

$$= \frac{0.064}{5.5}\left(15^{1.68} - 7.1^{1.68}\right)$$

$$= 0.0116\ (04.6 - 26.9) = 0.785\ \mu m$$

$$R = 0.79\ \mu m$$

Use Equation 23.2 to find the minimum suitable pixel density where $L = 100{,}000\ \mu m$, $M = 860x$, and $R = 0.79$ mm:

$$n > \frac{L}{M \cdot R}$$

$$n > \frac{100{,}000\ \mu m}{860 \times 0.79\ \mu m} = 147$$

Table A23.1. Dwell Time Per Pixel

Pixel density	Total number of pixels	Time per pixel ($t_f = 15$ min)
64 x 64	4096	220 ms
128 x 128	16384	55 ms
256 x 256	65536	14 ms
1024 x 1024	1,048,576	<1 ms

e.g., 147 x 147 pixel density. But the only choices available are 128 x 128 and 256 x 256. In this case, it would be better to raise the magnification to 1000x than to pick 256 x 256 since more counts would be generated.

The final criterion is that the average pixel defining a discrete phase in the x-ray image should have at least 8 counts. The dwell time per pixel for a 15 min frame time can be calculated for all common pixel densities (Table A23.1). Since the average pixel should contain at least 8 counts, this criterion would be more easily met for 128 x 128 than for 256 x 256. If not enough counts per pixel are obtained at either 256 x 256 or 128 x 128, the acquisition time should be increased beyond 15 min. For the examples shown, the pixel density was 128 x 128 and the dwell time per pixel was 55 ms.

(b) To set up the digital scan with an EDS system we must make a window file for the selected elements. Generally, windows are chosen to be slightly larger than full-width-at-half maximum (FWHM) although full-width windows are often used in cases with no overlap. Be sure to set up at least one background window in a region containing no peaks (low-energy side of iron K_α peak). Figure A23.3 shows an EDS spectrum for which the beam was scanned over an area of the basalt in an effort to detect all of the elements present. Elements

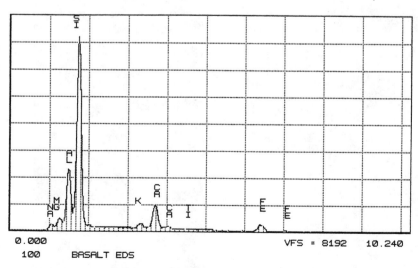

Figure A23.3. EDS spectrum of elements detected in basalt. This spectrum is used to determine which elements should be imaged in the digital x-ray image.

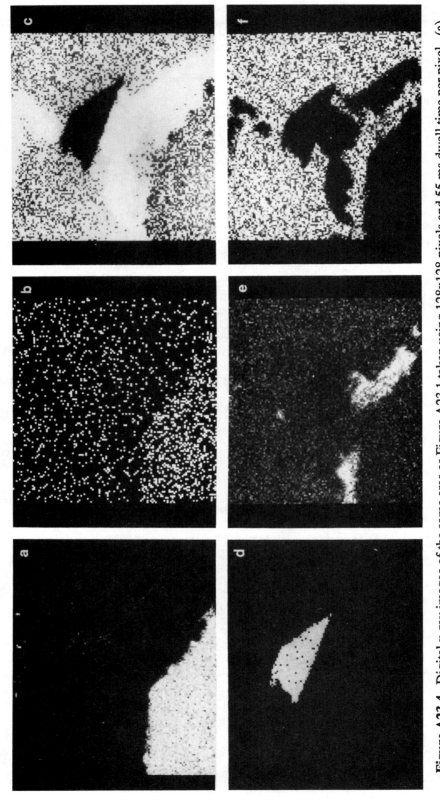

Figure A23.4. Digital x-ray images of the same area as Figure A23.1 taken using 128x128 pixels and 55 ms dwell time per pixel. (a) EDS iron image; (b) EDS background image; (c) EDS silicon image; (d) EDS calcium image; (e) EDS potassium image; (f) EDS aluminum image.

detected were sodium, magnesium, aluminum, silicon, potassium, calcium, titanium, and iron. The parameters selected in Experiment 23.2a produced the EDS digital x-ray images shown in Figure A23.4. While these images are not background-subtracted, they provide a good indication of the location of each element. However, the iron EDS image (Figure A23.4a) does not reveal the distribution of iron in the iron-poor regions as well as the analog WDS dot map iron image (Figure A23.1e,f) for which the peak-to-background is about 10 times greater.

(c) The WDS system will easily produce more than 8 counts per pixel in the iron-containing phase. Figure A23.5b shows the WDS digital iron image, which exhibits an even better signal-to-background ratio than the analog dot map (Figure A23.1e,f) because relative intensity values are displayed. The estimated average counts per pixel in the iron-rich phase (white) is about 800 in Figure A23.5b. In the regions of lower iron concentration near the middle of the frame the range is from 8 to 170 counts per pixel.

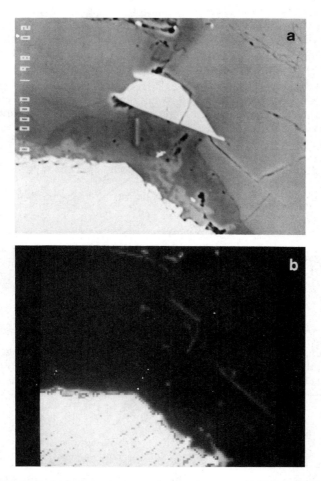

Figure A23.5. WDS digital x-ray image. (a) SE image; (b) WDS FeK_α image using 128x128 pixels and 55 ms dwell time per pixel.

23.3 Background Removal

Experiment 23.3: Background Subtraction. Background may either be subtracted on-the-fly for each pixel before the counts for that pixel are stored or subtracted after the element image is acquired by subtracting an image collected with a nearby background window. Figure A23.6 shows results from on-the-fly background subtraction for silicon, titanium, and iron. The chromium and background images (Figures A23.6e,f) appear similar. Both images show a slight enhancement of background signal in the area of high iron concentration. The background image is a good check to see if an element is really present. If present an element should show an x-ray image with contrast stronger than the background image.

23.4 Dead Time Considerations

Experiment 23.4: Test of the Dead Time Circuit. Early digital imaging systems employing EDS detectors did not correct for dead time variations across the image. Thus, when a region of the specimen was encountered containing a very high concentration of an element, the counting rate increased until the dead time reached 100% which effectively shut off the detector. This produced the annoying situation in which a round region of pure copper would be imaged as a copper ring rather than a filled circle. One remedy for this problem is to use a shorter time constant on the detector amplifier. Another solution is to extend the live time to compensate for the effect. However, even this correction may not be accurate, and the test described in this experiment is a check of this circuit.

(a) At 5% dead time the EDS iron image (Figure A23.7) should look similar to the digital WDS image (Figure A23.5b) in that it should reveal both high and low iron concentrations. In this case, a count rate yielding 5% dead time produced an image with too few iron counts. In order for this experiment to be statistically meaningful, the counting time should be extended to 1000 sec. Figure A23.7b shows an iron image taken at 20% dead time. This EDS image is similar to the WDS image (Figure A23.5b) except that there are still too few counts collected.

(b) At 50% dead time the iron image has more counts per pixel but it shows much less detail and has a blurred appearance (Figure A23.7c).

(c) The ratios of iron counts in the same pixels for the high- and low-iron regions at each dead time are shown in the following table:

Region	5% dead time	20% dead time	50% dead time
High iron	1	11	59
Low iron	0	5	8

Clearly, in this case the differences in iron count ratios are more likely due to insufficient counts in the pixels rather than an error in the dead time circuit. To establish that a problem exists in the dead time circuit more counts in each iron-containing phase should be taken.

23.5 Intensity Measurement

Experiment 23.5: X-Ray Line Scans. The x-ray line scan is useful for quantitative analysis since relative changes in concentration along a line in the image can be accurately

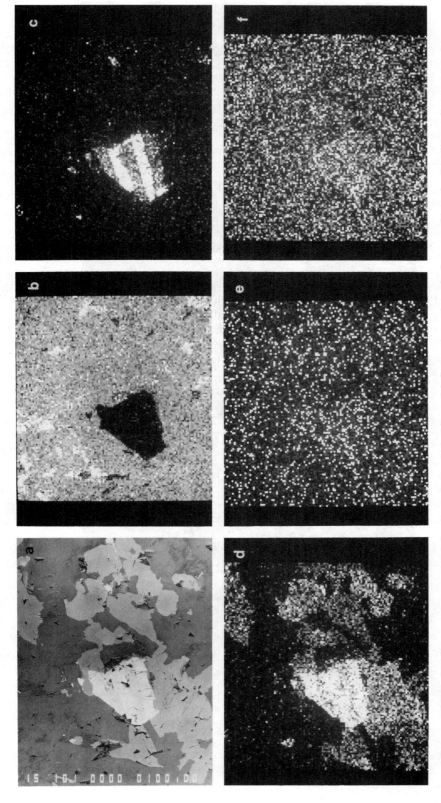

Figure A23.6. EDS x-ray images with background subtraction. (a) Secondary electron image of basalt; (b) silicon image; (c) titanium image; (d) iron image; (e) chromium image; (f) background image from a region of interest of lower energy than chromium. Background was subtracted on-the-fly for images (b)–(d).

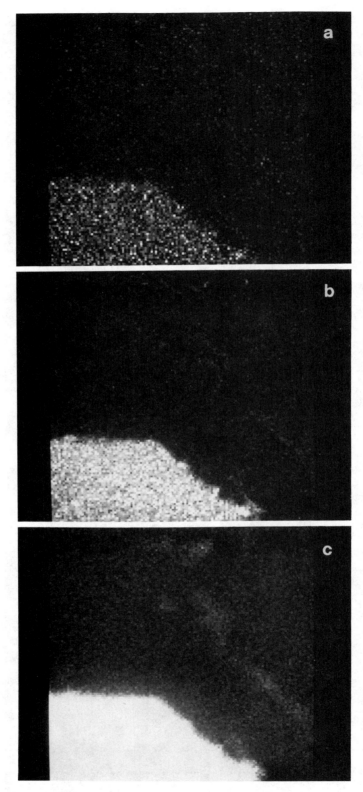

Figure A23.7. Effect of counting rate. (a) Iron image taken with 5% dead time; (b) iron image taken with 50% dead time; (c) iron image taken with 50% dead time.

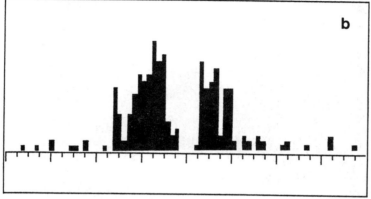

Figure A23.8. Line scan across a titanium region of basalt. (a) EDS titanium image from Figure A23.6; (b) titanium line scan.

determined without the nonlinear effects of human vision. Figure A23.8 shows a line scan across a portion of the titanium image from Figure A23.6. Line scans are especially important when there are not enough counts in a full range image to show a smooth variation in concentration. If the time devoted to a 128x128 digital image were used just to collect a digital line scan, the counts per pixel should increase by two orders of magnitude.

23.6   X-Ray Image Processing

Experiment 23.6: Image Smoothing. Image smoothing can be of great benefit for x-ray images particularly those collected with few counts. However, worm-like artifacts can appear after smoothing with a kernel that is too small. See the solution to Experiment 17.7 for examples of this.

Experiment 23.7: Primary Coloring. Color coding three elements with the primary colors allows different phases to be imaged clearly. For example, adding the red silicon image (Figure A23.4c) to the green potassium image (Figure A23.4e) clearly shows the regions containing both elements in yellow. For another example, combine a red iron image (Figure A23.6d) with a green titanium image (Figure A23.6c) and a blue silicon image (Figure A23.6b). Other examples are shown in *ADSEMXM*, Chapter 5, Figures 5.17A-D.

Part IV

ANALYTICAL ELECTRON MICROSCOPY

SOLUTIONS

Laboratory 24

Scanning Transmission Imaging in the AEM

24.1 Characteristics of the STEM Electron Probe

In a TEM/STEM the relationships demonstrated by Experiments 24.1-24.4 can be visualized directly on the TEM screen. In a dedicated STEM these effects can be inferred from effects on the signal profile across a sharp edge and the probe current [1].

Experiment 24.1: Effect of C_1 Lens Strength. In a TEM/STEM increasing C_1 lens strength decreases the probe size and decreases the probe current as shown in Figure A24.1. Typical STEM probe current values are shown in Figure A24.2. For a thermionic gun, decreasing the probe size d_p *drastically decreases* the probe current i_p since, neglecting lens abberations, $i_p = \text{const}\beta\,(d_p\,\alpha_p)^2$ where β is the gun brightness. In a DSTEM, changing C_1 also changes the probe size.

Experiment 24.2: Effect of C_2 Aperture Size. Decreasing the C_2 aperture size decreases the area of the gaussian probe and decreases the probe current. At the smallest aperture sizes (<70 μm) the shadow of the C_2 aperture is clearly visible around the probe (see Figure A24.3). If the C_2 aperture is misaligned the electron distribution in the probe becomes asymmetrical, which may give misleading microanalysis results (see Figure A24.4). In a DSTEM, changing the VOA changes both the probe size and probe current.

Experiment 24.3: Effect of Condenser and Objective Stigmators. If the stigmators are improperly adjusted the electron distribution in the probe will be asymmetrical giving an astigmatic probe image, and also an astigmatic scanning image.

Experiment 24.4: Effect of Objective Lens Focus. Defocusing the objective lens results in a spreading of the electron beam and a loss of sharpness in the probe image as the probe profile on a DSTEM. The focus of the objective lens should have no effect on the probe current.

24.2 STEM Image Formation

For forming conventional bright field (BF) and annular dark field (ADF) transmission

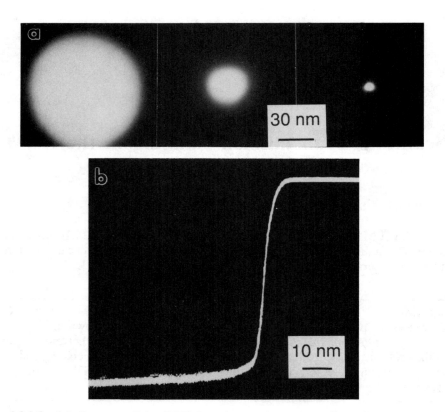

Figure A24.1. (a) Images of the STEM probe on the TEM screen showing how the probe size decreases with increasing condenser one lens strength (150-μm C_2 aperture). (b) Profile of a DSTEM probe scanned across the edge of a MgO cube, at a magnification of 1×10^6x. Probe size may be found by measuring the horizontal distance associated with the rise in signal.

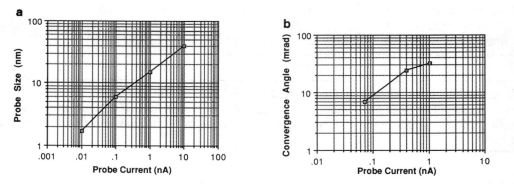

Figure A24.2. (a) Typical probe sizes for a STEM/STEM with a tungsten gun as a function of probe current. (b) Probe convergence angle (2α) versus probe current in a thermionic source TEM/STEM.

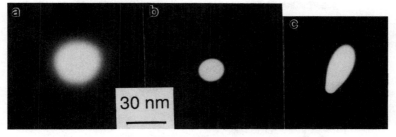

Figure A24.3. STEM probe image on the TEM screen outlined by (a) 150-µm aperture; (b) 70-µm C_2 aperture; (c) STEM probe image with misaligned C_2 aperture.

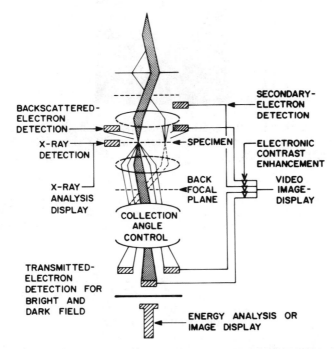

Figure A24.4. The optics of a TEM/STEM. Note that in a TEM/STEM the microscope may be in diffraction mode and the detectors intercept the diffraction pattern in a plane conjugate with the back focal plane of the objective lens.

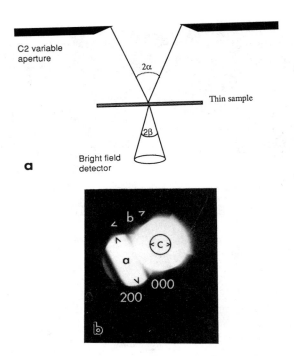

Figure A24.5. (a) Definition of the beam convergence angle (2α) and collection angle (2β) in a STEM. (b) The dimensions on a CBED pattern required to calculate 2α and 2β (see equations in answers).

images, the electron optics of a TEM/STEM are shown in Figure A24.5. Note that in this case the STEM system operates in the TEM diffraction mode, and the detectors are conjugate with the back focal plane of the objective lens.

Experiment 24.5: Image Relationships. The STEM image is often rotated and inverted with respect to the TEM image. To obtain an image in TEM, a parallel beam of electrons illuminates several square micrometers of the sample and the electron distribution in the image plane of the objective lens is projected onto a fluorescent screen. To obtain images in STEM, a fine convergent probe is scanned across the specimen. When the STEM is operated in diffraction mode, the electron distribution in the back focal plane of the objective lens is projected onto solid-state or scintillator/photomultiplier electron detectors. Signals from these detectors are used to modulate the brightness of a CRT producing an image as the beam scans across the specimen. There is no obvious spatial orientation relationship between TEM and STEM images.

24.3 Measurement of Convergence and Collection Angles

Experiment 24.6: Calculation of $2\alpha_s$ and $2\beta_s$. A schematic diagram defining $2\alpha_s$ and $2\beta_s$ in a TEM/STEM is shown in Figure A24.5. To calculate the values experimentally it is necessary to obtain a convergent beam diffraction pattern from a known sample under known diffracting conditions (Figure A24.5b). This permits determination of the Bragg angle $2\theta_B$ for the spot (from Bragg's law: $\lambda = 2d_{hkl} \sin\theta$), which is then used as a calibration to determine α_s (the size of the diffraction disks) and $2\beta_s$ (the projected width of the BF detector on the pattern). Thus, from Figure A24.5b we find that

$$2\alpha_s = \frac{a 2\theta_B}{b} \text{ and } 2\beta_s = \frac{c 2\theta_B}{b}$$

4.4 Contrast Effects in STEM Images

These experiments clearly demonstrate advantages and disadvantages of STEM imaging.

Experiment 24.7: Diffraction Contrast

(a) Under two-beam conditions with approximately parallel illumination the TEM BF image will show strong dynamical contrast effects such as bend contours or thickness fringes. These contrast effects will be less obvious in STEM BF since $2\beta_s$ is usually kept large to obtain an acceptable signal level. A comparison of TEM and STEM BF images is given in Figure A24.6. To reproduce true TEM contrast, $2\beta_s$ would have to be 10^{-4} rad according to the theorem of reciprocity. Such a small collection angle would severely limit the available signal. The annular DF STEM image can only approximate TEM DF contrast. In a two-beam condition, the annular detector receives electrons from all diffracted beams not just the strongly excited beam.

(b) Increasing the probe size by decreasing the C_1 lens excitation makes the image more intense (and less noisy) because the probe current increases. However, the contrast is not improved and the resolution degrades as the probe size increases.

(c) Increasing $2\alpha_s$ by using a larger C_2 aperture will increase the intensity because of the increased current in the probe. When $2\alpha_s > 2\theta_B$, the diffraction contrast will start to disappear because diffracted beams as well as the central beam will be incident on the BF detector.

(d) Increasing $2\beta_s$ by decreasing the camera length has a similar effect to increasing $2\alpha_s$. However, if $2\beta_s$ is decreased by going to very long camera lengths, the diffraction contrast at

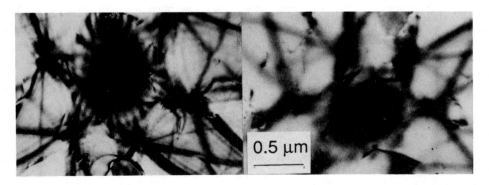

Figure A24.6. TEM (left) and STEM (right) BF image of bend center in stainless steel showing loss of diffraction contrast in the STEM image.

bend contours will improve. At the same time the image quality will suffer because the smaller signal falling on the detector will result in a lower signal/noise ratio in the image. Inserting a small objective aperture can further decrease $2\beta_s$ at the longest camera length. Again diffraction contrast and the noise level should both increase. Therefore, in general, STEM images of specimens showing diffraction contrast do not approach the combined contrast and quality of TEM images. Exceptions to this generalization are images formed in STEMs with field emission sources, such as those obtained in a DSTEM.

Experiment 24.8: Mass-Thickness Contrast

(a) Mass-thickness effects result in BF and DF images which show complementary contrast effects in both TEM and STEM images. Contrast is higher in STEM partly because the contrast may be expanded electronically and partly because of the efficient manner in which scattered electrons are collected on the ADF detector. Unlike diffraction scattering, mass-thickness scattering does not occur in specific directions. Therefore, the annular DF STEM detector is more efficient at collecting scattered electrons than the objective aperture in TEM. For this type of specimen, there is always a strong high-contrast signal on the STEM DF detector. This explains why STEM DF images only were observed in this part of the laboratory.

(b) Increasing the probe size will increase the image intensity but degrade the resolution.

(c) Increasing $2\alpha_s$ will increase the image intensity and eventually degrade the image quality as spherical aberration effects cause the electron tails around the probe to increase.

(d) Increasing $2\beta_s$ (decreasing the camera length) will increase the image intensity and also increase the contrast because of the increase in the number of scattered electrons picked up by the detector. This contrast increase means that STEM DF imaging of specimens showing mass-thickness contrast is the most efficient transmission imaging mode in a STEM. A comparison of TEM and STEM DF images is given in Figure A24.7.

Experiment 24.9: Phase Contrast Imaging. High-resolution lattice fringe images, or atomic structure images can also be formed in a STEM if more than one diffraction maximum allowed to hit the detector. In addition the probe size must be close to the lattice spacing (~0.3 nm). This is only possible in a FEG STEM because a thermionic source STEM cannot supply

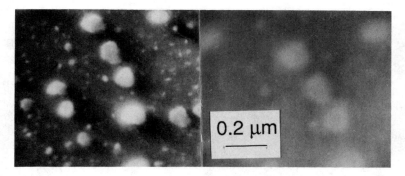

Figure A24.7. ADF STEM image (left) and TEM image (right) of stained two-phase polymer showing enhancement of mass-thickness contrast in STEM.

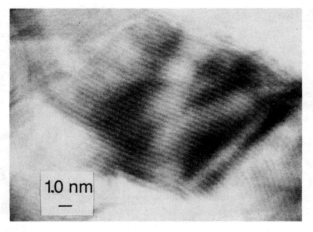

Figure A24.8. DSTEM phase contrast image showing the (0002) lattice planes in graphitized carbon.

enough current in a 0.3-nm probe to provide detectible contrast. A phase contrast image of graphitized carbon is shown in Figure A24.8.

24.5 Scanned Images Using Other Electron Signals

Experiment 24.10: Secondary Electron Imaging. SE images from thin specimens in the STEM are of higher resolution than SE images from bulk specimens in a standard SEM for several reasons:

(a) At 100-200 kV, the STEM electron gun is substantially brighter than the standard SEM gun at 20-30 kV. This means larger currents can be generated in smaller probes.

(b) The spherical aberration coefficient C_s is about 3 mm in a STEM versus 20-30 mm in an SEM. Again, this allows larger currents in smaller probes in the STEM.

(c) In a *thin* specimen there is less backscatter of electrons and therefore fewer remote sources of secondary electrons. This means the signal/noise ratio is higher.

(d) In the STEM the SE detector is in the upper objective polepiece, out of direct line-of-sight of any backscattered electrons. This also improves the SE signal/noise ratio in comparison with a standard SEM. A typical high-resolution SE image from a TEM/STEM is shown in Figure A24.9.

Experiment 24.11: Backscattered Electron Imaging. BSE images from a thin specimen in the STEM are of higher resolution than BSE images from bulk samples in the SEM for two of the same reasons as described in Experiment 10.9 above: (a) the gun is brighter in STEM and (b) the C_s value is smaller in STEM. In addition, more BS electrons come from near the probe since there is less remote scatter such as from the electron diffusion zone beneath the surface of a bulk specimen. However, the smaller number of BSEs collected from a thin specimen means that a somewhat larger probe size (>10 nm) must be used to

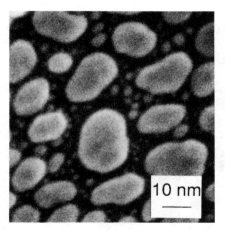 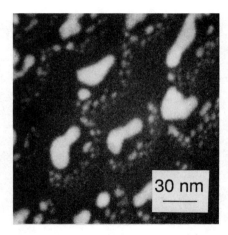

Figure A24.9. SE image of gold islands on a carbon film in a TEM/STEM showing 2 nm resolution.

Figure A24.10. BSE image of gold island on a carbon film in a TEM/STEM showing ? nm resolution.

generate sufficient signal for a reasonable image with an acceptable noise level. A typical BSE image from a STEM is shown in Figure A24.10.

Reference
[1]   J. R. Michael and D. B. Williams, *J. Micros.* **147** (1987) 289-303.

Laboratory 25

X-Ray Microanalysis in the AEM

25.1 Sources of Spurious X-Rays

Radiation emanating from regions of the specimen other than the beam-specimen interaction volume can limit the usefulness of x-ray microanalysis. Two major sources exist: the illumination system and postspecimen scatter. It is possible to observe the presence of these effects and determine if they will limit any proposed experiment.

Experiment 25.1: The "Hole Count." By placing the electron probe down a hole in the specimen (open grid square), spurious sources of radiation (both electrons and x-rays from the illumination system) can be detected because they will generate characteristic x-rays from regions of the specimen far from the beam location. Thus, superimposed upon the true spectrum from under the beam will be a small spectrum associated with the average composition of the entire specimen.

Traditionally, a disk specimen of silver or molybdenum has been used to detect the relative level of the "hole count." However, with a disk specimen the value of hole count measured varied with disk rim thickness and the thickness of foil used for the on-foil spectrum. To avoid this variation we need a thin film of constant thickness supported by a thick grid or aperture. Possible test samples are a chromium film on a gold grid (which we will use) or a chromium film on a molybdenum aperture. We will define the hole count as the ratio of the AuL_α intensity (in the hole) to the CrK_α intensity (on the uniform film). Of course, the values of hole count obtained with different specimens cannot be compared directly. Figure A25.1 shows an electron-generated chromium on gold spectrum and Figure A25.2 shows a spectrum generated down the hole, primarily by high energy x-rays or bremsstrahlung. Note that the CrK_α/K_β lines disappear but the AuL and M lines remain.

Using a thick C_2 aperture (0.5 mm platinum) cuts down the stray x-rays to reasonable levels at 100 kV, i.e., below the background intensity in the specimen spectrum. Compare the intensity scale in Figure A25.3 (thick aperture) with Figure A25.2 (conventional aperture). This remedy may not be sufficient at voltages above 100 kV and for specimens of high atomic number.

Experiment 25.2: Postspecimen Scatter. Even if the illumination system is "cleaned up" by using a thick C_2 aperture, x-rays can still be excited from regions of the sample remote from the beam-specimen interaction volume. This is illustrated by the presence of a strong AuL peak in the spectrum from the chromium thin film sample. See Figure A25.4 which shows a similar experiment with a thin film of gold on carbon supported on a copper grid. This copper peak was generated at the copper grid, which is many microns from the probe position. The peak arises because of electron scatter within the sample and from the objective aperture or other bulk material below the plane of the sample. Another possible source is specimen-generated bremsstrahlung, but this contribution is thought to be minor.

Figure A25.1. Electron-excited spectrum from chromium film on a gold grid showing the CrK_α/K_β lines at ~5.5 keV, the AuL line family at 10-15 keV, and the AuM_α line at ~2 keV.

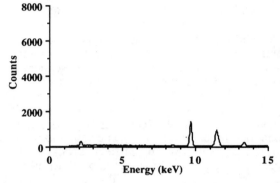

Figure A25.2. The spectrum from the chromium on gold sample, but with the beam down a hole in the sample. The spectrum is generated by high-energy bremsstrahlung from the illumination system which only fluoresces the gold grid.

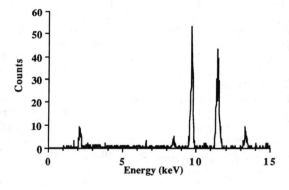

Figure A25.3. Same conditions as Figure A25.2 except a thick PtC_2 aperture was inserted to reduce the flux of bremsstrahlung from the illumination system. The AuL_α intensity is reduced from ~1500 to ~50 counts.

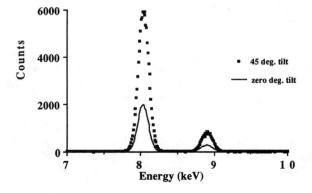

Figure A25.4. CuK_α/K_β lines generated from the support grid when the beam strikes a thin film, several micrometers from the grid bar. The copper lines increase in intensity with higher sample tilts, indicating that increased scattering from the sample is responsible for the CuK_α and K_β peaks.

Tilting the sample increases the gold (copper in Figure A25.4) peak intensity by increasing the amount of gold (copper) grid that interacts with electrons and x-rays scattered by the sample. Other elements in the specimen holder and stage region can also contribute x-rays to the spectrum. These x-rays cannot be distinguished from the x-rays generated by small amounts of the same element present in the sample. Therefore, there is a specific limitation to the analysis of a minor element in any sample for which the element is present as a major constituent in areas of the specimen (or specimen holder) away from the analysis point.

5.2 X-Ray Data Collection

Experiment 25.3: EDS Detector Position with Respect to the Image. It is important to know the direction of the EDS detector axis relative to the sample in order to maintain the smallest x-ray path length (see Figure 25A-5). Pushing on the side-entry goniometer specimen holder uniquely establishes the orientation of the sample to the EDS detector. Once the orientation is determined, always position the sample with the thinnest region towards the detector. Also, if the region around an interface is being analyzed, the interface should be parallel to the EDS axis, especially if x-ray absorption is likely to be a problem. Otherwise, differential absorption of x-rays will occur on one side of the interface.

Experiment 25.4: Collection of X-Rays. When collecting x-rays for quantification, the maximum number of x-ray counts should be obtained in the minimum time. At the same time the detector must not be overloaded. Dead time should be kept below 25%-30%. If the dead time is too large, artifacts such as sum peaks appear in the spectrum and peak shapes become distorted. It is convenient to monitor the count rate in a major peak (e.g., FeK_α) and the dead time, as a function of several variables:

(a) Maximum kV means maximum gun brightness and therefore maximum x-ray generation.

(b,c) Probe current i_p and probe size d_p for a thermionic source are related theoretically by $i_p \sim d_p^{8/3}$ if lens aberrations are the limiting factor. Generally, $i_p = \text{const}\beta\, d_p^2\, \alpha_p^2$ where β is the gun brightness, and so increasing d_p substantially increases i_p, thus increasing the x-ray count rate. In an FEG system the probe size may be changed by the virtual objective aperture (VOA) or by the C_1 lens, but only the VOA changes the current.

(d,e) A larger C_2 aperture increases the probe current with the possibility of an increase in probe size due to spherical aberration. A smaller C_2 aperture results in a well-defined probe, but with a much lower current.

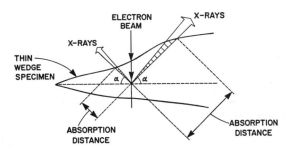

Figure A25.5. The variation in x-ray absorption path length in a nonuniform wedge specimen. If the EDS detector position with respect to the sample is known, then the sample can be oriented so that x-rays follow the shortest path length through the sample prior to detection. Here the EDS detector should be located on the left.

Table A25.1. Results of Quantification Using Different Spectral Manipulation Schemes

	wt% Fe	wt% Ni
Window method (1.2 FWHM) for peaks and averaged background	65.9 ± 1.1	34.1 ± 1.1
Gaussian peak fitting with Kramers' Law background fitting	66.1 ± 1.1	33.9 ± 1.0
Peak fitting using library standards, with digital filtering of background	63.9 ± 1.1	36.1 ± 1.0

Errors are ±3σ *(99% confidence limits using counting statistics only).*
Specimen is Fe-35% Ni (using electron microprobe bulk analysis).

Experiment 25.5: Background Subtraction. Several techniques may be used for subtracting the background (bremsstrahlung) x-ray intensity from beneath the characteristic peak; for example, computer curve fitting or mathematical filtering methods. However, for a relatively simple spectrum such as that from Fe-Ni neither technique has any advantage over the more primitive methods such as drawing a straight line under the peak (Figure A25.6) or using "windows" on either side of the characteristic peak (Figure A25.7). A comparison of quantification using different background subtraction routines is given in Table 25.1.

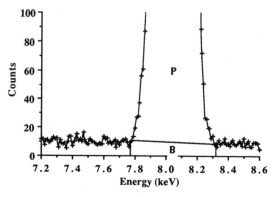

Figure A25.6. Background subtraction under an isolated K_α peak ("gross/net" approach).

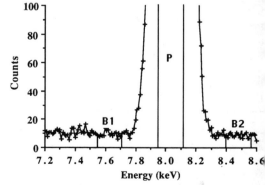

Figure 25.7. Background estimation under an isolated K_α peak by averaging the background intensities in "windows" B1, and B2 on either side of the peak.

25.3 Quantification of Data

Experiment 25.6: Measurement of the k-Factor. Thin foil microanalysis can be very straightforward and often requires only the determination of a k_{AB} factor, using an appropriate standard. Determination of the k_{AB} factor is quite simple if the specimen is sufficiently thin:
1. Use a standard with known amounts of elements in weight percent or atomic percent.
2. Use the same kV for standard and unknown.
3. Use the same detector for standard and unknown.
4. Collect intensity data in identical manner for standard and unknown. Then:

$$\frac{C_A}{C_B} = k_{AB} \frac{I_A}{I_B}$$

5. From the standard obtain the k_{AB} value: $k_{AB} = (C_A/C_B)_{standard} \times (I_B/I_A)_{standard}$. Thus, C_A/C_B may be determined for the unknown from $(I_A/I_B)_{unknown}$ and k_{AB}. The individual elemental fractions C_A and C_B for a two-element system may be calculated from $C_A + C_B = 1$. If no standard is available k_{AB} can be calculated (see *PAEM*, p. 160).

Experiment 25.7: Errors in k-Factor Determination. As in most analysis procedures the error is minimized if the statistics are good. Therefore, the more x-ray counts the better. Thus if there are $N = 10^4$ counts in I_A and I_B, then the standard deviation is

$$\sigma = \sqrt{N} = \sqrt{10^4} = 10^2$$

for both I_A and I_B. Assuming that one wants errors at the "$\pm 3\sigma$" or 99% confidence limit, then 3σ (or more appropriately s_n) = 3×10^2. Therefore, the relative error in I_A and I_B is $\pm 3s_n/N = \pm 300/10{,}000 = \pm 3\%$ relative. The total error in k_{AB} is the sum of the errors in I_A and I_B ($\pm 6\%$) plus the error in the standard. Thus, for a k_{AB} measured on a standard that is well characterized (better than $\pm 1\%$ relative accuracy), the total error in k_{AB} may be about $\pm 7\%$. If we collect 10^5 counts from the standard in I_A and I_B, a similar argument leads to an error in k_{AB} of $\pm 3\%$, which is a typical minimum error.

If either component A or B is present in small amounts, then the error in k_{AB} increases rapidly. For example, if there are 10^4 counts in I_A but only 10^3 counts in I_B, then using the same argument as above, $\pm 3s_n$ is $\pm 10\%$ in I_B, 3% in I_A, and the total error in k_{AB} is about $\pm 14\%$ relative.

It should be pointed out that the errors calculated above are overestimates. Since the measurements are statistical in nature, the standard deviations may be summed in quadrature which reduces the level of the relative error by about a third. Thus, the total relative error in a single AEM measurement, where I_A and I_B each contain more than 10,000 counts, is approximately $2/3(\pm 3\%$ in k_{AB} plus $\pm 6\%$ from $I_A/I_B) = \pm 6\%$. Thus, many AEM references in the literature refer to a relative error of about 5%-10% for this technique.

If several independent measurements of the k-factor are made a lower relative error may be calculated using the standard deviation and the "Student's *t*" distribution. Values for *t* may be found for several confidence levels in published statistical tables.

Experiment 25.8: Multielement Quantification. Since most materials of interest are not simple binary alloys the quantification can be more tedious. In the sample of biotite, several of the characteristic peaks are very close together in the low-energy region of the spectrum where the background x-ray intensity is changing rapidly (Figure A25.8a). Background subtraction in this case must be done by computer (Figure A25.8b) with appropriate standards and k-factors fed into the computer. Time does not permit all these data

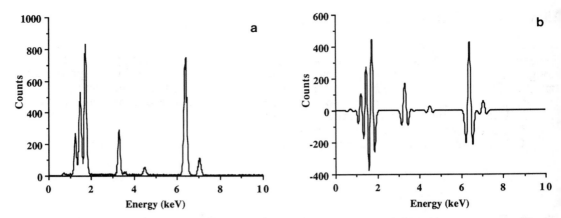

Figure A25.8. (a) EDS spectrum of biotite showing closely spaced peaks. To subtract the background, a mathematical method such as digital filtering is required as shown in (b), which is the result of convoluting spectrum (a) with a "top-hat"-shaped filter function. The background intensity is reduced to zero between the characteristic peaks.

to be gathered during the laboratory session. Quantification using calculated k_{AB} factors can be carried out to give an approximate answer. Compare with the nominal concentration values stated in the experiment. For any microanalysis situation, the "correctness" of our measurements can only be assessed by using various standards as unknowns to check the validity of our methods.

Experiment 25.9: Absorption Correction. The simple Cliff-Lorimer equation has to be corrected if characteristic x-rays are absorbed or fluoresced significantly (correction >10%). In the NiAl specimen, AlK_α x-rays are strongly absorbed and this should be observable as an increase in the Ni/Al intensity ratio as the thickness of the specimen increases.
  The equation to correct the collected spectrum is

$$\frac{C_A}{C_B} = k_{AB} \frac{I_A}{I_B} \text{ (A.C.F.)}$$

where

$$\text{A.C.F.} = \frac{\left.\frac{\mu}{\rho}\right|_{spec}^{A} \left[1-\exp\left(\left.\frac{\mu}{\rho}\right|_{spec}^{B} \rho\, t \cosec \alpha\right)\right]}{\left.\frac{\mu}{\rho}\right|_{spec}^{B} \left[1-\exp\left(\left.\frac{\mu}{\rho}\right|_{spec}^{A} \rho\, t \cosec \alpha\right)\right]}$$

where ρ is the specimen density, t is the specimen thickness, α is the x-ray take-off angle, and $\mu/\rho\; A_{spec}$ is the mass absorption coefficient for element A x-rays absorbed in the specimen. This factor may be determined for nickel in NiAl using a weighted average of mass absorption coefficients:

$$\left.\frac{\mu}{\rho}\right|_{spec}^{Ni} = \left.\frac{\mu}{\rho}\right|_{Ni}^{NiK_\alpha} C_{Ni} + \left.\frac{\mu}{\rho}\right|_{Al}^{NiK_\alpha} C_{Al}$$

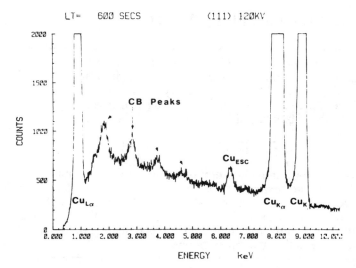

Figure A25.9. A series of copper spectra taken at 120 kV. The CuL, CuK_α, K_β, and copper escape peaks are labeled, but other small Gaussian peaks exist. These peaks are due to coherent bremsstrahlung production.

where C_i is the weight fraction of element i. Stoichiometric NiAl is 68.51 wt% Ni. A similar equation must be developed for aluminum in NiAl. The following values of μ/ρ are required and are taken from Heinrich's Tables [1]:

$$\left.\frac{\mu}{\rho}\right|_{Ni}^{Ni} = 58.9 \text{ cm}^2\text{gm}^{-1}$$

$$\left.\frac{\mu}{\rho}\right|_{Al}^{Ni} = 60.7 \text{ cm}^2\text{gm}^{-1}$$

$$\left.\frac{\mu}{\rho}\right|_{Ni}^{Al} = 4837.5 \text{ cm}^2\text{gm}^{-1}$$

$$\left.\frac{\mu}{\rho}\right|_{Al}^{Al} = 385.7 \text{ cm}^2\text{gm}^{-1}$$

In addition, the specimen thickness t has to be determined, and the take-off angle α and specimen density ρ have to be known. The contamination spot separation method often overestimates the thickness of the specimen and therefore results in overcorrection of the intensity data for absorption effects. In this case, the calculated result will show an excess of aluminum with respect to the known stoichiometric composition of the specimen. If the composition and/or density are unknown, it is necessary to follow an iterative procedure by first estimating C_A and C_B in order to determine an appropriate ρ, and then using this value of ρ in the correction factor equation to calculate a value of C_A and C_B. From these values redetermine ρ and redo the calculation until the result is self-consistent. Most computer programs allow for this iterative procedure.

25.4 Coherent Bremsstrahlung Effects

Experiment 25.10: Detection of Coherent Bremsstrahlung. Coherent bremsstrahlung peaks arise in thin foil specimens under high-voltage irradiation because the sample is usually a single crystal through the analyzed region (see *PAEM*, p. 147). Bremsstrahlung is therefore produced regularly through the periodic crystalline sample and the result is a coherent beam of bremsstrahlung of energy E given by

$$E(\text{kV}) = 12.4\,\beta/L[(1 - \beta \cos(90 + \theta)])$$

where $\beta = v/c$, θ is the take-off angle (assuming the specimen is normal to the beam) and L is the interplanar spacing in the beam direction. Specific gaussian peaks arising from different higher order Laue zone planes are visible at about the 1% intensity level. The peaks move as a function of kV and orientations as predicted by the above equation (see Figure A25.9). These peaks may cause difficulty in measuring grain boundary segregants such as S and P which have peaks of similar intensities at similar energies.

Reference
[1] K. F. J. Heinrich, *The Electron Microprobe*, ed. T. D. McKinley, K. F. J. Heinrich, and D. B. Wittry, Wiley and Sons, New York (1966) 351. Many mass-absorption coefficients from Heinrich's tables are quoted in Goldstein et al., *Scanning Electron Microscopy and X-Ray Microanalysis*, Plenum Press, New York (1981) 624.

Laboratory 26

Electron Energy Loss Spectrometry

26.1 Characteristics of the Energy Loss Spectrum

Experiment 26.1a: Serial Collection Conditions. There are many more variables related to spectrum acquisition in EELS than in EDS. These variables can greatly affect the nature and quality of the data. Often when a satisfactory EELS setup is obtained for a particular application, the same settings will be used for a similar application in future experiments so that the data can be compared. Because of the large dynamic range of the EELS spectrum (very high intensities and very low intensities), some method of scaling the two regions is required. This scaling may be obtained in three ways in serial EELS: by switching the PMT from analog current measurement to single electron pulse counting, by changing the gain of the PMT, or by changing the dwell time per channel of the MCA. Generally, data taken using a specific gain change are preferred; however, to reduce scintillator exposure and to speed data collection we used only the dwell time change in this experiment. The slit width adjustment is performed while observing the video monitor. For moderate energy resolution the slit is closed down until a small flat top remains on the zero loss peak. The three features in the spectrum are: the zero loss E_o peak at 0 eV, the carbon plasmon loss at 24 eV, and the carbon K edge at 284 eV. See Figure A26.1a.

Experiment 26.1b: Parallel Collection Conditions. Scaling the different intensities in PEELS collection is achieved by setting up difference acquisition modes for the intense low-loss and fainter high-loss regions of the spectrum. In the future, it is expected complete automatic acquisition will be feasible, thus allowing the complete spectrum to be viewed immediately.

26.2 Calibration of the Spectrometer

Experiment 26.2a: SEELS Energy Calibration. Since different MCAs employ different schemes for calibration of the EELS spectrum, we describe a typical calibration in this section. Generally, two parameters are stored in the MCA: the eV/channel, which should remain constant from run to run, and the zero offset which may change somewhat due to magnet hysteresis. It is useful to keep a record of the operational eV/channel and zero offset as calculated by the MCA so that these conditions may be reproduced.

Experiment 26.2b: PEELS Energy Calibration. For this case it is difficult to calibrate the energy-loss scale using a known edge and the zero-loss peak, as in SEELS. If the spectrum is intense enough to yield a detectable edge such as carbon K, above 100 eV, the zero-

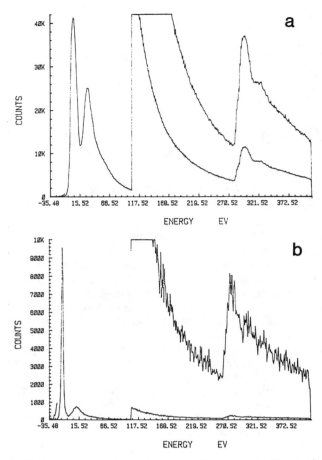

Figure A26.1. Typical electron energy loss spectra from a thin film of amorphous carbon. (a) Image mode, LaB$_6$, C$_2$ aperture = 150 µm, spectrometer entrance aperture = 3 mm; (b) same as above but entrance aperture = 1 mm.

loss peak will be too intense for the scintillator detector and the protection circuitry will prevent a spectrum from being acquired in this manner. The best way to calibrate the PEELS energy scale is therefore the drift tube voltage offset method. This method provides an independent measurement of energy loss, being based on a well-calibrated (~±0.3%) voltage reference. With this method, it is possible to calibrate the full energy spectrum by fixing the position of the zero-loss peak alone.

26.3 Measurement of the Energy Resolution of the Spectrometer

Experiment 26.3: Determination of the Energy Resolution from the Zero-Loss Peak. The resolution of the spectrum can be assessed from the FWHM of the zero loss peak. For the 3-mm aperture and the coarse (1-eV/channel) display the resolution should be several eV. For the 1-mm aperture and the fine display (0.1 eV/channel) the resolution should be <2 eV for a thermionic source and <0.5 eV for a FEG.

26.4 Spectrometer Variables

Experiment 26.4: Effect of Entrance Aperture Size on Resolution. By inserting the 1-mm spectrometer entrance aperture, the energy resolution should improve considerably. Limiting the collection angle even further by inserting a small objective aperture would further improve the resolution on the zero loss peak, but ionization-loss intensities would be drastically reduced (see Figure A26.1b).

Experiment 26.5: Effect of Spectrum Scan Direction. In PEELS, the whole spectrum is acquired simultaneously, so "scan direction" is not a variable. When scanning in the direction of low to high energy loss, the background near the zero loss peak contains a component due to the afterglow of the scintillator. By reversing the scan direction, the peak-to-background of features near the zero loss peak should improve.

Experiment 26.6: Exit Slit Width. Reducing the slit width to the point where the height of the zero loss peak on the video monitor is about 0.7x (maximum with flat top) gives the best resolution (about 1-2 eV when combined with the l mm or 2 mm entrance aperture). With this resolution the π^* transition fine structure at 284 eV on the amorphous carbon K-edge should be clearly resolved (see Figure A26.2b). In PEELS there is no exit slit, and the energy resolution is governed by the electron source and the spectrometer entrance aperture. The smallest aperture should easily insure resolution of the π^* in a PEELS spectrum.

26.5 Near-Edge Fine Structure

Experiment 26.7: Near-Edge Fine Structure of Carbon. Graphitic carbon has a near edge fine structure distinct from amorphous carbon (and also distinct from diamond) [1]. These differences sometimes can be used as a "fingerprint" to detect various forms of carbon when distinctly separate phases are analyzed as in this sample (see Figure A26.3).

26.6 Specimen Thickness Effects

Experiment 26.8: Shape of the Carbon Edge. Figure A26.4 shows the change in shape of the carbon edge for a modest increase in specimen thickness. The second hump on the edge is caused by the excitation of a carbon plasmon loss by electrons that also excited the K-ionization.

Experiment 26.9: Plasmon Losses in Aluminum. Plasmon losses, due to collective oscillations of free electrons, can occur in multiples as the thickness of the specimen increases. Thus, in thicker regions of foil aluminum plasmon losses occur at 15 eV, 30 eV, 45 eV, etc. Observation of ionization edges for losses less than 200 eV can become difficult when the specimen thickness is much greater than 20 nm (at 100 kV). The aluminum $L_{2,3}$ edge at 73 eV can be most easily observed in thin areas of the specimen because the increased background caused by plasmon and single electron scattering in thicker areas (see Figure A26.5) tends to obscure this relatively low intensity energy loss.

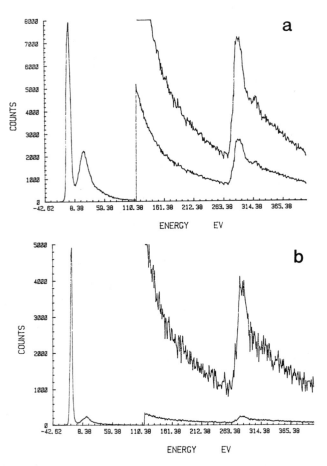

Figure A26.2. Amorphous carbon. (a) Image mode, LaB$_6$, C$_2$ aperture = 150 μm, entrance aperture = 3 mm; (b) same as above but slit width reduced.

Experiment 26.10: Measurement of Specimen Thickness. For reasonably thin specimens yielding plasmon excitations, the approximate equation given can be used to estimate the specimen thickness. Thus, for aluminum, if $I_p/I_o = 0.1$, then $t \sim 14$ nm at 120 kV. This method is useful for relative thickness estimates since the other methods of thickness measurement are difficult to apply to very thin, clean specimens.

Experiment 26.11: Thickness Required for Quantitative Microanalysis. Multiple scattering in thicker particles of BN will cause the boron and nitrogen K edges to become less distinct as they are swamped by a higher background. Electrons in the edges may lose energy to plasmon losses in thicker specimens. Specimen thicknesses where $I_p/I_o \leq 0.1$ are generally required for accurate EELS quantitation of boron nitride as demonstrated in the next section.

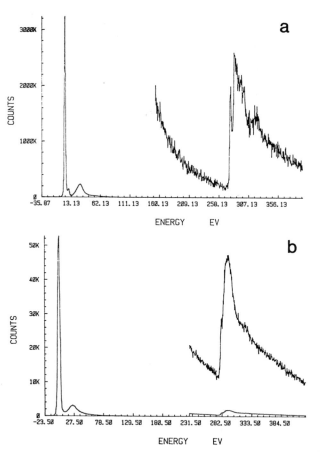

Figure A26.3. Two allotropes of carbon. (a) graphite flake (diffraction mode, $L = 190$ mm); (b) amorphous carbon film. Note that the specimen thickness is approximately the same (similar plasmon peak height), but that the shapes of the spectral features are different in both the low-loss and the K-edge regions.

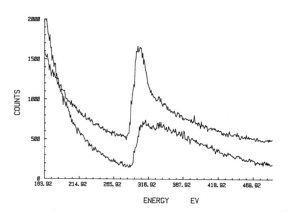

Figure A26.4. Carbon K-edge shapes for thin (top) and thick (bottom) carbon layers. Note that the slope of the background before the edge also changes.

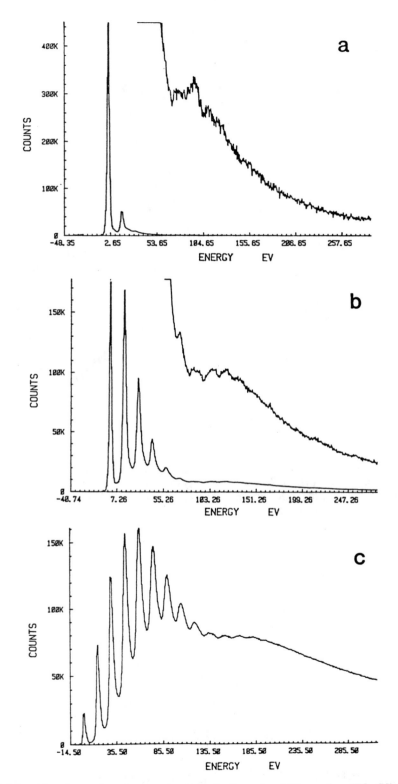

Figure A26.5. Aluminum low-loss spectra including the Al$L_{2,3}$ edge (73 eV) for (a) thin, (b) thicker, and (c) very thick specimens.

26.7 Quantification of the Ionization Loss Spectrum

Experiment 26.12: The Atomic Ratio of Boron to Nitrogen in BN. Note that EELS quantification involves setup variables such as the collection angle $2\beta_s$ and the energy window Δ. The fact that these two setup variables affect the quantitation is a major difference between EDS x-ray analysis and EELS.

(a) The "thin" flake should have $I_p/I_o \leq 0.1$.

(b) Only moderate energy resolution is required for this experiment. Much more important is collection of a statistically significant number of counts in both the boron and nitrogen edges. The spectrometer resolution should now be about 5-10 eV.

(c,d) Note that $2\beta_s$ should be determined experimentally for each microscope (see Laboratory 24) to obtain accurate results. Spectra with and without the background subtracted are shown in Figure 26.6.

(e,f) Note that, ideally, B/N = 1 in stoichiometric boron nitride. However, with different $2\beta_s$ and Δ, B/N may depart markedly from unity.

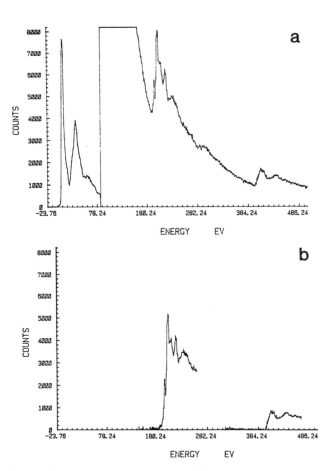

Figure A26.6. Energy loss spectra of boron nitride. (a) Spectrum as collected; (b) background-subtracted boron and nitrogen edges ($\Delta = 80$ eV).
of the background before the edge also changes.

26.8 Energy Loss Imaging (Optional)

Experiment 26.13: False EELS Imaging. The point here is that without proper background subtraction under the edges for each point in the image, the image intensity for each element will be a function of specimen thickness, not chemical composition [2] (see also *PAEM*, pp. 272-274).

Reference
[1] R. F. Egerton and Whelan, *Journal of Electron Spectroscopy and Related Phenomena* 3 (1974) 232.
[2] R. F. Egerton, *Electron Energy Loss Spectroscopy in the Electron Microscope*, Plenum Press, New York, 1986, 122.

Laboratory 27
Convergent Beam Electron Diffraction

27.1 CBED Pattern Formation

Experiment 27.1: Setup in STEM Mode. The ray diagram showing CBED formation is given in Figure A27.1 and is compared with SAD formation with a parallel beam of electrons. The CBED is present whether the probe is scanning or stationary. However, the most useful information is obtained in stationary mode.

Ideally the probe should be focused at the eucentric plane. Under these conditions, having obtained a focused STEM image, a focused CBED pattern should be present on the TEM screen. The results of focusing the pattern to either side of the correct focus for CBED are shown in Figure A27.2. Note the inversion of the shadow image on either side of exact focus. Refocusing may again be necessary if the specimen is moved or tilted out of the eucentric plane using the second tilt axis.

Experiment 27.2: Effect of Electron Optical Variables.
(a) Changing the Region Contributing to the Pattern. The region contributing to the CBED is controlled either by the mechanical specimen traverses or the electronic probe traverses. These adjustments change the diffraction pattern if the specimen orientation or thickness changes. It is often easier to adjust the zone axis pattern by electronically shifting the probe since backlash may limit the sensitivity of the mechanical specimen traverses.

(b) Effect of C2/VOA Aperture Alignment. If the C_2 or VOA aperture is off-axis, it can be difficult to obtain exactly symmetrical zone axis patterns. Occasionally it may be necessary to adjust the C_2 aperture in conjunction with the probe/specimen traverses to obtain a perfectly symmetrical zone axis pattern.

(c,d) Changing Probe Convergence. Changing the C_2/VOA aperture size and/or the C_2 lens strength changes $2\alpha_s$. At very small aperture sizes and strong C_2 lens strengths, $2\alpha_s$ is very small and the CBED pattern consists of very small discs looking very much like a standard SAD pattern (although coming from a much smaller region). As $2\alpha_s$ is increased (discretely by changing the C_2 aperture or continuously by changing the C_2 lens strength) the CBED discs expand. As long as $2\alpha_s < 2\theta_B$ for the particular diffracting conditions, a K-M patten is formed as shown in Figure 27.3 (upper pattern). Such a pattern is essential for performing conventional BF/DF STEM imaging. This pattern may be indexed by the same methods as conventional SAED patterns. As $2\alpha_s$ is increased to values greater than $> 2\theta_B$, the CBED discs overlap and Kossel conditions are obtained as shown in Figure 27.3. Under these conditions multibeam STEM images would be formed. Kossel patterns are discussed more in Experiment 27.7.

Maximum probe convergence is obtained in STEM mode or TEM mode with the C_2 lens turned off, and this is the preferred operating condition. Activating the C_2 lens in TEM mode invariably reduces $2\alpha_s$.

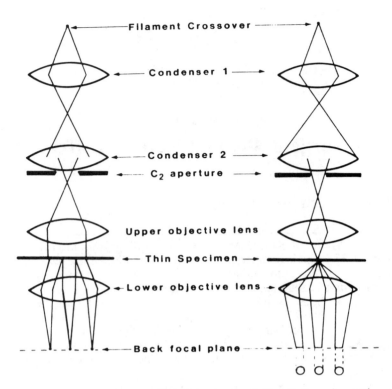

Figure A27. 1. STEM optics to form a CBDP in the back focal plane in stationary convergent probe mode (right). Selected areas of electron diffraction formation with parallel incident beam (left).

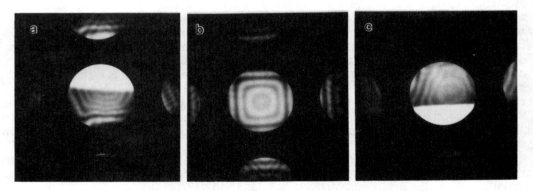

Figure A27. 2. Through focus series for CBDP formation. (a) Underfocus; (b) focus; (c) overfocus. Note reversal of spatial information in central 000 disc between (a) and (c).

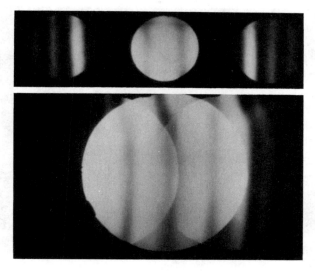

Figure A27.3. Kossel-Möllenstedt (upper) and Kossel (lower) conditons.

27.2 Kossel-Möllenstedt Fringes for Thickness Determination

Experiment 27.3: Thin and Thick Foils. In a very thin region of the specimen (< 50 nm) dynamical diffraction effects do not occur and under these kinematical conditions the CBED discs contain no intensity variations inside (uniform intensity). Such a pattern contains no more information than an SAD pattern (see Figure A27.4). With increasing thickness, dynamical diffraction is more likely and more information becomes available within the CBED discs as shown in Figure A27.4 (bottom). This extra information will be discussed in subsequent sections. When the specimen is very thick, absorption will dominate and the information in the CBED discs will become "washed out" and less intense.

Experiment 27.4: K-M Fringes for Thickness Measurement. To determine specimen thickness, the angular spacing of the K-M fringes $\Delta\theta_i$ is measured as shown in

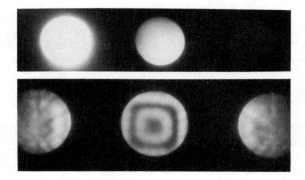

Figure A27.4. CBDPs from thin (upper) and thick (lower) samples.

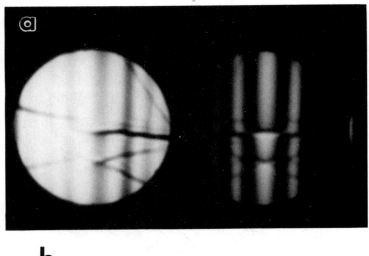

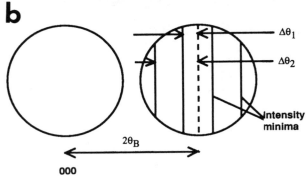

Figure A27.5. Measurement of specimen thickness from K-M fringes. (a) Experiment pattern taken under two-beam conditions; (b) schematic measurements required from (a).

Figure 27.5 and used to calculate s_i from the following equation [1]:

$$s_i = \frac{\lambda \Delta \theta_i}{d_{hkl}^2 2\theta_B}$$

where λ is the electron wavelength, d_{hkl} is the interplanar spacing giving rise to the *hkl* diffraction disc, and $2\theta_B$ is the Bragg angle for that *hkl* plane. When the values of s_i are plotted as shown in Figure A27.6, both the extinction distance $\xi_g(hkl)$ and t can be determined.

27.3 Higher-Order Laue Zone Lines (Defect HOLZ Lines)

Experiment 27.5: Observation of HOLZ Lines. Observation of HOLZ lines in the 000 disc is not always straightforward. There is a critical thickness range over which they are visible. However, local strain, surface films, beam heating, and local disorder can cause the

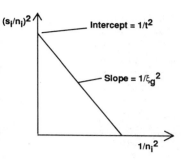

Figure A27.6. Plot to determine thickness (t) and extinction distance (ξ_g) from K-M fringe spacing.

lines to be very weak or absent. Experimental conditions to minimize the causes of absence involve (a) cooling to liquid N_2 temperatures; (b) choosing (UVW) such that the lattice plane spacing is large; (c) using a large $2\alpha_s$ (STEM mode); and (d) selecting a reasonably small probe (20 nm). The following observations may be made:

(i) By varying the specimen tilt away from the exact zone axis, the HOLZ lines will move in exactly the same manner as Kikuchi lines (of which they are the elastic scattering analogue). Therefore, the symmetry information they contain is most useful under exact zone axis conditions.

(ii) If the specimen thickness is increased, the dynamical fringes will increase in number and the HOLZ lines will eventually fade away as inelastic scattering dominates.

(iii) Changing the voltage of the instrument results in a shifting of the HOLZ lines. If the microscope has a continuous voltage control, the HOLZ line positions can be used to calibrate the operating voltage very accurately, and, given that the voltage has been calibrated, very small lattice parameter changes (10^{-4} nm) can be detected. Typical HOLZ lines in the 000 disc are shown in Figure A27.7.

Figure A27.7. HOLZ lines in 000 disc of CBED pattern taken down <114> zone axis.

27.4 Higher-Order Laue Zone Diffraction Maxima (HOLZ Rings)

Experiment 27.6: Observation of HOLZ Rings. HOLZ reflections arise at large scattering angles (>5°) and therefore can only be seen at small camera lengths (< 500 mm). Electron scattering is weak at such angles, therefore cooling the sample or using large probes and larger $2\alpha_s$ values can help. The HOLZ reflections result from the Ewald sphere intercepting reciprocal lattice points of planes not parallel to the beam direction (UVW) as shown in Figure A27.8. The radius of the first HOLZ ring G is related to the spacing between reciprocal lattice planes H through:

$$H_m = \frac{\lambda G^2}{2}$$

where H_m = the experimentally measured lattice spacing along the beam direction. The expected value of H, calculated from the reciprocal lattice can be determined from

$$H^{-1} = |\vec{B}| \quad \text{(in units of nm)}$$

where $\vec{B}$ is the direction of the electron beam (UVW). This gives rise to the following general equations:

For lattices with orthogonal coordinates,

$$H^{-1}_{UVW} = \left[(aU)^2 + (bV)^2 + (cW)^2\right]^{1/2}$$

For a hexagonal or rhombohedral lattice using a three index system

$$H^{-1}_{UVW} = \left[(U^2 + V^2 - UV)a^2 \,(Wc)^2\right]^{1/2}$$

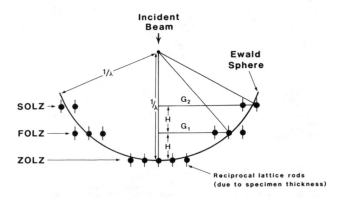

Figure A27.8. Origin of HOLZ reflections. Schematic diagram showing Ewald sphere intersecting the zero order Lane zone (ZOLZ), first-order Lane zone, (FOLZ), and the second-order Lane zone (SOLZ).

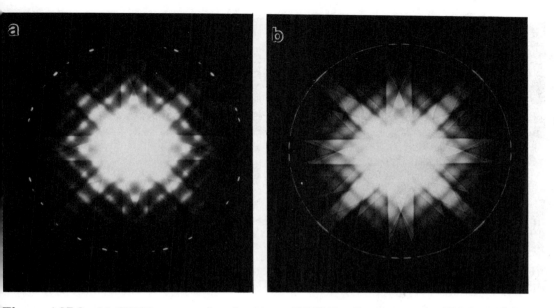

Figure A27.9. (a) CBED pattern showing ring of HOLZ reflections under K-M conditions with small $2\alpha_s$; (b) same pattern under Kossel conditions with lage $2\alpha_s$.

For a hexagonal or rhombohedral lattice using a four index system

$$H^{-1}_{UVW} = \left[3(U^2 + V^2 + UV)^2 + (Wc)^2 \right]^{1/2}$$

For a monoclinic lattice (*b* axis unique)

$$H^{-1}_{UVW} = \left[(Ua)^2 + (Vb)^2 + (Wc)^2 + 2UWac \cos\beta \right]^{1/2}$$

Experimentally H_m^{-1} should be equal to H^{-1}, or differ by an integral multiple if extinctions have caused the absence of a complete HOLZ ring [2]. Thus, direct comparison can be made between experimental values of H_m^{-1} and the expected values of H^{-1} for particular crystal structures.

Experimentally it is sometimes easiest to measure G by using a large value of $2\alpha_s$, since the individual HOLZ maxima merge into a ring of intensity containing bright HOLZ lines. Several bright lines may constitute the ring and generally it is considered that the value of G should be measured from the innermost ring. A typical pattern containing HOLZ reflections is shown in Figure A27.9a (small $2\alpha_s$) and A27.9b (large $2\alpha_s$).

27.5 Kossel Patterns

Experiment 27.7: Observation of a Kossel Pattern. The use of a large $2\alpha_s$ to give a ring of HOLZ lines as in Figure A27.10 is also the condition necessary for generation of Kossel patterns. If $2\alpha_s > 2\theta_B$, then all the scattering information is elastic in nature and the pattern symmetry contains three-dimensional information.

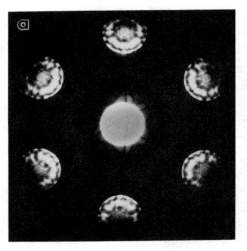 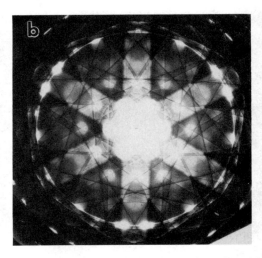

Figure A27.10. <111> zone axis of austenitic stainless steel. (a) Symmetry $3m$ within 000 disc; (b) whole Kossel pattern symmetry 3m.

27.6 Determination of Point Group Symmetry

Experiment 27.9: Point Group of Austenitic Stainless Steel. Determination of the plane group symmetry (two-dimensional in the plane of the pattern) of the details in the bright field 000 spot and the whole Kossel pattern along a low-index zone axis allows one or two diffraction group candidates to be assigned to that crystal direction using Table A27.1 [3]. For example, Figures A27.10a and A27.10b show the <111> CBED patterns at high and low

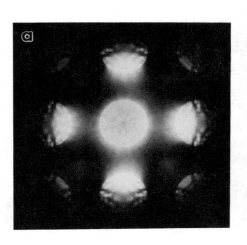 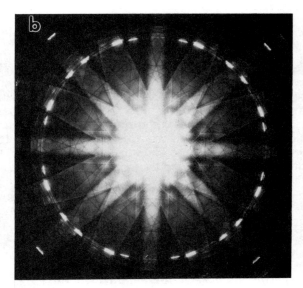

Figure A27.11. <100> zone axis of austenitic stainless steel. (a) BF symmetry $4mm$ within 000 disc; (b) whole Kossel pattern symmetry $4mm$.

Table A27.1. Convergent Beam Diffraction Pattern Symmetries

Diffraction Group	Bright Field	Whole Pattern	Dark Field General	Dark Field Special	±G General	±G Special	Projection Diffraction Group
1	1	1	1	none	1	none	1_R
1_R	2	1	2	none	1	none	
2	2	2	1	none	2	none	
2_R	1	1	1	none	2_R	none	21_R
21_R	2	2	2	none	21_R	none	
m_R	m	1	1	m	1	m_R	
m	m	m	1	m	1	m	$m1_R$
$m1_R$	2mm	m	2	2mm	1	$m1_R$	
$2m_R m_R$	2mm	2	1	m	2	—	
2mm	2mm	2mm	1	m	2	—	$2mm1_R$
$2_R mm_R$	m	m	1	m	2_R	—	
$2mm1_R$	2mm	2mm	2	2mm	21_R	—	
4	4	4	1	none	2	none	
4_R	4	2	1	none	2	none	41_R
41_R	4	4	2	none	21_R	none	
$4m_R m_R$	4mm	4	1	m	2	—	
4mm	4mm	4mm	1	m	2	—	$4mm1_R$
$4_R mm_R$	4mm	2mm	1	m	2	—	
$4mm1_R$	4mm	4mm	2	2mm	21_R	—	
3	3	3	1	none	1	none	31_R
31_R	6	3	2	none	1	none	
$3m_R$	3m	3	1	m	1	m_R	
3m	3m	3m	1	m	1	m	$3m1_R$
$3m1_R$	6mm	3m	2	2mm	1	$m1_R$	
6	6	6	1	none	2	none	
6_R	3	3	1	none	2_R	none	61_R
61_R	6	6	2	none	21_R	none	
$6m_R m_R$	6mm	6	1	m	2	—	
6mm	6mm	6mm	1	m	2	—	$6mm1_R$
$6_R mm_R$	3m	3m	1	m	2_R	—	
$6mm1_R$	6mm	6mm	2	2mm	21_R	—	

*After Buxton et al 1976. Courtesy B. F. Buxton. Reprinted by permission of The Royal Society.

Table A27.2. Relation Between the Diffraction Groups and the Crystal Point Groups

Diffraction Groups	1	1̄	2	m	2/m	222	mm2	mmm	4	4̄	4/m	422	4mm	4̄2m	4/mmm	3	3̄	32	3m	3̄m	6	6̄	6/m	622	6mm	6̄m2	6/mmm	23	m3	432	4̄3m	m3m
6mm1$_R$																											×					
3m1$_R$																										×						
6mm																									×							
6m$_R$m$_R$																								×								
6$_{1R}$																							×									
3$_{1R}$																						×										
6																					×											
6$_R$mm$_R$																				×												×
3m																			×											×		
3m$_R$																		×													×	
6$_R$																	×															
3																×												×				
4mm1$_R$															×																	×
4$_R$mm$_R$														×																	×	
4mm													×																			
4m$_R$m$_R$												×																		×		
4$_{1R}$											×																					
4$_R$										×																						
4									×																							
2mm1$_R$							×								×										×	×	×					×
2$_R$mm$_R$			×			×						×		×										×	×	×						×
2mm					×																					×						
2m$_R$m$_R$			×				×	×													×					×		×				
m1$_R$				×			×	×													×	×									×	
m		×		×			×	×			×		×								×	×									×	
m$_R$			×		×	×			× ×			× ×		× × ×			×		×		×		× × ×					×		× ×	
2$_{1R}$			×												×																	
2$_R$	×			×			×				×		×		×									×			×					×
2		×						×							×																	
1$_R$		×									×																					
1	×		× ×		× ×		× ×	×	×		× ×		× ×	×	× × ×		×		× ×													
Point Groups	1	1̄	2	m	2/m	222	mm2	mmm	4	4̄	4/m	422	4mm	4̄2m	4/mmm	3	3̄	32	3m	3̄m	6	6̄	6/m	622	6mm	6̄m2	6/mmm	23	m3	432	4̄3m	m3m

*After Buxton et al (1976) courtesy B. F. Buxton, reproduced by permission of the Royal Society.

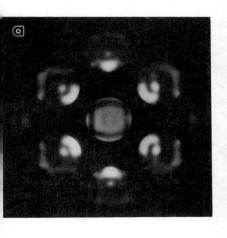 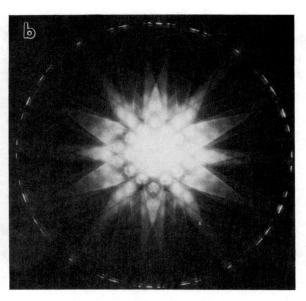

Figure A27.12. <110> zone axis of austenitic stainless steel. (a) BF symmetry of $2mm$ within 000 disc; (b) whole Kossel pattern symmetry $2mm$.

camera lengths, respectively. The BF symmetry (i.e., that of the HOLZ line distribution in the 000 disc) is $3m$ (a threefold rotation axis combined with a parallel mirror plane). The whole pattern symmetry in Figure 27.10b is also $3m$. Reference to Table A27.1 shows that possible diffraction groups consistent with this symmetry are $3m$ and $6_R m m_R$. For each of these diffraction groups, Table A27.2 gives possible point groups [3]. Diffraction group $3m$ would be consistent with a point group of $3m$ or $\bar{4}3m$, while $6_R m m_R$ is consistent with a point group of $3m$ or $m3m$.

Similar analysis from Figures A27.11a and A27.11b (<100> axis) and Figures A27.12a and A27.12b (<110> zone axis) reveal the possible point groups summarized in Table A27.3. From this table the only point group with a match to the symmetry in each of the three zone axes is $m3m$, otherwise known as $\frac{4}{m}\bar{3}\frac{2}{m}$. Thus, the point group of austenitic stainless steel is $m3m$.

Table 27.3. Symmetries along Cubic Electron Beam Directions and Related Point Groups

Point group	Zone axis		
	<111>	<100>	<110>
$m3m$	$6_R m m_R$	$4 m m 1_R$	$2 m m 1_R$
$\bar{4}3m$	$3m$	$4_R m m_R$	$m 1_R$
432	$3 m_R$	$4 m_R m_R$	$2 m_R m_R$
$m3$	6_R	$2 m m 1_R$	$2 m_R m_R$
23	3	$2 m_R m_R$	

References
[1] P. Kelly, A. Jostons, R. Blake, and J. Napier, *Physica Status Solidi* **A31** (1975) 771.
[2] M. Raghavan et al., *Metallurgical Transactions* **15A** (1984) 1299.
[3] B. F. Buxton, J. A. Eades, J. Steeds, and G. Rackman, *Philosophical Transactions of the Royal Society of London* **281** (1976) 181.

Index

Absorption correction, 146-147, 378-379
Alloy data, 125, 339-340
Aluminum plasmon losses, 151, 383
Analog image processing, 20-21, 202-203
Analog x-ray dot maps, 132, 352
Analog x-ray images, 132-133, 352
Analytical electron microscopy, *see* specific types
Angular scale calibration, 75, 263
Apertures, *see also* Objective apertures
  depth-of-field and, 14, 192-193
  in EELS, 150, 383
  in STEM, 140, 365
Artifact peaks, 36, 38, *see also* specific types
Astigmatism, 7
Atomic number contrast, 16-17, 197
  in BSE imaging, 52, 53, 54, 227, 228, 229, 230, 231
  in computer-aided imaging, 299
  in high-resolution SEM, 65, 244-245
  in SE signal components, 69
Atomic ratios, 152, 387
Automatic qualitative analysis, 216

Background intensity, 118-119, 330-333
Background removal, 133-134, 358
Background subtraction, 134, 145, 358, 376
Backscattered electron (BSE) imaging, 16, 51-54, 227-231
  collection efficiency of, 53
  contrast and, 16, 17, 18, 197
    topographic, 52, 53, 54 198, 199, 227, 228, 229, 230
  in difference mode, 53-54, 229
  in electron channeling contrast, 74, 76
  in environmental SEM, 86, 89, 287, 288, 289, 290, 291, 292, 293
  E-T detector in, 51-52, 227-228, 230
  in high-resolution SEM, 61, 62, 63, 64, 65, 66, 244-245, 247
  SC signal in, 54, 230, 231
  SE signal components and, 67, 69, 70, 251, 254
  STEM and, 142, 371-372
  in sum mode, 52-53, 228
  x-rays and, 133
Basic imaging, 3–7
Beam current, *see* Electron beam current
Be window Si(Li) detectors, 27, 33
Biological tissues
  bulk specimen preparation of, 166-168
  imaging of, 89, 291-293, 294
  thin specimen preparation, 177-179

Block trimming, 177, 178
Bremsstrahlung effects, 147, 380
Brightness, 6, 13, 19, 192
BSE imaging, *see* Backscattered electron imaging
Bulk specimen preparation, 159-170
  of biological tissues, 166-168
  of ceramics, 159-164
  of fibers, 168-170
  of metals, 159-164
  of minerals, 159-164
  of particles, 168-170
  of polymers, 164-166

Calibration curve method, 121, 334
Carbon K peak shape, 118, 330
Carbon structure, 150-151, 383
Cathode ray tubes (CRTs), 4, 5
  stereo images and, 18
  stereo microscopy and, 22, 23
  x-ray images and, 132, 133, 352
CBED, *see* Convergent beam electron diffraction
Ceramics
  bulk specimen preparation of, 159-164
  thin specimen preparation of, 172-176
Cleavage, 175
Coatings, 176, 178, 180-186
  for high-magnification imaging, 183-184
  for low-magnification imaging, 181-182
  for medium-magnification imaging, 182-183
  metal, 180-181
  thin film, *see* Thin film coatings
  in x-ray microanalysis, 184-186
Coherent bremsstrahlung effects, 147, 380
Collection angles, 140-141, 368-369
Collection efficiency
  in BSE imaging, 52, 53
  in SE signal components, 68-69, 251
Complex spectrum analysis, 39, 215, 216
Computer-aided imaging, 90-95, 296-305
  contrast in, 92, 298, 299
  digital images in, 90-92, 296
  intensity measurements in, 92-94, 298-300
  kernels in, 94-95, 302-305
  stereo images in, 18
Condenser lenses, 4
  beam current and, 9, 10–11, 190
  focusing and, 6
Conductive specimens, 5
Contrast, 16-18, *see also* specific types
  atomic number, *see* Atomic number contrast

401

Contrast (*cont.*)
  in computer-aided imaging, 92, 298, 299
  diffraction, 141, 369-370
  edge brightness, *see* Edge brightness contrast
  electron beam induced, *see* Electron beam induced contrast
  electron channeling, *see* Electron channeling contrast
  in high-resolution SEM, *see under* High-resolution SEM
  magnetic, *see* Magnetic contrast
  mass-thickness, *see* Mass-thickness contrast
  obtaining of, 6
  particle, 65, 242-243
  relief, 64, 65, 242
  in STEM, 141, 369-371
  topographic, *see* Topographic contrast
  voltage, *see* Voltage contrast
Contrast enhancement, 92, 297, 298
Convergence, *see* Electron beam convergence
Convergence angles, 140-141, 368-369
Convergent beam electron diffraction (CBED), 140, 153-156, 368, 389-399
  HOLZs in, *see* Higher–order Laue zones
  pattern formation in, 153-154, 389, *see also* Kossel-Mollenstedt patterns; Kossel patterns
  pattern recording in, 155-156
  point group symmetry in, 156, 396-399
Converter plates, 69, 72, 254, 257, 261
CRTs, *see* Cathode ray tubes

Dead time
  in EDS analysis, 28-31, 207-210
  in WDS analysis and, 44
  in x-ray images, 135, 358, 360
Density slicing, 300, 302
Depth-of-field, 8, 13–15, 192-194
Detector resolution, 30, 31, 210
Difference mode, 53-54, 229
Diffraction contrast, 141, 369-370
Digital images, 20, 203
  acquisition of, 90-92, 296
  intensity levels of, 91-92, 296-297
  parameters of, 134, 352-357
  x-ray, 133-134, 352-357
Digital scanning, 91, 296
Drying, 177, 178
Dynamic environmental experiments, 89, 293-295
Dynamic voltage contrast, 83

EBIC, *see* Electron beam induced contrast
EBIV, *see* Electron beam induced voltage
Edge brightness contrast
  in high-resolution SEM, 64-65, 242
  in SE signal components, 68
Edge detection kernels, 304-305
Edge energy, 108-110
Edge enhancement, 95, 304-305
EDS, *see* Energy-dispersive spectrometers
EELS, *see* Electron energy loss spectrometry
Electron beam convergence, 8, 12-13
Electron beam current, 3, 4, 5, 8, 9–11, 189-190

Electron beam current (*cont.*)
  beam size and, 194-196
  channeling contrast and, 76
  damage from, 59-60, 235
  in EDS analysis, 28-31, 41, 207-210, 217
  effect of on EBIC, 85, 282, 284
  in electron channeling contrast, 263, 269
  focusing and, 6
  image quality and, 19, 201
  in low-voltage SEM, 57-58, 59-60, 234, 235
  in quantitative EDS analysis, 108, 110, 115
  in quantitative WDS analysis, 99-100
  specimen insertion and, 5
  in trace element microanalysis, 122
  vs. topographic contrast, 57-58, 234
  in WDS analysis, 42-43, 219
Electron beam induced contrast (EBIC), 81, 85, 279, 281-286
  observation of devices with, 84-85, 281-282
  observation of materials with, 85, 284-285
  specimen preparation for, 163-164
Electron beam induced voltage (EBIV), 85
Electron beam parameters, 8–15, 189-196
  beam convergence, 8, 12–13, *see also* Electron beam convergence
  beam current, 8, 9–11, 189-190, *see also* Electron beam current
  beam size, 8, 11–12, 15, 190-191, *see also* Electron beam size
  brightness, 8, 13, 192, *see also* Brightness
  depth-of-field, 8, 13–15, 192-194
  gun alignment and saturation, *see under* Electron guns
Electron beam size, 8, 11–12, 190-191
  current and, 194-196
  image quality and, 19
  small vs. large, 15
Electron channeling contrast, 73-77, 263-274
  measurements in, 75-76, 263
  microstructure imaging and, 76-77, 270, 271
  specimen preparation for, 163
  wide-area patterns in, 74-75, 263, 264
Electron channeling pattern (ECP)
  measurements with, 75-76, 263
  selected area, 77, 273-274
  wide-area, 74-75, 263, 264
Electron column, 3
Electron energy loss spectrometry (EELS), 148-152, 381-388
  calibration in, 149, 381-382
  characteristics of, 148-149, 381
  energy resolution in, 149, 382
  false imaging in, 152, 388
  near-edge fine structure in, 150-151, 383
  parallel, *see* Parallel EELS
  quantification in, 152, 387
  serial, *see* Serial EELS
  specimen preparation for, 172
  specimen thickness effects and, 151, 383
  spectrometer variables in, 150, 383
Electron guns, 3, *see also* specific types
  alignment of, 8–9, 189

lectron guns (*cont.*)
　brightness of, *see* Brightness
　digital images and, 203
　field-emission, *see* Field-emission guns
　in high-resolution SEM, 61
　LaB6, *see* LaB6 guns
　saturation of, 6, 8–9, 189
　SE signal components and, 67
　thermionic, 203
　tungsten hairpin, *see* Tungsten hairpin guns
lectronics console, 3
lectron lenses, *see* Lenses
lectron optical variables, 154, 389
lectron probe, 139-140, 365
lectron probe microanalyzer (EMPA)
　in light element microanalysis, 117
　in quantitative EDS analysis, 108
　in quantitative WDS analysis, 99
　in trace element microanalysis, 122
　in WDS analysis, 42, 44
lectron range, 57-58
lectropolishing, 173
mbedding, 177, 178
mission current, 9–10, 189-190
MPA, *see* Electron probe microanalyzer
nergy calibration, 28, 149, 207, 381
nergy-dispersive microanalysis, 33-41, 213-217,
　　see also Energy-dispersive spectrometry
　qualitative analysis in, 36-39, 214-216
　quantitative analysis in, 39-41, 216-217
　spectra families in, 33-36, 213
nergy-dispersive spectrometry (EDS), 27, 27-32,
　　207-212, *see also* Energy-dispersive
　　microanalysis
　beam current and, 28-31, 207-210
　compared to EELS, 387
　compared to WDS analysis, 42, 44, 45, 46, 47, 219,
　　220, 221, 222
　in computer-aided imaging, 90
　dead time in, 28-31, 207-210
　in light element microanalysis, 117, 118, 330
　in particle microanalysis, 127, 128, 129, 131, 343,
　　345, 346, 350
　spectral artifacts in, 31-32, 210-212
　spectrometer setup in, 27-28, 207
　in trace element microanalysis, 122, 123, 124, 125,
　　126, 335, 336, 337, 338, 339, 340, 341
　x-ray images and, 132, 134, 135-136, 355-357,
　　358, 359, 361
　in x-ray microanalysis, 143, 144, 145, 146, 147,
　　375, 378
nvironmental secondary electrons (ESEs), 87,
　　287, 288, 289, 291, 292, 293
nvironmental SEM, 86-89, 287-295
　dynamic experiments in, 89, 293-295
　imaging of insulators in, 88-89, 291-293
　imaging of water in, 87-88, 289-290
　signal quality in, 87, 287-289
scape-depth contrast, 63-64, 68
scape peaks, 31, 36, 38, 210-211
SEs, *see* Environmental secondary electrons
tching, 342

E-T detectors, *see* Everhart-Thornley detectors
Everhart-Thornley (E-T) detectors, 5, 8, 16
　in BSE imaging, 51-52, 227-228, 230
　collection efficiency of, 52
　contrast and, 17, 18, 197, 198, 199
　determining saturation with, 9
　in electron channeling contrast, 74
　in high-resolution SEM, 61
　in magnetic contrast, 78, 79
　obtaining an image with, 6
　in particle microanalysis, 128
　SE signal components and, 67, 68, 69, 254
　in STEM, 55, 232
　in stereo microscopy, 22, 23, 24
　voltage contrast and, 279
Exit slit width, 150, 383

False peaks, 31, 211
Families of x-ray spectra, 33-36, 213, *see also*
　　specific spectra
Faraday cup imaging, 71, 254, 259, 261
FEGs, *see* Field-emission guns
Fiber specimen preparation, 168-170
Field-emission guns (FEGs), 6
　brightness of, 192
　digital images and, 203
　in high-resolution SEM, 61
　image quality and, 19
　saturation of, 9, 189
　SE signal components and, 67
Final lenses, *see* Objective lenses
Flat specimen biasing, 69, 251, 253
Focusing, 6–7
　WDS analysis and, 42-43, 219
Fracturing, 164

Golden dictum, 36

Hard tissues, 166-167
Higher-order Laue zones (HOLZs), 154-155, 156,
　　392-395, 399
High-intensity peaks, 36
High-phosphorus standard, 123, 335
High-resolution SEM, 61-66, 242-250
　contrast mechanisms in, 61-64
　imaging requirements in, 66, 245-248
　nontopographic contrast in, 65, 243-245, 247
　topographic contrast in, 63, 64-65, 242-243, 248
Histograms, 94, 299-300, 301
Hole count, 143-144, 373
Homogeneous specimens, 40

Image analysis, 92
Image contrast, *see* Contrast
Image processing, 92
Image quality, 16, 19-20, 201-202
Image rotation, 15, 194
Image smoothing, 136, 361
Incomplete charge collection, 32, 212
Insulators
　imaging of, 88-89, 291-293
　minimizing charging effects on, 58-59, 235

Intensity measurements, 92-94, 298-300
Ion-beam milling, 173
Ionization loss spectrum, 152, 387

Kernels, 302-305, *see also* specific types
  in computer-aided imaging, 94-95
  defined, 94
  smoothing, 136, 303-304, 361
  in x-ray images, 303, 361
K-factor, 146, 377
K-M patterns, *see* Kossel-Mollenstedt patterns
Kossel-Mollenstedt (K-M) patterns, 153, 154, 395, *see also* Kossel patterns
  for thickness determination, 154, 391-392
Kossel patterns, 155, 156, 395-396, *see also* Kossel-Mollenstedt patterns
K spectra, 33-34, 37, 213

LaB_6 guns, 6
  brightness of, 192
  in high-resolution SEM, 61
  saturation of, 9
  SE signal components and, 67
Laplacian kernels, 304, 305
Lattice parameters, 76, 263
Lenses, 4, 140, 365
  beam current and, 10–11, 190
  condenser, *see* Condenser lenses
  objective, *see* Objective lenses
Light element microanalysis, 117-121, 330-334
  data collection for, 117-119
  measurement of concentrations in, 119-121, 333–334
Linear tubes, 4
Line scans
  in computer-aided imaging, 93, 298
  in x-ray images, 220-221, 358-361
Live time, 28-29, 207
Low-intensity peaks, 37
Low phosphorus concentration, 124, 337-339
Low-voltage SEM, 57-60, 234-241
  beam damage in, 59-60, 235
  beam energy in, 57-58, 234
  minimizing charging effects on insulators in, 58–59, 235
  topographic contrast in, 57-58, 234
L spectra, 33-34, 36, 37, 213
LVSEM, *see* Low-voltage SEM

Magnetic contrast, 78-80, 275-278
  type I, 78-79, 275-278
  type II, 79-80, 277-278
Mass-thickness contrast
  in high-resolution SEM, 63, 65, 244-245
  in STEM, 141, 370
Maximum output count rate, 30, 208
MCAs, *see* Multichannel analyzers
Metal coating, 180-181
Metallography, 160-162
Metals
  bulk specimen preparation of, 159-164
  thin specimen preparation of, 172-176

Meteorites, 126, 340, 341
Microscope-detector relationship, 128-129, 343
Milling, 173
Minerals
  bulk specimen preparation of, 159-164
  thin specimen preparation of, 172-176
Minimum detectability limits, 125, 339-340
Mirror imaging
  BSE, 256
  SE, 255
  of specimen chamber, 69, 254
Morphology
  microstructual, 160-162
  surface, 164-165
M spectra, 33-34, 36, 37, 213
Multichannel analyzers (MCAs)
  in EDS analysis, 33-41, 213-217, *see also* Energy-dispersive microanalysis
  in quantitative EDS analysis, 108
  stereo images and, 18
Multielement quantification, 146, 377-378

Near-edge fine structure, 150-151, 383
Noisy images, 94, 302-304
Nontopographic contrast, 65, 243-245, 247

Objective apertures, 4–5
  beam convergence and, 12
  beam current and, 9, 10
  real, 9, 10
  virtual, 9, 10, 12
Objective lenses, 4, 5
  beam convergence and, 12
  beam current and, 10–11, 190
  focusing and, 6
  in STEM, 140, 365, 367

Parallel EELS, 148
  collection conditions in, 149, 381
  energy calibration in, 149, 381-382
  spectrometer variables in, 150, 383
Particle contrast, 65, 242-243
Particle microanalysis, 127-131, 343-351
  of fractured surfaces, 128-129, 343-348
  of irregular or rough particles, 129-131, 348-351
  of spherical particles, 128-129, 343-348
Particle specimen preparation, 168-170, 175
Particulates, 168-169, 176
Peak intensity
  in quantitative EDS analysis, 115
  in quantitative WDS analysis, 106-107
Peak overlaps
  in EDS analysis, 38, 39, 40
  in light element microanalysis, 118, 333
  in quantitative WDS analysis, 105
  in WDS analysis, 46, 221
Peaks, *see also* Peak intensity; Peak overlaps; Peak shape; Peak-to-background ratios
  artifact, 36, 38
  in EDS analysis, 31, 36-41, 210-211, 214-217
  escape, *see* Escape peaks
  false, 31, 211

peaks (cont.)
  high-intensity, 36
  low-intensity, 37
  in quantitative EDS analysis, 110
  in quantitative WDS analysis, 100, 105
  sum, see Sum peaks
  system, 31, 210-211
Peak shape, 118, 119, 330
Peak-to-background ratios
  in particle microanalysis, 128, 129, 131, 333-344, 350
  in quantitative WDS analysis, 314
  in WDS analysis, 45, 221
PEELS, see Parallel EELS
PEs, see Primary electrons
Petrographic thin sections, 163
Phase contrast imaging, 141, 370-371
Phosphorus
  high-standard, 123, 335
  low concentration, 124, 337-339
  in meteorites, 126, 340, 341
Photomultiplier tubes (PMTs), 5
Plasmon losses, 151, 383
PMTs, see Photomultiplier tubes
Point beam analysis, 129, 131, 343, 347, 350
Point group symmetry, 156
Polished flat specimens, 40
Polishing, 164, 173
Polymers
  beam damage of, 60, 235
  bulk specimen preparation of, 164-166
  imaging of, 88-89
  thin specimen preparation of, 176-177
Postspecimen scatter, 144, 373-375
Primary coloring, 136, 361-362
Primary electrons (PEs)
  in environmental SEM, 287
  in high-resolution SEM, 62
  in SE signal components, 69
Pulse pile-up, 31-32, 211
Pure element standards, 125-126, 340

Qualitative analysis
  automatic, 216
  of a complex spectrum, 39, 215, 216
  in EDS analysis, 36-39, 214-216
  of a simple spectrum, 38, 214, 215
Qualitative stereo microscopy, 22-23
Quantitative analysis
  in EDS analysis, 39-41, 216-217
  in EELS, 152, 387
  in quantitative EDS analysis, 116, 324-329
  in WDS analysis, 47, 221
  in x-ray microanalysis, 145-146, 377-379
Quantitative energy-dispersive microanalysis, 108–116, 316-329
  calculation of composition in, 114-116, 321-324
  operating conditions for, 108-110, 316
  quantitative analysis in, 116, 324-329
  standardless analysis in, 116, 324-329
Quantitative measurements, 71-72, 254-262
  in computer-aided imaging, 92-94, 298-300

Quantitative stereo microscopy, 25-26, 204-206
Quantitative wavelength-dispersive microanalysis, 99-107, 309-315
  calculation of composition in, 106-107, 309, 315
  operating conditions for, 99-105, 309

Rastered beam analysis, 129, 131, 343, 347
  utilizing particle shape, 348, 349
Real objective apertures, 9, 10
Real time, 28-29, 207
Relief contrast, 64-65, 242
Resolution, 8
  of channeling contrast images, 77, 270, 272
  in EELS, 149, 382
  in WDS analysis, 44-45, 219
Rough specimen biasing, 68, 251, 252
Rough surface microanalysis, see Particle microanalysis

SACP, see Selected area electron channeling pattern
SAD, see Selected area diffraction
Salt crystallization, 89, 293-295
Sample current, 43-44, 219
Scanning electron microscopy (SEM), see specific types
Scanning electron microscopy (SEM) parts, 3–5
Scanning system, 4
Scanning transmission electron microscopy (STEM), 55-56, 139-142, 232-233, 365-372
  in CBED, 153-154, 389, 390, 393
  characteristics of probe, 139-140, 365
  compared with SEM, 56, 232, 233
  contrast in, 141, 369-371
  convergence and collection angles in, 140-141, 368-369
  in EELS, 148
  image formation in, 140, 365-368
  with other electron signals, 142, 371-372
  relationship with TEM, 140, 368
  in x-ray microanalysis, 143, 144-145
Scanning transmission imaging, see Scanning transmission electron microscopy
Scan rate, 20, 201-202
Schottky barriers, 85, 163-164, 285
Schottky diodes, 85, 285-286
SC signals, see Specimen current signals
Secondary electron (SE) imaging, see also Secondary electron signal components
  contrast in, 17, 197
  in environmental SEM, 86, 87-88, 89, 287, 288, 290, 291, 292, 293
  in high-resolution SEM, see High-resolution SEM
  STEM and, 142, 371
  x-rays and, 133
Secondary electron (SE) signal components, 67–72, 251-262
  collection efficiency of, 68-69, 251
  contrast in, 68
  contributions of, 69, 254
  quantitative measurement of, 71-72, 254-262

Sectioning, 164
SEELS, *see* Serial EELS
SE imaging, *see* Secondary electron imaging
Selected area diffraction (SAD), 153, 155, 389
Selected area electron channeling pattern (SACP), 77, 273-274
SEM, *see* Scanning electron microscopy
Semiconductors, 163-164
Serial EELS, 148
  collection conditions in, 148-149, 381
  energy calibration in, 149, 381
  false imaging in, 152
  spectrometer variables in, 150
Signal processing, 16, 20-21, 202-203, *see also* specific types
Silicate analysis, 116, 324, 325, 326, 327, 328, 329
Simple spectrum analysis, 38, 214, 215
Single pixel contrast enhancement, 92, 297, 298
Skeletal tissues, 166-167
Smoothing kernels, 136, 303-304, 361
Soft tissues, 167-168
Spatial resolution, 79, 80, 278
Specimen chambers, 5, 6, 69, 254
Specimen current (SC) signals
  as BSE detector, 54, 230-231
  contrast and, 16, 17, 18, 197, 199
Specimen grids, 72, 261
Specimen insertion, 5-6
Specimen rotation, 79, 275, 276, 277
Specimen stage motions, 5-6
Specimen thickness
  in CBED, 154, 391-392
  in EELS, 151, 383
Spectral artifacts, 31-32, 210-212
Spectrum acquisition, 28, 207
Stage motions, 5-6
Staining, 177
Standardless analysis
  in EDS analysis, 39, 40, 41, 216-217
  in quantitative EDS analysis, 116, 324-329
Static voltage contrast, 83, 279-280
STEM, *see* Scanning transmission electron microscopy
Stereo effect, 25, 204
Stereo images, 18, 199-201
Stereo microscopy, 22-26, 204-206
  qualitative, 22-23
  quantitative, 25-26, 204-205
Stereo photography, 24-25, 204
Stigmators, 7, 140, 365
Stray radiation, 30, 210
Sum mode, 52-53, 228
Sum peaks, 25, 36, 38, 210-211
Surface imaging, 87-88, 291
Surface topography, 159-160
System peaks, 31, 210-211

TEM, *see* Transmission electron microscopy
Tent kernels, 302-303
Thermionic electron guns, 203
Thin film coatings, 176
  SE yields from, 66, 245-248, 249

Thin films, 175, *see also* Thin film coatings
Thin specimen preparation, 172-179
  of biological tissue, 177-179
  of ceramics, 172-176
  of metals, 172-176
  of minerals, 172-176
  of polymers, 176-177
Topographic contrast, 197-201
  in BSE imaging, 52, 53, 54, 198, 199, 227, 228, 229, 230
  in high-resolution SEM, 63, 64-65, 242-243, 248
  in low-voltage SEM, 57-58, 234
  SC signal and, 54, 231
  in SE imaging, 17
  in SE signal components, 68
  vs. beam energy, 57-58, 234
Trace element microanalysis, 122-126, 335-342
  data collection for, 122-124, 335-339
  measurement of concentrations in, 126, 341-342
  minimum detectability limits in, 125-126, 339-340
Transmission electron microscopy (TEM), 139, 365
  in CBED, 153, 154, 389
  contrast in, 141, 369-370
  in EELS, 148
  image formation in, 140, 366, 367, 368
  relationship with STEM, 140, 368
  in specimen preparation, 170
  specimen preparation for, *see* Thin specimen preparation
  in x-ray microanalysis, 145
Tungsten hairpin guns, 6, 8, 9, 192

Ultramicrotomy, 175, 176-177, 178

Vacuum system, 5, 6
Virtual objective apertures, 9, 10, 12
Voltage contrast, 81, 82-83, 279-280

Water imaging, 87-88, 289-290
Water pressure, 88, 289
Wavelength-dispersive microanalysis, *see* Wavelength-dispersive spectrometers
  quantitative, *see* Quantitative wavelength-dispersive microanalysis
Wavelength-dispersive spectrometers (WDS), 42-47, 219-223
  analysis situations of, 46-47, 221-223
  characteristics of, 44-46, 219-221
  in light element microanalysis, 117, 118, 330
  operating conditions for, 42-44, 219
  in particle microanalysis, 127, 128, 129, 131, 350
  in trace element microanalysis, 122, 122, 123, 125, 126, 335, 337, 340, 341
  in x-ray images, 132, 133, 134, 352, 353, 354, 357, 358
WDS, *see* Wavelength-dispersive spectrometers
Wide-area electron channeling pattern (ECP), 74-75, 263, 264

Working distance
  depth-of-field and, 14, 192, 193-194
  vs. x-ray intensity, 46, 221

X-ray collection, 145
X-ray dot maps, 47, 132, 223, 352
X-ray emission imaging, 147
X-ray images, 132-136, 352-362
  analog, 132-133, 352
  background removal in, 134, 135, 358
  dead time in, 135, 358, 360
  digital, 133-134, 352-357
  dot mapping technique of, *see* X-ray dot maps
  intensity measurement in, 135-136, 358–361
  processing of, 136, 361-362
X-ray intensity
  in quantitative EDS analysis, 110, 316
  in quantitative WDS analysis, 100, 105, 309
  vs. sample current, 43-44, 219
  vs. working distance, 46, 221
X-ray microanalysis, 143-147, 373-380 *see also* specific types

X-ray microanalysis (*cont.*)
  bremsstrahlung effects in, 147, 380
  data collection in, 144-145, 375-376
  data quantification in, 145-147, 377-379
  emission imaging in, 147
  specimen preparation for, 372-374, *see also* Bulk specimen preparation
  spurious x-rays in, 143-144, 373-375
X-ray spectra families, 33-36, 213, *see also* specific spectra

ZAF method, 47
  defined, 114
  in EDS analysis, 39, 40, 41, 216-217
  in light element microanalysis, 120, 333
  in particle microanalysis, 127, 128, 129, 131, 343, 344, 345, 346, 348
  in quantitative EDS analysis, 114-115, 321-323
  in quantitative WDS analysis, 106-107, 309-315
Zero-loss peak, 149, 382
Zero-order Laue zone (ZOLZ), 154, 155
ZOLZ, *see* Zero-order Laue zone